GENES, BRAINS, AND HUMAN POTENTIAL

—

KEN RICHARDSON

GENES, BRAINS, AND HUMAN POTENTIAL

—

The Science and
Ideology of
Intelligence

COLUMBIA UNIVERSITY PRESS
NEW YORK

Columbia University Press
Publishers Since 1893
New York   Chichester, West Sussex
cup.columbia.edu
Copyright © 2017 Columbia University Press

Cataloging-in-Publication Data available from the Library of Congress
ISBN 978-0-231-17842-6 (cloth)
ISBN 978-0-231-54376-7 (electronic)

Columbia University Press books are printed on permanent
and durable acid-free paper.

Printed in the United States of America

Cover design: Noah Arlow

CONTENTS

PREFACE

The causes of variation in human potential—for attributes like intelligence, and educational and occupational achievement—are always interesting. The expression "fulfilling our potential" is widely used and implies destinies already laid down, with variable, but definite, limits. Moreover, the causes of variation really matter: perceptions of someone's—a child's—potential can seriously prejudice how the individual might be treated by others and by our institutions. As everyone knows, that is why a nature-nurture debate has smoldered around the issue for hundreds, if not thousands, of years.

Thank goodness, then—you might think—that modern scientists are, according to widespread reports, settling the issue at last. They can tell us about the real nature of intelligence, and even measure it with IQ tests. They have revealed, in remarkably exact proportions, how individual differences are due to different genes. Thanks to the brilliant technology that sequenced the human genome, we are told that scientists are now even identifying the genes responsible for that variation. They are also showing how those genes—in interaction with environments—shape our brains to determine our levels of intelligence (and differences in them). What is more, those scientists might soon be able to design genetically informed interventions in schools to help those not so well endowed, or even target specific genes to boost IQ and give the world more geniuses. So now, at last, we can put that heated nature-nurture debate behind us once and for all.

However, there is something wrong with that scenario. None of it is true. The advances turn out to be more hype than reality. The findings are crude, based on assumptions that are decades old and long since criticized as deeply flawed. There is not even an agreed-on theory of how to describe potential or intelligence. And as for discovering "genes for" intelligence or other potential, none have been found (in spite of the mind-boggling costs and continuing promissory notes).

Nor will they, it is now clear, because the endeavor is based on several misconceptions of the nature of genes, the nature of human potential, the nature of development and of brain functions, and even the nature of the environment. The problems behind the hype are conceptual, not the need for more data to add to the (inconclusive) piles we already have. Whatever powerful new technologies are applied, we will still only get slightly more sophisticated expressions of essentially the same message. That is because the concepts themselves are really only veneered expressions of a very old—albeit often unconscious—ideology, rooted in the class, gender, and ethnic structure of society: a ladder view of a social order imposed on our genes and brains.

This book seeks to reveal, describe, and explain all of that. But it is also aims to do much more. Nearly all the debate and discussion about human potential takes place in a fuzzy atmosphere of hunch and informal preconception. The joke is well known: ask a dozen psychologists what intelligence is, and you will get a dozen different answers. Behind the exaggerated claims there remains little scientific theory and definition of potential, in fact—little understanding of intelligence, how it evolved, how it develops, how we can promote it. These are prime conditions for ideological "infill"— the real obstacle, I shall argue, to understanding human potential.

The book seeks to remedy that fog and deficit. It requires a conceptual, not an empirical, revolution, and it so happens that one is now looming in biological systems research. It is beginning to permit, for the first time, an integration of findings and theory, from the molecular ensembles of the single cell to the amazing creativity of human social cognition. It is already showing, for example, that the classic, but elusive, "gene" is a conceptual phantom with deep ideological roots. The reappraisal puts us, as Evelyn Fox Keller explained recently, at "a critical turning point in the history of genetics," where "recent work . . . obliges us to critically reexamine many of our most basic concepts."[1]

Similar conceptual advances are revolutionizing our understanding of brain, cognitive systems, the engagement of these in social systems from ants to humans, and, finally, that transformative evolution to human culture and social cognition. They are nothing if not far reaching. Imagine that genes are not the "blueprints" or "recipes" we have been told they are; that living things existed before genes; that a child's potential is not pre-limited, but is created in the course of development; that the environment is vastly more complex—yet more providential—than it looks; that forms of "intelligence" exist even in single cells; that the brain and human intelligence develop throughout life; that, in humans, they are shaped by the "social tools" they have access to rather than innate programs.

Above all, imagine that, far from the gene-based ladder-view of people with graded brains, the vast majority of us, and our children, will be constitutionally "good enough" for participation at all levels of social activity and democratic institutions. This is the view that the new biology and psychology, and even many experts in human resources, in commerce and industry, are coming around to. It might just be bringing humans out of a long period of ideological gloom in which only the few are really "bright," into a new enlightenment for everyone. It all suggests a far better, more hopeful, story to be told about human potential.

That is the story to be told in this book, but it involves a long route through much novel territory. To give you some sense of direction, here is a rough route map. In the first chapter, I explain why a new look at the whole field covered by the title is badly needed. I illustrate how ideology has (even unwittingly) perfused much of what passes for a science of potential through a key weakness—the vagueness of its basic concepts. The rest of the chapter illustrates, at some length, such weakness in the recent hype-ridden "advances" about genes and brains and intelligence.

What follows deconstructs the current edifice around those basic concepts—and then slowly builds a new one on sounder foundations. Chapter 2 is about the peculiar, and largely mythical, model of the gene at the roots of the edifice; it exposes the flawed methods of inquiry (and results) developed around it. Chapter 3 shows how the IQ test—the basis of nearly all that is said about genes, brains, and intelligence—is the opposite of an objective measure. There is no agreed-on theory of intelligence: test constructors have simply decided in advance who is more or less "intelligent" and then built the test around that decision.

The construction of a genuine biological model of intelligence, and of a new vision of potential, starts in chapter 4. It goes right back to basics: to molecular networks, and life before genes; to evolution and cells; the true nature of complex environments; the kind of information living things really need to survive in them; and the "intelligent systems" that use it. It also explains why the new concepts of "dynamical systems" are needed to understand them. It will be, for most readers, the most challenging chapter, but is crucial to the reconstruction that follows. There are a number of summaries, though; and some of the more complex parts can be skipped, anyway.

Chapter 5 applies these ideas to the explanation of development: the transformation of an original "speck" of matter into bodies and brains of dazzling variety and competencies, utilizing the same genome. It starts to tell us how potential and variation are actively created through the system dynamics, rather than passively received in genes. A dynamical, "intelligent" physiology, coordinating activities in disparate tissues, also has much to tell us about the nature of individual differences.

Chapters 4 and 5 begin to show how intelligent systems have evolved at many different levels, corresponding with more changeable environments. Chapter 6 describes how a "neural" system of intelligence emerged as more changeable environments were encountered. It contrasts the traditional mechanical and computational metaphors of brain functions with the emerging concepts of dynamical processes. Only the latter can deal with unpredictable environments, and I will show how brains based on dynamical processes are—even in nonhuman animals—far cleverer than we think.

Scientists' views of brain functions, though, have been much informed by models of cognition. In chapter 7, I summarize these models, show their inadequacies, and offer the new perspective now emerging in dynamical systems research. For the first time, that perspective clearly describes how cognitive intelligence both transcends, yet *emerges from*, that in brain networks.

Chapter 8 puts cognition in the context of the evolution of social groupings from ants to apes. Even in ants, it has entailed a further leap of intelligent functions, *between* brains, which is even more complex than those within them (I call it "epicognition"). And that, in turn, sheds new light on the origins and nature of individual differences in social groups, some of which were hinted at by Charles Darwin himself.

Such new perspective is especially important for understanding human evolution. Chapter 9 describes how humans evolved especially close forms of group dynamics, resulting in a new "layer" of regulation, namely, human culture. Much of the chapter goes on to describe the epoch-making fecundity of that unique intelligent system, explaining why humans adapt the world to themselves, when all other species are locked in to specific niches.

Throughout these chapters, I draw out implications for the understanding of individual differences. That is particularly the case when I turn, in chapter 10, to considering how to promote human potential, and contrast the emerging dynamical framework with traditional "input-output" models of causes and interventions (and why those have been disappointing). As with chapter 11, where I bring the same perspective to bear on schooling and education, the implications for policy are stark and far reaching.

ACKNOWLEDGMENTS

Integrating such a wide spectrum of research has been a necessary, but also a thrilling, trail to follow, and I hope readers will share some of that excitement. It would not have been possible without the thoughts and inspiration of the many ideas-makers I have had the pleasure of encountering along the way, either in person, or by other means. They are too numerous to list, here, but I hope their presence in these pages and the nature of the product will reflect my gratitude. The following, however, took time out from their own busy schedules to read drafts of the present work and offer extremely helpful comment and feedback: Claudia Chaufan, Jonathan Latham, Robert Lickliter, Mike Jones, Jay Joseph, Richard Lerner, David Moore, Sarah Norgate, Steven Rose, and Allison Wilson. My partner Susan Richardson has been a patient and supportive ally throughout the project, and Brian Richardson helped with some of the diagrams.

I am grateful to all of them for helping me turn a rough draft into a more readable and coherent product. If it still is not, the responsibility is entirely my own.

1

PINNING DOWN POTENTIAL

SCIENCE AND IDEOLOGY

Science is a systematic enterprise that builds and organizes knowledge in the form of testable explanations and predictions about the universe.

—Wikipedia

Ideology: The body of ideas reflecting the social needs and aspirations of an individual, a group, a class, or a culture.

—Dictionary.com

For millennia—probably at least since the emergence of class-structured societies—scholars, philosopher-psychologists, and state authorities have told people that social inequalities are inevitable, the consequence of immutable differences in mental potential in people themselves. Those messages have always formed powerful ideologies, making inequality seem just and natural, preempting protest and coaxing compliance.

Until relatively recently, legitimation took the form of appeals to supernatural powers as a kind of ultimate authority. In ancient Greece, in Plato's *Republic*, it was God that made men of Gold, Silver, or Bronze; women frivolous; and slaves subhuman. The medieval period in Europe invented the Divine Right of Kings. The poor in Victorian Britain sung in chapel of how God made them "high and lowly" and "ordered their estate." In the Imperial colonies, subjugation of natives was excused through allusions to innate inferiority and the "white man's burden."

Charles Darwin changed all that, as everyone knows. He introduced a new "power"; an objective, impartial one that takes authority from material reality and transparent reason rather than from the supernatural. Darwin proved, indeed, that biological differences are an important part of the evolution of species. But Social Darwinists and psychologists ran away with aspects of a far more circumspect Darwin and used them to legitimize the wealth and power of the strong, and the poverty of the weak in the human species. Wrapped in the convincing language, concepts, and gravitas of objective science, the old message was given an even greater—and at times more deadly—effectiveness.

Over the past century or so, essentially the same message has been honed, impressed on us, and turned into an ever-stronger scientific orthodoxy. Instead of Plato's myth of the metals, this science tells us about lucky or unlucky permutations of genes that determine levels of potential in our brains and are more or less fulfilled through child rearing and education (otherwise called the "environment"). Such science, it has been claimed, supplants the old, now redundant, ideologies. It lays the foundations for a more just, meritocratic society in which equal opportunity replaces equality of outcome.

That science of human potential has played a powerful role in human affairs over the past century or so. It has influenced policy makers and social institutions, and has reinforced class structure in so many ways. Education, employment, immigration, and related policies have been forged around it. In families, it warns that expectations for our children need to be cautious, because outcomes will be uncertain but inevitable. In individuals, it has instilled self-images of potential, thereby inducing willingness to accept inequalities in society, the limited extent to which we share the fruits of our labors, and, with that, unequal power and privilege.

Of course, this science of human potential has not developed entirely without challenge. All the while, critics have balked at its brashness, its fatalism, and the implied limits to individual possibilities. There have been constant attempts to moderate the message in an awkward nature-nurture debate: individual differences depend on environments as well as genes. Underneath, though, the hereditarians and environmentalists have really shared much of the basic conceptual furniture: the nature of genes, of environment, of development, and even the assumption that potential can be measured, as in intelligence quotient (IQ) testing. So the debate has never moved beyond questions of emphasis over what makes the most difference—genes or environments.

As a consequence, the history of the science seems to have depended more on how much of the genetic logic the social body could stomach than on fundamental scholarly challenge (of which there has been much). As is now well known, enthusiastic acceptance of the logic in the 1920s and 1930s was followed by disgust over the consequences in Nazi Germany and elsewhere. A period of benign environmentalism followed in the postwar period. But then the hereditarians came back with even more hard-hitting claims: with, they said, new methods and new data, a smart new title—behavioral genetics—but still with some skepticism to overcome. More recently, that wave has turned into a mass assault, confronting us, they have claimed, with even stronger genetic credentials; but also with kinder, even emollient, messages of benign interventionism, with benefits for all. The science of human potential has certainly never been boring.

This book suggests that now is a good time to be looking at that science afresh. New technologies and exciting visions have brought a new outpouring of scientific claims about genes and brains and human potential. The old skepticism is waning, and behavioral genetics is riding the tide of more receptive social conditions. Faced with widening social inequalities and increasing social tensions, governments are looking for biological and psychological rationales with which to appease their populations. Massive funds have become available for shedding light on causes of social inequality that will not threaten the status quo.

So every day, it seems, press releases and news headlines inform us of the latest sensational discoveries. They bombard the public mind, but really say much the same as before: inequalities are in our genes, which are in our brains, determining our level of potential as seen in our intelligence.

However, now something else is also going on; the message is rebounding with unpleasant dissonance. It is as if the very intensity, scale, and exuberance of output are revealing a science overstretched. Inconsistencies are coming to light, exposing deeper fault lines in the science itself. Put simply, the more we get, the less reliable the science seems to be: the stronger the claims, the more patently improbable the results. It suggests, what many have long suspected, that something is—and has been—going on other than pure science. Many human scientists are now speaking of a rising crisis in this and related domains.

They started to become concerned at least ten years ago. In 2005, John Ioannidis, in *PLoS Medicine*, summarized the "increasing concern that most current published research findings are false." Ten years later, an editorial in the journal *BioMed Central* (September 2, 2015) states: "In recent decades, the reproducibility of a shocking number of scientific studies has been called into question . . . with the increasing number of studies revealing that much of science cannot be reproduced or replicated." A paper in *BMC Neuroscience* (July 23, 2015), concurs that "hallmark papers . . . have been flagged as largely unreproducible."

More sensational has been a "Reproducibility Project" in psychology, in which investigators asked scientists to attempt to replicate published results of a hundred key projects. They found that only 39 percent of them were successful.[1] Some dispute these findings,[2] but it is now generally accepted that many, if not most, findings about genes and brains are not as strong as originally thought. I will have much more to say about this later in this and other chapters. The point is that we are now being asked to take a harder look at this science, and its deeper, social, preconceptions. It is dawning on critics that, although claiming to supplant the old ideological authority, the new science of human potential may have simply become another tool of that ideology.

Some readers may be surprised that such a thing is even possible in science. So I want to be clear about what I mean. We now know that ideology is not just the bombastic roar or blatant self-interest of the ideologue; it can also arise in the more quiet output of the scholar. In science generally, it tends to flow in more subtle, usually unconscious, currents, shaped by the social and political landscape from which it springs. Like everyone else, scientists tend to absorb and reflect the prejudices, social structures, and institutions of their time. They build models that, like the pre-Copernican model of the sun and planets circling the earth, fit everyday social experience and what seems obvious. So the ideology gets wrapped up in the paraphernalia of science and all its trimmings.

That, at least, is the gist of an article in the science journal *Nature* (September 9, 2015).[3] It first reminds us of the "fuzzy boundaries between science and ideology." Regarding recent trends, it warns us that "a culture of science focused on rewarding eye-catching and positive findings may have resulted in major bodies of knowledge that cannot be reproduced." Expressing concern that more and more research is tackling questions that are relevant to society and politics, it urges scientists "to recognize and openly acknowledge the relationship."

So that is one of the things I invite you to do in this book. But there is another reason this is a good time for a fresh look. The fault lines now being exposed also suggest something wrong with fundamental preconceptions in the science. The perpetuation of the nature-nurture debate is continuing evidence of that. The history of science suggests that, when scientists are locked in such disputes, it usually takes a radically new conceptual framework to break out of the deadlock. "The important thing in science," said William Bragg, Nobel laureate in physics, "is not so much to obtain new facts as to discover new ways of thinking about them."

It so happens that just such a new way of thinking about human potential is now emerging from work in many allied sciences: in genetics and molecular biology, in evolutionary theory, in brain sciences, in a deeper understanding of the environment, in new revelations about development, and (at last) in a genuine theory of intelligence. Nature and nurture in the nature-nurture debate have simply been opposite sides of the same coin, when what is really needed is a change of currency. Moreover, it is a new way of thinking that is also being galvanized by masses of new facts and findings. What I offer in this book is such an alternative.

It seems appropriate to start showing how the fuzziness of the boundaries between science and ideology, in the area of human potential, really lies in the haziness of its key underlying concepts. So I illustrate that in what immediately follows. Then in the rest of the chapter, I show how that haziness is being fatally glossed over in the contemporary exhuberance, pointing the way as I do so to aspects to be picked up in later chapters.

Chapters 2 and 3 attempt to lift the fog around those (incredibly perva-sive) ideas. The rest of the book will present the steps to the new way of thinking.

THE USEFULNESS OF OBSCURITY

The popular concept of human potential is, in fact, the perfect vehicle for turning ideology into science. The terminology that abounds today—such as "genes for" or "brain networks related to"—can make conclusions about potential sound quite convincing. But the scientific definitions of "poten-tial" are no clearer than those of a standard dictionary: a "something" capable of becoming "something else," a "capacity" for developing into something, the "possibility" of being or becoming, a "latent" quality, and so on.

It is precisely that ambiguity that makes the concept of human poten-tial so prone to ideological infill and political rhetoric. It subtly weaves hope and fatalism into our unequal societies. It suggests that each of us *might* become more than we are now; but also that nature—the luck of the genetic dice—ensures there will be strict limits to it. The concept, in other words, is an ideological convenience that perfectly maintains the notion of innate differences and limits while framing the contemporary rhetoric about "equal opportunities" and "fulfilling our children's poten-tial." This is reflected in the very concepts of intelligence, of the gene, and of the brain. A quick look should illustrate the point.

THE "g" PHANTOM

The contemporary scientific concept of intelligence is an offspring of that convenient vagueness: a hunch—something "obvious"—clothed in the precise language of science but with a murkier history.

Although Charles Darwin considered himself to be "rather below the common intelligence" (and had an undistinguished school and univer-sity career), he revealed to us the role of biological variation in heredity and evolution. His cousin, Francis Galton, who had inherited a fortune, and his place in the British upper class, seized on the idea to scientifically vindicate what, to him, was already quite obvious: that the class system so favoring him was a reflection of inherited natural ability. His interpretation of Darwin's theory was that it finally quashed all "pretensions of natural equality" (as he put it in his book *Hereditary Genius* in 1876).

Galton soon turned this new biology of heredity into a social and political mission. He founded the Eugenics Society, with the aim of restricting reproduction to those with the most potential (as he perceived it). To make it scientific, and seemingly objective, though, he needed a scientific measure of potential. Such measure would serve, he argued, "for the indication of superior strains or races, and in so favouring them that their progeny shall outnumber and gradually replace that of the old one."[4]

Galton thus defined intelligence as whatever it is the upper class has more of, and he invented the first intelligence test with a political purpose in mind. He was the first, in what has been a long line of scientists, to exploit the notions of potential and intelligence in that way (I will have much more to say about that in chapter 3).

By the turn of the twentieth century, Galton's ideas were taken up by many others, including renowned statistician Karl Pearson. Like Galton, Pearson argued that social reforms would not eradicate "feeble-mindedness." Only programs for selective breeding could prevent the degeneration of society, he said. He, too, used rather crude intelligence tests to promote the idea. After administering them to Jews and other immigrants of East London, he warned that they are "inferior physically and mentally to the native population."

Darcie Delzell and Cathy Poliok have shown how Pearson's class background and ideology encouraged errors in methodology, data quality, and interpretation, and influenced his "often careless and far-fetched inferences." With a warning that expertise and esteem offer no immunity to such errors, they note how "the influence of personal opinions and biases on scientific conclusions is a threat to the advancement of knowledge."[5]

That ideological perversion of science was, however, only the start of much that was to follow on both sides of the Atlantic. After the turn of the century, genes had been identified as the hereditary agents, and Galton's followers had revised his test to become the modern IQ test. The essential message was that social inequalities are really just smudged expressions of genetic inequalities, and the test proves it.

A vigorous testing movement grew, clearly nourished by those beliefs. Lewis Terman, the founder of the first English-speaking test in the United States, declared how the test would help "preserve our state for a class of people worthy to possess it."[6] Certainly, turning social classes into numerical ranks, with the upper class at the top, made the enterprise *look* scientific. It furnished the first scientific-seeming tool of social policy.

The political spin-off was huge, as were the social consequences. The IQ testing movement became highly influential in the United States in the adoption of eugenic and anti-immigration policies in the 1920s.[7] Likewise, the test became used to help justify education selection policies on both sides of the Atlantic. The issue went quiet during and after the Second World War, following the horrors of Nazi Germany. But it soon took off again. The racist undertones of Arthur Jensen's counterblast to compensatory education in the 1970s, and of Richard Herrnstein and Charles Murray's *The Bell Curve* in 1994, became notorious.

Nearly all of what is said scientifically about human potential today is still derived from IQ testing. But to this day, its supporters are not sure what IQ tests measure. In the absence of a clear theory, they can only resort to metaphors from common experience—a kind of generalized mental "strength" or "energy" that they call "*g*"; a "power" or "capacity." As Mark Fox and Ainsley Mitchum explained in a paper in 2014, "Virtually all psychometric models . . . take for granted that a score on a test can be accurately interpreted as placement along one and only one dimension."[8]

Indeed, such a proposition put by Charles Spearman in 1915 has been described as the single biggest discovery in psychology. He introduced the concept of "general intelligence," or "*g*," although he was later to admit that the idea is poorly defined. In their standard textbook *Behavioral Genetics*—basic fodder for generations of psychology students—Robert Plomin and colleagues candidly admit, albeit with some understatement, that "it is rather less certain what *g* is." And, in a popular introductory book on the subject, Ian Deary also admitted that "there is no such thing as a theory of human intelligence differences—not in the way that grown-up sciences like physics or chemistry have theories."[9]

Any glance at the literature confirms this. Behavioral geneticists themselves are inclined to use terms like "bright," "smart," "talent," and so on, which are hardly scientific concepts. Take, for example, contributions to the *Cambridge Handbook of Intelligence* (2011). Janet Davidson and Iris Kemp note that "few constructs are as mysterious and controversial as human intelligence," and that "there is little consensus on what exactly it means . . . for one person to be more intelligent than another." Susana Urbina reviews some of the "excessive and unjustified meanings that the IQ label has acquired." Robert Sternberg and Barry Kaufman simply say "there has never been much agreement on what intelligence is."[10]

I discuss this pretend intelligence at great length in chapter 3, and show that IQ is not a measure of any general intelligence. It is the deceptive tool of an inexact science looking for rather exact consequences. But it does that by trying to match that ideological intelligence to an equally ideological gene.

THE IDEOLOGICAL GENE

The gene that is now conceived as the very basis of human potential is an incredibly powerful entity. It is difficult to exaggerate its dominion. Potential is conceived as "genetic potential and the way that is expressed phenotypically," states an October 2014 editorial in the journal *Neuroscience and Biobehavioral Reviews*. The genome—an individual's complete set of genes—is viewed everywhere as the primary repository of potential and the origins of form and variation. So individual differences are largely conceived as genetic differences.

Indeed, leading scientific journals now regularly print statements like "gene expression controls and dictates everything from development and plasticity to on-going neurogenesis in the brain" (statement of the editors of the journal *Science*, August 30, 2011). Psychologists like Steven Pinker (*How the Mind Works*) claim that "a linear string of DNA can direct the assembly of an intricate three-dimensional organ that lets us think, feel and learn." In *The Selfish Gene*, Richard Dawkins told us that "they created us body and mind." Even Nessa Carey, in her book *The Epigenetics Revolution* (intended to rewrite our understanding of genetics, the cover says), describes DNA—the chemical of the genes—as "like a script," and goes on, "There's no debate that the DNA blueprint is a starting point." (Well, there is a debate, from many directions, as we shall see in chapters 2 and 4).

This all-powerful gene chimes in with contemporary individualism, and people are being urged to have their DNA sequenced in order to learn who they "really" are. There is constant speculation about designer babies, obtained through genetic engineering, and of enhanced human intelligence through gene editing. Meanwhile, sperm banks introduce more exclusion criteria for supposed genetic disorders in cognition (creating worries about a new eugenics), and the *Royal Society* of London assures us that "human learning abilities vary, in the same way that human height and blood pressure vary."[11]

Many seem to think, in fact, that it is the genome that actually develops, like an internal homunculus with a self-directing program of self-realization, more or less assisted by the environment. So the genes are idealized as "little brains" with wide executive functions, as if they were active commands and instructions for development, instead of the purely passive templates for protein construction that they really are (as I shall explain in chapter 4). In the statistical models used to estimate their effects, they are reduced to being independent charges, like the batteries in a flashlight, that just add together to produce a more or less bright light.

In her book, *The Ontogeny of Information*, Susan Oyama described how the gene has become institutionalized as a quasi-cognitive power, an almost conscious agent. She sees such superstitious attributions as part of a Western cultural tradition: "Just as traditional thought placed biological forms in the mind of God, so modern thought finds ways of endowing the genes with ultimate formative power."[12]

Others have agreed that there may be something quasi-religious in the contemporary vision of the gene. In 2001, Alex Mauron published an article in the journal *Science* titled "Is the Genome the Secular Equivalent of the Soul?" He described how a scriptural metaphor of an overarching, controlling being has been attached to genes. Perhaps it is hardly surprising, therefore, that Bill Clinton described the laboratory sequencing of the human genome as "learning the language in which God created life." Indeed, many biologists still refer to the genes as "The Book of Life." I have just read on the BBC website that "the genome is the instruction booklet for building a human." And when referring to people's durable qualities, nearly everyone now uses clichés like "It's in their DNA."

Perhaps this vision of the all-powerful gene arises from some deep psychological need in people, perhaps due to being raised in patriarchal, hierarchical societies and living in the cult of leadership and dependency relations. So genes are themselves now revered as paternalistic agents with controlling intentions, as if serving a deep need for an ultimate authority.

Accordingly, these ideological genes don't merely contain codes or blueprints; they "act," "behave," "control," "direct," are "selfish," and so on. The developmental psychologist Jean Piaget described such attribution of purpose to inert materials as a form of animism. It is common among young children, who tend to endow inanimate objects with the qualities of consciousness, free will, and intention. It is also common among pre-scientific cultures and refers to an inborn essence containing and manifesting the properties we observe.

AN IDEOLOGICAL BRAIN

Historically, too, an ideological brain has been created. A brain-centered view of human intelligence long ago produced the obvious, if crude, hypothesis of a relationship between brain size and intelligence. Victorian scientists, already keen on anthropometry (body measurement) soon turned to craniometry (head measurement) to estimate brain sizes, first from simple skull circumferences in the living and then from cranial volumes in the dead. They duly reported on the expected brain-size differences between social classes and between males and females.

Meanwhile, explorer/anthropologists and colonialists were doing the same on natives in the colonies; they reached equally reassuring conclusions about the brains of different "races." In addition, there have been postmortem inspections of brains of high-achieving individuals, Einstein's among them. Even recently (2015), scientists claim to have discovered another little fold in the great man's brain not present in others. It has, of course, been pointed out many times that those methods are hopelessly unreliable and meaningless. But even in recent times, psychologists like Philippe Rushton have published strong opinions on brain size and race.

In the twentieth century, understanding of the brain has relied heavily on allusions to technological metaphors. So brain functions have been likened to hydraulic and electric machines, clocks, telephone switchboards and networks, computers, factories and their command systems, and many other such systems. Metaphors are, of course, common and useful in science. I quibble only when they are recruited for ideological functions.

In the brain, all those metaphors mentioned have been taken to suggest origins in genes with individual differences expressed in brains. "High cognitive ability," says James Flynn in his book *Intelligence and Human Progress* (2013), "begins with genetic potential for a better-engineered brain." He doesn't tell us, though, how genes can actually *be* such "engineers." In fact, the computer or computational metaphor has been predominant since the 1970s.[13] But other accounts appeal to social-structural metaphors. In her book, *The Executive Brain* (2009), Elkhonon Goldberg tells us that "the frontal lobes are to the brain what a conductor is to an orchestra, a general to the army, the chief executive officer to a corporation."

So the ladder view of society and its institutions is imposed on the brain as on genes and intelligence. These metaphors are still prevalent in discourse about human potential, even though there is little evidence for their real counterparts in brains. I will have more to say about them all later in this chapter, and in chapter 6. More generally, whenever brain research is being done today in the contexts of social (including gender and ethnic) classes, education, employment, economic competition, commercialism, and marketing, then there is a danger that we are being presented with an ideological brain.

A CONFUSING HERITABILITY

The selfish, independent gene has been the vehicle of another ideological tool—a statistical device that many have heard of, but few seem to be clear about. This is the concept of heritability. It was devised in the 1920s as a guide to selective breeding, essentially as a simple statistical concept: the proportion of variation in a trait, such as egg laying in hens, that is statistically associated with genetic variation (and expressed as a fraction from

0 to 1, or as a percentage). It is estimated from breeding experiments under controlled environmental conditions.

In the hands of psychologists, however, heritability has—like the concept of potential itself—been consciously or carelessly obfuscated. Instead of a statistical ratio, it has come to be understood—and presented to the public and media—as the deterministic concept of *in*heritability: that is, telling us the degree to which a trait value, such as IQ, for any individual is *determined by* heredity. Ideologically transmogrified in that way, heritability is read as the degree to which differences among individuals, social classes, and ethnic groups are scientifically proven to be genetic, or immutable.

In humans, the heritability of "intelligence" has been estimated by comparing resemblances in, say, IQ scores among pairs of twins. Identical twins share all the genes involved in variation, whereas nonidentical twins share only half of those genes. If the identical twins are more alike, then it is concluded that genetic variation is important. Formulas have been devised for estimating the heritability from those resemblances, and the estimates have usually been quite high.

This logic overlooks the simple (and, indeed, demonstrable) possibility that identical twins will also share more environment, in the sense of being treated more alike by parents, teachers, and so on (an Achilles' heel of twin studies to be more fully discussed in chapter 2). In the heady announcements of heritabilities from twin studies, such possibilities are virtually swept aside. But there are many other flaws and assumptions in twin studies that I discuss at length in chapter 2.

The more important truth, though, is that even a perfectly accurate heritability estimate of human potential or intelligence—which we certainly do not have—tells us nothing about the genetic makeup of individuals or subgroups reared in different environments. To suggest otherwise is an elementary falsehood (for reasons to be made clear in chapters 2 and 4). Outside agricultural breeding programs, where environment is carefully controlled, it is a completely pointless statistic—except, that is, for its ideological thrust.

In spite of those problems, the past two or three decades have seen an intense new focus on twin studies and heritability estimates, backed by streams of funding and turning out new waves of results enthusiastically disseminated to the unwary. "We can't ignore the evidence," says Jill Boucher, in *Prospect Magazine* (November 13, 2013), that "genes affect social mobility." And evolutionary theorist Oliver Curry warns us to expect "a genetic upper class and a dim-witted underclass to emerge" (as reported on the BBC website, April 11, 2014). Linda Gottfredson says that, "given what we know about *g*'s nature and practical importance, Black–White genetic differences in *g* render the goal of full parity in either IQ or achievement unrealistic," even though we know nothing about "Black-White genetic differences in *g*" (whatever that is).[14] Similar misunderstandings of heritability and IQ are found in Nicholas Wade's 2014 book, *A Troublesome Inheritance*.

However, the fatalistic logic had already been put succinctly by Arthur W. Toga and Paul M. Thompson in *Annual Review of Neuroscience* (2005).[15] "Nature is not democratic," they said. "Enriched environments will help everyone achieve their potential, but not to equality. Our potential seems largely pre-determined."

Apparently, then, we have to conclude from this hard science that democracy among humans is an *un*natural state. More broadly, of course, none of these remarks is socially benign. Individuals, and whole groups, so blighted by "scientific" doubts about their own abilities, are deterred from social and political participation, decision making, and actions they may otherwise have taken. That affects personal development and impairs democracy, as I explain in chapters 10 and 11.

These conclusions about genes and brains and potential have been based almost entirely on "old" methods, like twin studies and measures of cranial volumes. Over the past two decades, however, new weapons have been unleashed in the pursuit of the "final" proof. They have dazzled the media, the public, and policy makers more than ever. This new wave of findings and claims takes up most of the rest of this chapter.

DNA SEQUENCING FOR POTENTIAL

Over most of the past century, these claims about genes determining differences in potential were made without anyone actually "seeing" and describing them. Conclusions about genes have been inferred almost entirely from twin studies that (even if valid) impute sources of statistical variance, not gene identities.

Over the past two decades, however, brilliant advances in molecular biology have made it possible to describe more or less directly the specific genes, or different versions of them, that different individuals actually have. This is what the Human Genome Project has been about. Immediately it occurred to behavioral geneticists that the limitations of twin studies could at last be transcended. Now we could potentially describe direct associations between variation in genes and variation in ability on a one-to-one basis—or so it has been thought.

These days, of course, most people know that genes derive their individual identity from the different ways that four molecules (called *nucleotides*) are combined in different sequences to form the genes on the DNA strands that make up the chromosomes. The new methods enable the identification not only of the whole genes (the different "words" on the DNA strand) but also of the nucleotides (the "letters" that make up the words) at each location in each gene. Variations in these, from person to person, are called *single nucleotide polymorphisms*, or SNPs. Such sequencing was the process that made the Human Genome Project possible, culminating in first publication of a representative sequence in 2000.

The sequencing process has rapidly developed into an industrial-scale enterprise done by machines and computers. It has been accelerated by the commercial production of DNA "microarrays" or "chips." These are templates of known nucleotide components against which unknown sequences can be compared and matched for identification. They have permitted the sequencing of large samples of individual genomes at rapid speed and rapidly reducing cost.

Matching differences in these SNPs to differences in IQ seems a simple, and obviously hugely appealing, step. The DNA can be extracted from a few drops of blood or a few cells scraped from the lining of the cheek. Little in-depth knowledge or technical expertise is required, because the sequencing is done by machine, with any associations printed out by a computer program. Surely this will identify "genes for intelligence," or other aspects of potential, it was immediately reasoned.

So large-scale, international consortia of gene sequencers and psychologists have formed, backed by huge funds from traditional funding bodies like the National Institutes of Health in the United States, and the U.K. Medical Research Council. The rationale is that, by identifying the "true" causes of problems, it may be possible to devise interventions more fruitful than those used in the past.

Such genome-wide association studies, or GWASs, have been primarily aimed at the genetics of various diseases and disorders. But the appeal, and its simple logic, took an early hold on psychologists in the field of intelligence. Promises of a new dawn for the understanding of differences in human potential were soon being issued. As early as 2000, Robert Plomin was claiming that genes could be found in infancy to predict adult cognitive ability, enabling parents to recognize genetic limits to their children's ability. Thus primed for breathtaking discoveries, the BBC Radio 4 reported (August 8, 2000) that at least one research team is close to identifying "genes for IQ," and that "scientists could soon test the potential intelligence of new-born babies." The report was duly echoed in the *Times* and other newspapers on both sides of the Atlantic—even though no such identification has taken place either then or since!

A decade and a half later, several hugely expensive screens for genes for human potential continue, hunting for correlations between SNP (or other genetic) variants and IQ or school performance. A typical example of the logic is in the research of the team assembled by Steve Hsu and others in China, and involving Plomin's group in Britain. They are scanning DNA from blood samples from the one in a thousand individuals with the highest IQ scores. They hope to find the "special" genes prominent in such individuals. Hsu has hinted that genes might then be engineered in others to produce more superintelligent individuals (see chapter 10). It has been suggested that such studies could one day help parents select embryos with genetic predispositions for high intelligence.[16]

Another example is the Mind Genes project at Imperial College, London. Its website states that "Mind Genes is a research project using next generation genome sequencing technology to identify genetic variants associated with cognitive abilities." It says further that "the reasons why people differ in their cognitive abilities are complicated," but nevertheless suggests that "genetic factors may account for 70% of the variation in cognitive abilities." "However," it admits, "it has been very difficult to identify these genes."

But splashed around the world's media just before Christmas 2015 was the declaration that Imperial College researchers have identified two "intelligence gene networks." Lead author of the paper (published online in *Nature Neuroscience* on December 21, 2015) Michael Johnson is quoted as saying, "This research highlights some of genes involved in human intelligence, and how they interact with each other." He goes on, "What's exciting about this is that . . . potentially we can manipulate a whole set of genes whose activity is linked to human intelligence," and "our research suggests that it might be possible to work with these genes to modify intelligence, but that is only a theoretical possibility at the moment."[17]

Numerous other gene-hunting projects have been springing up in various parts of the world. Many are extending this logic to education. In chapter 11, I have much more to say about the ideology of schooling as a "test" of potential. But one example is the large group of researchers forming the Social Science Genetic Association Consortium. Their remit has been to find associations between genetic SNPs and educational achievement. They are funded by a number of prestigious bodies and have been published in the leading journal *Science* in 2013. Across the billions of SNPs scanned, they claimed to find a small number statistically associated with school test scores.[18]

What wasn't emphasized in the press releases or excited media reports was that associations were found with only 2 percent of the variance in educational performance. That is, even if the associations are real (and there are doubts), 98 percent of individual differences in school achievement were *not* associated with genetic variation. Ironically, the group's press release declared this to be a genetic "advance," claiming that the "findings will eventually be useful for understanding biological processes underlying learning, memory, reading disabilities and cognitive decline in the elderly."

In a circumspect scientific environment, such a minuscule correlation (and a correlation is not a cause) would ordinarily have been dismissed as a chance or, at best, indirect effect. Jonathan Latham, in the *Independent Science News* (August 3, 2013) described this publicity to be "as spectacular a mis-description of a scientific finding as is to be found anywhere in the scientific literature." He wrote that the claims to find genes, even when we haven't, have more to do with the ideological and political gains to be found in biological fatalism. But scientific meetings and seminars with titles like "The Molecular Genetic Architecture of Educational Attainment" are now springing up.

As I write, yet another study has appeared in *Nature*. With such a handy methodology now available, there will no doubt be many more. And, no doubt, associations will be seized on as the discovery of "genes for education," with far-reaching conclusions. The main conclusion is that we can soon start identifying groups of children for special genetically informed treatment in schools.

This seems to be the position, for example, of Kathryn Asbury and Robert Plomin, as reported in their book *G Is for Genes*. It conjures up a startling vision, where the DNA of all our children will be available on bio-data banks, from which we can read their true fates from a very early age. In schools, they tell us that "the technology will soon be available . . . to use DNA 'chips' to predict strengths and weaknesses for individual pupils and to use this information to put personalized strategies in place for them." Each child, it seems, is to enter school with a genetic barcode, to which teachers will respond with tailored treatments, like an optician's prescription after an eye test.[19]

Over two thousand years ago, Plato urged us in his *Republic* to "discover with accuracy the peculiar bent of the genius of each." Today, Asbury and Plomin claim to be on the verge of such discovery by identifying children's genes. Although fundamentally flawed, as we shall see in chapters 2 and 4, the idea nevertheless set politicians' antennae twitching. Plomin was called to give evidence to the United Kingdom's Parliamentary Select Committee on Education in 2013. This has been followed by several sole-interview TV programs on the BBC to discuss that vision of gene-based education.

What also stands out is the self-confidence of this new wave of genetic determinists. In one paper, for example, Ian Deary and colleagues reported results that "unequivocally confirm that a substantial proportion of individual differences in human intelligence is due to genetic variation."[20] "Unequivocal" is a word rarely used in research reports, even in the advanced sciences. Since Deary has already told us that there is no "grown-up theory of intelligence differences," we're entitled to ask: What exactly is being confirmed?

The worrying way in which the hyperbole has penetrated the public mind, however, is illustrated in the *Hastings Center Report* (September–October 2015). It is a well-meaning publication that discusses ethical matters in areas of science. The journal issue in question considers ethical dilemmas in the area of genetics and intelligence. It does this based on the assumptions that (a) IQ is a measure of intelligence, (b) we know what intelligence is, (c) genes for intelligence have been found (albeit with tiny effects), and (d) we can use that knowledge for therapeutic purposes.

None of these assumptions is true. So there is no dilemma. Yet authors of the report, somehow, prefer to believe that there is.

Unfortunately, gene hunting has also created an aura of "real science" in a discipline conspicuously short of it, and a wide range of psychologists are scampering to join in. Those who urge caution are condemned as "science deniers."[21] Others refrain from criticism with some trepidation about funding and careers. This, too, suggests something afoot other than pure science.

THE GENE BUBBLE BURSTING

The truth is that the trail is now littered with unsuccessful attempts to identify such genes. In project after expensive project, no associations have materialized, or a marginal "finding" has not been replicated in repeat investigations. To date, no gene or SNP has been reliably associated with the normal range of human cognitive ability (let alone shown to actually cause differences in it).

Studies grow bigger and bigger, dozens of them now combining into ever-more-expensive consortia, and papers get published with indecent haste in a few journals. But still they come up with null or minuscule results, with correlations interpreted as causes, almost invariably not replicated in follow-up studies.

Even those firmly of the faith have been expressing their disappointment. Erik Turkheimer was prominent among those trumpeting the forthcoming revolution in the discovery of genes for IQ. More recently, he has realized how, "to the great surprise of almost everyone, the molecular genetic project has foundered on the . . . shoals of developmental complexity." And in 2015, he admits that "scientists have not identified a single gene that would meet any reasonable standard as a 'gene for' schizophrenia, intelligence, depression, or extraversion."[22]

The dilemma is now widely discussed as the "missing heritability" problem. Twin studies have estimated IQ and other aspects of potential to be at least 50 percent heritable. To the behavioral geneticist, this indicates that there must be many variable genes underlying individual differences in it. So where are they? As we shall see in chapter 2, given the flaws in twin studies and the true nature of intelligence, the heritability estimates are probably inaccurate in the first place.

In spite of these disappointments, the gene hunters continue to lace reports with such terms as "exciting," "breathtaking," and "momentous shifts," telling us that the fruits "will soon be available" and so on. These are the sort of subjective terms we would not normally expect to find in scientific papers. And, increasingly, we see truth being merely asserted through wishful thinking rather than empirical demonstration. "We now *know* that many genes of very small effect are responsible for the heritability of intelligence," say Nicholas Shakeshaft and colleagues in a paper in the journal *Intelligence* (February 2015, emphasis added)—even though no such responsibility has been shown. There are many biological reasons why not.

For a start, consider the scale of it all. Humans possess, in the original fertilized egg and in nearly every cell of their bodies, about twenty thousand of the gene "words." Each word is made up from different sequences of four nucleotide "letters"—the chemicals adenine (A), thymine (T), guanine (G) and cytosine (C). Each of these letters occupies a locus in a sequence in the DNA strand. And each locus can, in different people, have a different nucleotide (hence the term "single nucleotide polymorphism," or SNP).

For example, a DNA sequence . . . AAGGCTAA . . . , as part of a sequence in one person, may occur as . . . ATGGCTAA . . . in another. The second nucleotide has undergone a substitution. The trouble is that there are more than six billion of these nucleotides in the twenty-three pairs of chromosomes in each human cell. On average, an SNP occurs only once in every three hundred nucleotides, which means there are at least ten million SNPs in the human genome.

In other words, humans share the vast majority of their genes and their nucleotide sequences: any two humans taken at random from anywhere in the world will be more than 99 percent identical. On one hand, ten million SNPs still represent lots of variation. On the other hand, we also know that nearly all these variations are functionally neutral: it doesn't matter which version you have, they work equally well (as I explain in chapter 4). Trying to segregate those relatively few SNPs that supposedly make a difference from those that do not is difficult enough for a well-defined medical condition. But for traits as poorly defined as intelligence and using only statistical correlations as evidence, making firm conclusions already seems highly naïve.

In fact, the method of associating such tiny pieces of DNA with crude and imprecise mental scores—often obtained from shortened tests administered over the telephone, online, or by parents to their children, and usually statistically compressed (see chapter 2)—is almost designed for churning out reams of spurious correlations. And that is what is happening. The statistical associations seem to be scintillations in a fog.

An important part of the fog is that complex physiological and mental functions do not vary like eye or hair color, through a single gene or very few genes. Educational and cognitive "phenotypes" are not simple physical traits, like height and weight, milk yield in cows, or egg laying in hens. They involve thousands or even tens of thousands of genes. And they are not utilized like electrical charges being added together. Instead, individual genes are taken up as resources into complex biochemical networks that create variation out of intense interactions.

Most of the fog, however, is around what is said to be measured. Most reported correlations are with some sort of IQ scores. But there is little agreement about what they really measure except in equally obscure terms. The Imperial College group suggests that they have turned cutting-edge gene-identification technology to finding genes related to memory, attention, processing speed, reasoning, and executive function. The claim is made with great confidence, as if those were clearly defined and characterized functions, like (some) medical conditions. They are emphatically not, as I explain in chapters 3 and 7. The best analogy I can think of is using super high-tech surveillance equipment to capture a creature in the woods in the dark when we aren't sure what it looks like—and then, when we've captured something, not being at all sure what we've got.

Even more critical is the problem of demonstrating that such statistical associations are causal. That requires satisfying a phenomenal array of conditions. This problem is sidestepped in the imperative to find the genes. So reports are littered with terms like "gene for," "gene effects," "due to," "accounts for," "explains," "influences," "underlies," and so on. I wonder why journal referees and editors allow authors to get away with this. "Common Genetic Variants Influence Human Subcortical Brain Structures" is the title of a 2015 paper in *Nature*, when, of course, the paper demonstrates only statistical association covering less than 1 percent of variation.

Teachers and professors constantly remind naïve students of the dangers of interpreting correlations as causes. Correlations—or related association measures—are honeytraps for investigators; they are also the cheapest weapon of ideology. A correlation merely measures the extent to which two entities co-vary, or vary together, not of how one causes the other. Ice cream sales may well correlate with the extent of sun tans in white people: it doesn't mean that ice cream is causing sun tans. Stephen Jay Gould, in his 1981 book *The Mismeasure of Man*, described such misinterpretation of correlations as "probably among the two or three most serious and common errors of human reasoning."[23] But inquiry and interpretation in the field of human potential and genes for IQ thrives on it.

In view of this, it is perhaps not surprising that many scientists are now worried about relations between this kind of science and the general public. A yawning gulf is opening up between hype and reality. The speed with which results find an esteemed journal, and a press statement is issued, is beginning to turn science into something akin to show business.

So Michael Hiltzik remarked in the *Los Angeles Times* (October 27, 2013) how "the demand for sexy results, combined with indifferent follow-up, means that billions of dollars in worldwide resources . . . is being thrown down a rathole." An investigation reported in *Independent Science News* (August 8, 2013) states that "human genome research has now reached a crisis point. Very little in the way of genetic predispositions have been found yet the public has become convinced that genetics is a key factor in human disease, mental health and social inequality."

But there are now deeper ripples of concern. At last, it is beginning to dawn on people that the genome might not actually contain the information being sought. As a lead article in the journal *Nature* in 2009 complained, "Despite the successes of genomics, little is known about how genetic information produces complex organisms." The appalling answer now emerging from molecular biology is that it doesn't—genes contain no such information!

This is the huge conceptual hurdle now to be overcome. Researchers like the Imperial College group may talk of gene products forming gene regulatory networks (GRNs) that govern development in bottom-up fashion, as per the standard model. But data now pouring out of molecular genetics labs are revealing how GRNs are themselves regulated by layers of other networks in and far beyond the cell, in top-down organization. Chapters 4 and 5 describe how organisms, and their variations, are constructed during development, using genes but not programmed by them.

Accordingly, modern molecular biology reveals that, for normal ranges of complex functions, little if any relationship exists between genes on the "inside" and variation on the "outside." We all want to understand the detail of developmental pathways. But that requires sensitivity to the dynamic multi-level system that has actually evolved, in which regulation is much more top-down than bottom-up. It has been demonstrated how "phantom" heritability can arise out of such interactions—an artifact of the statistics, but not there in reality.[24]

Of course, all kinds of rationalizations and obfuscations are now created to explain the missing genes. These are most commonly along the lines that the candidate genes are too numerous, with individual effects on variation too tiny, to actually detect. So now we seem to have another phantom—an imaginary cloud of tiny-effect genes, for which there is no causal scientific evidence. The idea falls into the same category as the other phantoms: the concepts of potential, intelligence, *g*, independent genes, heritability, and the causal correlation coefficient. We are presented with another fuzzy vehicle of ideology with which scientists confirm the social prejudices of the faithful without in fact proving anything scientifically.

This will not, of course, deter those gene hunters who now find themselves looking, not for single genes, but for (very large) packs of them.[25]

What really worries me is that attempts to so "chemicalize" human potential in this century could have even more dire consequences than the medicalization of "madness" in the last.

THE ENVIRONMENT AND NATURE-NURTURE KNOTS

Of course, everyone also acknowledges the role of the environment in creating individual differences and in realizing potential. Everybody knows that nutrition promotes growth, exercise will increase muscle, and dieting will reduce fat. But are there such analogous causes of individual differences in intelligence?

In fact, in this field, models of the environment are almost as simplistic as those of the gene. The environment seems to be mostly conceived as a kind of maternal counterpart to the paternal role of the genes. Apart from supplying resources (nutrients), the environment, in the dominant view, merely nurtures, supports, limits, or attenuates what's in the genes. Many authors genuinely want to stress that environmental experiences "influence the functioning of our genes," modify "how genes work," and even talk about gene-environment interaction or interplay. But, always, like powerful despots, the genes still remain in charge.

As with the genes, understanding of the environments that make a difference to human intelligence remains backward and piecemeal. Pioneering efforts have been made to identify aspects of the home, parenting, social class, and so on, that are statistically associated with IQ or school test scores. These have involved sending questionnaires to parents, and observations and interviews by researchers visiting homes and schools. Correlations reported are with broad factors like parental teaching style, discipline style, numbers of toys and books in the home, family income, and neighborhood and housing characteristics.

Deducing from the correlations how these really cause individual differences in intelligence is a different matter. When they have been used to guide intervention programs for young children, the effects tend to be small, labile, or nonexistent (see chapter 10). The differences in experience that make even children in the same family so different from one another are particularly difficult to pin down. In the commentary mentioned above, Eric Turkheimer pointed out that "attempts to [identify] systematic environmental causes that produce systematic differences in outcome almost always end in disappointment."[26]

In other words, investigators do not seem to be coming to grips with what the environment of human potential really is. The field is dominated by nominal, unanalyzed, and impressionistic accounts that do not specify what it is about environments that is important for individual differences. This point has been put repeatedly. As Alan Love said, we have encouraged "the simplification of environmental causal factors in favor of isolating causal import from [internal] components."[27]

In other words, compared to the billions spent on chasing improbable genes, relatively little attention has been given to this key issue. In *The Cambridge Handbook of Environment in Human Development*, Linda C. Mayes and Michael Lewis state: "Indeed, the features of the environment and their various outcomes are poorly understood . . . it is surprising how little systematic work has gone into their study." Likewise, Dale Goldhaber argues that "in fact, it is this lack of a definable environmental perspective . . . that has made it possible for the nature side to become increasingly visible and influential both within the discipline of human development and more broadly across the culture. Nativists take on the empiricists all the time, but the reverse is rare."[28]

As with the genes and intelligence, investigators fall back onto simplistic metaphors of the environment. One of the most popular metaphors reflects the way that farmers identify aspects of soil quality, feed, and fertilizer for boosting the yield potential of crops and animals. In the 1960s, London sociologist Basil Bernstein referred to the metaphor as the horticultural view of the child. It forms the basic conception of "the environment" for many baby books. In their book *G Is for Genes*, Kathryn Asbury and Robert Plomin tell us that knowledge of children's genes will suggest "ways of planting them in soil that will help them to grow as fully as their natures allow," and of using "the environment (of the school) to maximize genetic potential."[29]

Another metaphorical view of the environment is that of provision of exercise, in the sense that physical exercise helps develop the body generally. In an article titled "Brightening Up" (another metaphor), Guy Claxton and Sara Meadows downplay notions of innate potential. They nevertheless say that "most researchers now believe that young minds are better thought of as developing muscles. . . . They are made up of a lot of interwoven strands, and they get stronger with exercise. Like musculature, minds have a genetic element to them. Different people are born with different physical 'potential', different ranges and aptitudes."[30]

In chapter 4 and beyond, I present a quite different notion of the environment. I show how, compared with the attention given to genes, the nature of the environment has been sadly neglected (and how that has led to so many misunderstandings about the nature of genes). There is a need for an "enviromics" every bit as detailed and pored over as the genomics the public have been blitzed with over the past decade.

That will also mean understanding the environment at many different levels, including the social context in which human cognition evolved and operates. For example, it means realizing that theories of genetic determinism themselves become a deleterious part of the environment in which people develop and come to think of their potential. Suggestions that they are deficient in genes and brains really do affect children's cognitive development and their performance as adults (see chapter 10).

THE BRAIN AND POTENTIAL

The potential said to be coded in different ranks in the genes is thought to be reflected in the varying quality of neural networks in the brain. As mentioned above, James Flynn states that intelligence "begins with genetic potential for a better-engineered brain." Just as psychologists and educators have turned themselves into quasi-geneticists in recent years, many others have sought to become brain scientists. Taken together, these trends have produced a general conviction that this is the way to understand individual differences in potential and in human behavior generally.

In principle, there is nothing wrong with an awareness of how the brain supports mental functions. But its usefulness depends on the model of the brain that emerges and the nature of its relationship with those functions. I simply challenge the conceptual foundations on which yet more hype, and inappropriate models of the brain, are springing up. As with genetics, scholars and practitioners, as well as the media and general public, have been subjected to a stream of promissory notes laced with hyperbole. So an article in *The Psychologist* (March 2013), a journal of the British Psychological Society, tells us about "powerful methods" yielding "powerful insights" from a "bright new approach" through which "substantial insights may not be far away."

As with the genetics of intelligence, big promises about cracking the mystery of the brain, with great implications for intervention, have duly led to funds being poured into ambitious projects. The European Union's Human Brain Project, for example, has an estimated budget of €1.2 billion. Its remit is to build a computer that, by emulating the human brain, will reveal the secrets of intelligence. This, it is said by the project leaders, will reveal "fundamental insights into what it means to be human."

Again, over the past twenty years, the area has been boosted by brilliant technological advances. The most popular tool has been the brain scan: more specifically, the functional magnetic resonance imaging (fMRI) scan. This technique puts participants into a huge cylindrical chamber crossed with magnetic fields. The individual can remain at rest or be asked to engage in some problem-solving activity. The relative amount of oxygenated blood flowing to different parts of the brain is then indicated in X-ray-type pictures and is assumed to reflect localized differences in neural activity.

No one can doubt the extra dimension the technique has brought to studies of the brain, especially in medicine, and in providing evocative pictures of its activities and connectivities. But serious misgivings arise when the technology is applied to questions of human potential. The usual aim is to correlate levels of activity or anatomical volume in different individuals with different levels of IQ. Such correlations are taken to indicate how differences in IQ are caused by differences in brain tissue quality, in turn due to differences in the quality of genes. Then follows—as with the genes for IQ research—the next step in the argument: that telling teachers what kind of "brain" a child has will help them improve their teaching of that child.

Typical of this approach is that of Richard J. Haier. In an interview with the National Institute for Early Education Research, he was described as "a cutting-edge researcher on human intelligence whose work with neuro-imaging technology points to a future where knowing how individual children's brains function may help teachers tailor their approaches to educating each child."[31] His website tells us that "Richard has found that the density of gray and white matter in regions of the brain is related to differences in how people score on intelligence tests and other cognitive measures." It also includes a fascinating discussion room for topics like "Brain Scans May Improve Careers Advice" and "Why Do So Few Women Reach Top Ranks in Science?"

Another illustration is research by Jin-Ju Yang and colleagues, who attempted to associate a composite measure of brain network volume—thickness, degree of folding, and so on—with IQ.[32] And splashed across the world's media was the report that comparisons of scans of male and female brains "supported old stereotypes, with men's brains apparently wired more for perception and co-ordinated actions, and women's for social skills and memory" (reported in the *Guardian*, December 3, 2013).

There has been an upsurge of studies looking for such associations. These, too, have set antennae twitching in various policy-related groups, again looking for esoteric solutions in brain structures to what may well be much deeper problems in social structures.

Among these groups is the well-meaning Education Endowment Foundation, which is "dedicated to breaking the link between family income and educational achievement, ensuring that children from all backgrounds can fulfill their potential and make the most of their talents." In January 2014, it announced, in conjunction with the Wellcome Trust, "the launch of a fund supporting the use of neuroscience in classrooms . . . to develop and evaluate the effectiveness of neuroscience-based educational interventions."

Specialist journals, like *Mind, Brain & Education* have also proliferated, devoted to "transforming through neuroscience." As early as 2003, an editorial in *Nature* wrote about "Bringing Neuroscience to the Classroom." The journal produced a news feature called "Big Plans for Little Brains" on the emerging relations between education and cognitive neuroscience. Writing in the journal *Nature Neuroscience* in 2014, Marian Sigman and colleagues argue that "it's prime time to build the bridge." Paul Howard-Jones warns against "brain scan to lesson plan" approaches, but nevertheless advocates the "development of learning technology informed by neuroscience." As an example, he reports the design of a web app known as "Zondle Team Play, that allows teachers to teach whole classes using a games-based approach and which draws on concepts from neuroscience."[33]

Not to be outdone is the London Royal Society, with its series of *Brain Waves* modules starting in 2011. These modules are designed to "present important developments in neuroscience that have the potential to contribute to education." The Society makes big claims about connections between brains and genes and the implications for education and policy: "Neuroscience is shedding light on the influence of our genetic make-up on learning over our life span, in addition to environmental factors. This enables us to identify key indicators for educational outcomes, and provides a scientific basis for evaluating different teaching approaches."[34] Science transformed medicine about a century ago, so neuroscience can help transform education today, the module claims.

Just as Plomin and colleagues urge teachers to be geneticists, so well-meaning educational psychologists like the late John Geake (author of *The Brain at School*) had already spoken of "teachers as brain scientists." And with a similar Platonic idealism, we are told how differences in gene-brain make-up demand different "class groupings." Again, parents are being swept along on a tide of scientific optimism, often promulgated through popular media before due scientific process. And teachers are now receiving dozens of unsolicited ads in the mail each year selling "brain-based" learning schemes.

Again, I do not question motives—only the conceptual furniture on which these aspirations and claims are based. In the rush to join the bandwagon, that furniture is demonstrably flimsy. Take, for instance, the workshop held by the British Psychological Society on the "Neuroscience of Coaching" in July 2013. Its declared purpose was to develop a "brain-based approach" to teaching and learning. It claims, on the website, that much of this depends "upon the same brain networks to maximize reward and minimize threat as the brain networks used for primary survival needs. . . . In other words, social needs are treated in much the same way in the brain as the need for food and water."[35] This prompts the obvious question of why, then, do humans have such big brains?

Governments are, of course, interested in genes, brains, and potential for more utilitarian reasons: to enhance the labor supply for our economic machine; to sustain the competitiveness of business corporations in international markets; and also, as we shall see, to frame an ideology of "equal opportunities." In those contexts, and among human resources executives, potential is referred to as "intellectual capital," again rather vaguely and idealistically. "Neuroeconomics" is itself a growth area, with at least one journal of that name.

However, the corporations are also interested in the possibility of manipulating genes and brains to make us good consumers. For example, I have just received an email from an international organization called Neuromarketing. Its website invites us (with clumsy grammar) to "learn about how is the brain processing information" and "how do we decide to buy what we buy?" It goes on (amid many colorful pictures of brains, nerve networks, and cortical wiring) to claim that "researchers are studying reactions in people's brains to examine their reactions to advertising," and other related topics.

Again, I stress, it is not the genuine discovery and progress in these areas that worries me, but the hasty and reductionist interpretation that is forced on them (for example, that brains, or genes, are the basis of our potentials and actions, rather than resources for them). A neuroscience input to education is widely considered to be a good thing. But I will show in later chapters that the model of human brain functions on which most proposals seem to be based is seriously misconceived.

BRAIN BUBBLE SKEPTICISM

Whatever the impact on teachers and the public, large numbers of genuine neuroscientists are already reacting to what are seen as exaggerated and/or premature claims. Much brilliant work has been done, and much credit for that is due. But as so often happens, technology and method have far outstripped perspective, theory, and interpretation. Our understanding of the brain functions is not yet advanced enough. And, in any case, real understanding must be found at a variety of levels (as I argue throughout the second half of the book).

A simple illustration is the interpretations of brain-volume differences between males and females and explanation for the (formerly) huge attainment gap in math achievement. On the basis of "biological" (hormonal) and "brain" studies, scientists have told educators to pay more attention to biological differences between boys and girls. Over the past thirty years, however, that attainment gap has virtually disappeared in schools. In the United States, fully 30 percent of math PhDs are now awarded to women, obviously for reasons to be understood at levels other than neural networks.

Much of the problem lies in the tendency to resort to simple metaphors or mechanistic models of brain, as mentioned above. This is why, in their book *Brainwashed: The Seductive Appeal of Mindless Neuroscience*, Sally Satel and Scott Lilienfeld offer a stern warning about overly deterministic interpretations of scientific findings, especially when those findings gloss over limitations and complexities. And I may be skeptical about Paul Howard-Jones's apps (see above), but he is quite critical, too, about the way that "myths about the brain—neuromyths—have persisted in schools and colleges, often being used to justify ineffective approaches to teaching."[36]

A major problem is that, however impressive the technicolor pictures of brain scans may be, they are particularly difficult to interpret. In chapter 6, I discuss a number of problems, such as noise in the system, the experience of having a scan while confined in a large cylindrical enclosure, the difficulty of presenting realistic cognitive tasks and evoking meaningful responses, and so on.

Brain imaging is rather like trying to work out what goes on in a complex factory by observing how brightly lit different rooms become when raw materials go in or products come out. Except that, in the case of MRI scans, the light is averaged across thousands of different "windows" (or microscopic blocks of brain tissue). This says little about what really goes on behind them, on what bases, and with what results. As Matteo Carandini reminds us in *The Future of the Brain*, cognition does not reside in isolated circuits but in the computations that emerge between them.[37]

Perhaps it is not surprising that, in an amusing aside, Craig Bennett and colleagues discovered signs of apparent activity in an MRI scan of the brain of a dead salmon! They were duly awarded the 2012 IgNobel

Prize for Neuroscience "for demonstrating that brain researchers, by using complicated instruments and simple statistics, can see meaningful brain activity anywhere—even in a dead salmon."[38]

It's good fun, of course, and possibly a little unkind to the many benefits of the technology, especially in medical fields. However, it is because of such worries that the journal *Cognitive and Affective Behavioral Research* published a special issue on reliability and replication in cognitive neuroscience research. In their introduction, Deanna M. Barch and Tal Yarkoni stress the need to "to step back and develop new procedures and methods for tackling at least some of the problems contributing to the crisis of replication, whether real or perceived."[39] They particularly refer to choices of analysis in fMRI interpretations that can artificially *produce* the results desired. A special issue of the journal *Perspectives on Psychological Science* (January 2013) made similar points. That problem might particularly influence attempts to correlate IQ with aspects of brain function and structure.

However, we have already been warned of such problems in the report by a subgroup of the American Psychological Association that was specially convened in 2012 to review progress in understanding intelligence. "Consistency across studies in brain areas associated with reasoning . . . is limited," they said. "Patterns of activation in response to various fluid reasoning tasks [i.e., IQ tests] are diverse, and brain regions activated in response to ostensibly similar types of reasoning . . . appear to be closely associated with task content and context." They point out, for example, how "two different intelligence test batteries revealed only limited overlap in brain regions identified." So they question attempts to show that "high-ability individuals are more efficient problem solvers at the neural level," because "the results of these studies provide a somewhat disjointed picture of the neural basis for intelligence."[40] Michael Rutter and Andrew Pickles have also warned that, although "brain imaging constitutes a valuable tool . . . so far, its achievements do not live up to the claims and its promise."[41]

Much the same can be said about the promises surrounding neuroscience and education. A major review states that "that there are no current examples of neuroscience motivating new and effective teaching methods," and argues that "neuroscience is unlikely to improve teaching in the future." It further declares that "the theoretical motivations underpinning educational neuroscience are misguided," and indeed, that "neuroscientists cannot help educators, but educators can help neuroscientists."[42]

Such comments show that there is still little consensus about how human potential is related to the brain, except in very general terms. This is probably because, in spite of abundant specific findings, these have not yet been woven into an integrated view of what the brain is for, nor why, in the course of evolution, it has become so much more complex. As Steven Rose explained in an article in the *Lancet* titled "50 Years of Neuroscience," "Many of the problems that had beset the early days remain unresolved . . . for all the outpouring of data from the huge industry that neuroscience has become, [hopes] for bridging theories are still in short supply."[43]

Jonathon Roiser likewise complained in the *Psychologist* (April 2015) how "we lack a generally accepted neuroscientific explanation of how brains make minds." An article by Gary Marcus in the *New York Times* (June 27, 2015) is titled "Face It, Your Brain Is a Computer"—but then goes on to point out how neuroscientists tend to focus "on understanding narrow, measurable phenomena . . . without addressing the larger conceptual question of what it is that the brain does."[44]

As with genetics, then, we have brilliant accounts of some specific structures and processes in the brain. But how these figure in the higher functions of the brain and intelligence in real environmental and social contexts is another matter. In consequence, there is now great confusion in psychology about how to relate to neuroscience, even entertaining the notion that the neural level of analysis will eventually render psychology superfluous.[45]

In chapter 6, I also describe a new view now unfolding. It is telling us how the brain is a dynamic and interwoven network of connections and interactions. The brain operates at many levels and scales, from subcellular molecular ensembles, to synaptic connections, to local and regional circuits, and, ultimately, to large-scale networks interconnecting brain regions. Also, in humans, these hyper-networks are further embedded in social networks involving relations between brains. As we shall see, in the dynamic activity itself, potential is created rather than merely realized.

A MORE POSITIVE FUTURE

The purpose of this book is not merely to urge caution about questionable data and ebullient claims made public. It is first to warn about the possible ideological roots of much that has passed for the science of human potential, and then to describe a completely different view. Putting the nature-nurture debate into social and ideological perspective is an important part of that.

It will be important, as the *Nature* article mentioned earlier states, for scientists to take more care about how they assess scientific evidence. This will mean identifying the preconceptions beneath the science and checking their validity and objectivity: for example, in the "genes for IQ" area, proving the validity of tests and establishing that correlations are actually causes. Until we are clear about many such doubts, the safest position—surely appropriate in a democratic society—would be to adopt the null hypothesis. In other words, the default position of no relationship cannot be rejected.[46] Another response is that of Richard Lerner's proposal "for the major organizations in developmental science, and the major journals in our field, to collaborate in writing and broadly disseminating a consensus document about the bad science associated with past and contemporary genetic reductionist ideas."[47]

However, we are now in an increasingly strong position to make more positive, optimistic proposals about human potential. Many scholars are now talking about a sea change, or critical turning point, in genetics and the need to critically assess our basic concepts. Likewise with the brain: there is a shift from the metaphor of a machine or computer executing built-in programs to an interactive dynamical system that doesn't merely express potential but actively creates it.

My democratic null hypothesis, then, is as follows. There are disabilities stemming from biological causes, affecting a small number of people. But the vast majority of humans have biological constitutions "good enough" for development to allow full participation in social life, whatever that demands. Potential depends on how that development is helped or hindered. I show that this is, biologically and psychologically, a perfectly justifiable position to adopt. A far more exciting and hopeful story is now emerging about genes, brains, and human potential.

2

PRETEND GENES

A PSYCHOLOGICAL WITNESS

Individual differences in human intelligence may be obvious to most people; but behavioral geneticists have claimed to measure and explain them *scientifically*, meaning objectively and authoritatively. The conclusion, that at least half of the variation in measured intelligence is due to differences in genes, is also rigorously scientific, they argue. It is also unfortunate, but true, that, by claiming pristine scientific status, behavioral geneticists have had immense, and sometimes damaging, effects on the lives of millions of people over a long period of time.

Such influence now goes back a long way. As early as 1909, the British psychologist Cyril Burt was advising policy makers that differences in intelligence are innate, and that differences in mental potential between groups and social classes are largely immutable. As a young man, Burt had visited the eugenicist Francis Galton and was strongly drawn to his ideas. Later, he was taken on as advisor to the Consultative Commission on Education in 1938. It reported how "intellectual development" is "governed by a single central factor, usually known as 'general intelligence.'" They went on, "Our psychological witnesses assured us that . . . it is possible at a very early age to predict with accuracy the ultimate level of a child's intellectual powers." The verdict we have heard many times since, and the "new" recommendations of today's research on genes and brains, was that "it is accordingly evident that different children . . . require types of education varying in certain important respects."

Burt's ideas were shared with a number of psychologists in America and elsewhere in Europe. They have run like a river delta through social, family, and educational policy on both sides of the Atlantic ever since. The idea that psychologists can measure innate potential is widespread today. One example, among many, is the advice to parents given by the TheSchoolRun .com, a private company selling school preparation material to families. "Most secondary schools use Cognitive Abilities Tests, CATs, to test general intelligence and to stream overall or set for certain subjects," it says. It goes on, "CATs are used to give a snapshot of a child's potential, what they could achieve and how they learn best." Moreover, there is no point in trying to coach your child for them. "The tests are designed to be taken without any revision or preparation so they can assess a child's potential in his or her ability to reason. CATs are not testing children's knowledge and understanding as a math or English exam might, so they can't 'learn' how to answer the questions." Their psychological witness explains, "I think helping your child will not have anything other than an insignificant impact on the actual performance in the CATs test."

Such sites are, of course, well meaning. But they are misguided by what they have heard in some streams of the psychological literature and are na-ïve about their ideological roots. They are echoes of Cyril Burt's warning: "The facts of genetic inequality are something that we cannot escape."[1]

Today such interest tends to be expressed in the more benign terms of "personalized learning." But it remains rooted in the same underlying genetic model. Plomin's advice to the United Kingdom's Parliamentary Select Committee on Education (December 4, 2013) was that "50% of the variance is due to genetic differences . . . everybody knows that." Parents, teachers, and others do not usually ask how psychologists *can* know these things. But that is what this chapter is about—in some detail.

KNOWING ABOUT GENES

The statements about IQ by Burt and today's behavioral geneticists are decades apart. But they are based on essentially the same narrow concept of genes and the same flawed assumptions about how they enter into the development of form and variation. The basic problem is that their inferences are not based on real genes (or environments), but on imaginary models of them. No one can actually "see" the genes implicated by Burt and his heirs. We cannot count them or "weigh" them, measure their "charge," or otherwise grade them. Nor can we work out who has what permutations, with what effects (except in rare disorders). No one is able to prove the causal consequences of having these genes in properly controlled experiments (as the best science, in other fields, does).

Instead, behavioral geneticists resort to statistical models. Of course, such models can be useful in virtually every field of science (I have used them myself many times). But the conclusions we wring from such models are only as good as the assumptions that go into them. There is a well-worn joke about the "spherical cow" or "spherical horse" problem (it is told in many versions). Here is one: A physicist loudly claimed he could predict the winner of any horse race with great accuracy; but, he said more quietly, only if it was a perfectly elastic, spherical horse, carrying a fixed weight, and moving through a vacuum at fixed speed over an even surface.

The models used by virtually all behavioral geneticists to tell us about the importance of genes in human potential are highly debatable in that sense. They are based on highly simplified assumptions about genes, about environments, about the nature of intelligence and its development, and about describing and measuring individual differences in it. Since it is not difficult to show that the assumptions are false, it follows that the models are far removed from the reality of the development of form and individual differences in human intelligence.

It is important to examine all the assumptions and expectations of the model in turn. Let us start with the peculiar model of the gene that is used to produce such important claims, with the socially profound implications just mentioned.

MENDELIAN GENETICS

Charles Darwin emphasized the role of heredity in human variation and evolution. But he still did not know what the hereditary material, passing from parents to offspring and influencing their variation, consisted of. He rather vaguely thought it was some kind of germplasm that became blended in the meeting of eggs and sperm. That meant that most variation would be continuous in nature, like height and weight.

Subsequent discoveries in genetics, from Gregor Mendel's experiments on pea plants in the 1860s to others around 1915, suggested otherwise. They indicated that the hereditary material passed across generations as particles rather than in "plasm" or quasi-liquid form. At first, too, those studies appeared to implicate one gene governing the development of each bodily trait. For example, different versions of one gene would determine whether eyes were blue or brown; those of another gene would determine whether hair was dark or blonde; and so on. But things turned out to be much more complicated.

Mendel had chosen to study pea plants for special reasons. They present a number of traits with categorically distinct variations in which the frequencies of each variety can easily be counted. Moreover, pea seeds were readily available from merchants and they reproduce quickly with great fecundity. More importantly, the plants can be experimentally self-pollinated or cross-pollinated across or within categories. In fact, Mendel spent the first two years of his study establishing pure lines that faithfully reproduced with the chosen traits. For example, there were purple flowers or white flowers, smooth or wrinkled seeds, green or yellow pods, or long or short stems. These were the different *phenotypes* (Greek for "the form that is shown") showing *phenotypic variation*.

This is what Mendel found. Crossing purple flowers with white flowers produced offspring with only purple flowers: there was no blending, or production of intermediate colors. However, when the flowers of the first generation (F1) were allowed to *self* (were self-fertilized), they produced offspring (the F2 generation) with a mixture of purple and white flowers in the ratio of 3:1. Breeding with the other phenotypes produced closely similar results, with 3:1 ratios in the F2.

Thus was posed the obvious question: how can it be that the one (the *recessive*) phenotype appeared to be totally eclipsed by the other (the *dominant*) phenotype in the F1 generation, but then reappear in the F2? Although all the F1 flowers were purple, the plants evidently still carried the potential to produce white flowers. The theory of blending inheritance cannot explain such results.

Mendel went on to do other experiments of breeding and counting in F2 generations using crosses and backcrosses. Of course, we now know that offspring inherit two forms (or alleles) for each gene, one from each parent. One form tends to be dominant over the other (recessive) form, and the pairs of alleles may be identical or different. Taken together, those conclusions helped explain the peculiar phenotypic ratios and established the basic principles of Mendelian genetics. Note that the environment did not enter into consideration in Mendel's experiments, because, above a very basic threshold common to all plants, it did not make any difference to the result.

NON-MENDELIAN TRAITS

Such Mendelian traits have been identified in humans. Think of eye color; hair color; blood type; having (or not having) a widow's peak; being color blind; and some diseases, such as Huntington's disease or muscular dystrophy. It was eventually realized, however, that most traits are not like those specially selected by Mendel, with their phenotypes falling into neat categories (and even those just mentioned are now known to be influenced by other factors).

For example, in 1908, Swedish plant breeder Herman Nilsson-Ehle worked with the red and white kernels of wheat seeds. Instead of offspring falling into one or other of the same discrete (red or white) categories, he found that they exhibited a range of intermediate colors, from deep red to pure white. The ratios of these hues did not support the idea of single genes for each trait. Instead, they suggested that variants (or alleles) of three genes were involved as if their different effects just added together in different proportions. It was subsequently found that environmental factors also influenced color, so that in practice a continuous range of hues could be observed.

In fact, it was already obvious that most bodily traits and functions are, like height and weight, expressed in continuous values, not discrete ones. And far greater (usually unknown) numbers of variant genes are involved. Trait variation is more obviously affected by the environment, too. So the distinction was made between *qualitative* traits, for which individuals can be discretely categorized (Mendelian traits), and *quantitative* traits, for which individuals have to be measured and/or ranked (non-Mendelian traits). The latter are also referred to as biometric or *polygenic* traits. It soon became commonplace among psychologists to assume that human potential or intelligence is a continuous trait and is distributed like one.

IDENTIFYING CAUSES OF VARIATION

By the 1920s, great interest in genetics had developed in agricultural contexts. It was realized that knowing how much variation in particular traits could be attributed to genetic variation could help guide crop and animal breeding to maximize yields. If variation in milk yield in cows, say, is largely due to genetic variation (there is a high heritability), then selecting for breeding those individuals who already exhibit high milk yield may boost the average yield in offspring. If the variation is estimated to be mainly due to environmental variation (there is low heritability) selective breeding will not make much difference.

However, estimating heritability is much more difficult for continuous traits than for Mendelian traits. If we look at a group of individuals varying in a Mendelian trait, we can observe who has what gene. We cannot do this with polygenic traits because (a) there are many genes involved, (b) the variation is continuous, and (c) part of it will be due to different genes and part of it to different environments. We cannot distinguish one kind of effect from the other.

Then along came the statistician Ronald Fisher, who envisaged what he thought would be a possible solution. Fisher developed an interest in heredity in part because of his interest in eugenics in humans. His solution to modeling polygenic, continuous traits was published in 1918. In it he introduced new methods of statistical analysis of variation. Using these methods, he proposed to work out the relative contributions of nature and nurture to measured variation.

This was his suggestion. There are many genes, he acknowledged. But let's assume that their effects on individuals are independent of one another (as if they were random combinations of Mendelian genes) and just add together. Then we can take it that individual differences are just differences in the sums of such genes in those individuals. That is, the overall genetic effect on variation—the heritability—can be treated as if it were random variations of such sums, so long as the size of the genetic effect can be separated from the environmental effect. And that can be achieved because it will produce certain observable patterns of resemblance among known relatives. If the size of the effect is large, then identical twins should be closely alike, nonidentical twins or ordinary siblings less so, cousins still less so, and random individuals least of all.

So the heritability could be calculated after all. (The paper was actually titled "On the Correlation Between Relatives on the Supposition of Mendelian Inheritance"). Or so it was thought. As for humans, Fisher reached the remarkable conclusion that "an examination of the best available figures for human measurements shows that there is little or no indication of non-genetic causes," and that "the hypothesis of cumulative Mendelian factors seems to fit the facts very accurately."[2]

Fisher, however, noted in the paper that "throughout this work it has been necessary not to introduce any avoidable complications." One of the complications he mentioned was the possibility that factors might not just add together like independent weights or forces; that they might actually have effects on one another (called *gene-gene interaction*). Also, their effects may be different in different environments—what is called *gene-environment interaction*. Either complication screws up the formula.

Another complication is that environmental effects can be controlled in animal experiments. In practice, this usually means attempting to randomize environments, so that every individual genotype has an equal chance of experiencing every environment. That may be reasonable for cows in a field, or hens in a chicken shed, but it is clearly not possible in humans. The equality of environmental effects across different kinds of relatives just has to be assumed, and I return to that matter below. Such simplifying assumptions are common and often useful in science. In later life, however, Fisher seemed to realize he had devised a spherical horse solution when he admitted that "[heritability] is one of those unfortunate short-cuts which have emerged from biometry for lack of a more thorough analysis of the data."[3]

At any rate, the 1918 paper became hugely influential and soon was applied by psychologists or behavioral geneticists to human mental traits—but only by adopting additional assumptions. As we shall see, today's researchers into the genetics of intelligence are still neglecting "complications" and employing "unfortunate short-cuts."

THE RISE OF BEHAVIORAL GENETICS

It was the determined Cyril Burt who brought Fisher's "solution" into the domain of human mental abilities. As Burt explained in a paper in 1956, his aim was to reveal the true causes of variation in traits like human intelligence. He defined intelligence as innate cognitive ability and assumed it to be well measured by IQ scores. His problem, again, was that of separating the genetic from the environmental effects on individual differences in IQ scores.

Controlled breeding experiments, especially for complex functions like cognitive ability, cannot be achieved in humans. Apart from not knowing the genetic backgrounds of individuals, or what environmental factors to control, they would be unethical. However, as noted by Fisher (and even by Galton), the known genetic resemblances among relatives, especially twins, suggested a natural experiment. In particular, identical twins will share all the genes that make a difference in a trait, because they arise from a single egg—they are monozygotic (MZ); nonidentical twins share only half of them, on average, because they arise from two different eggs—they are dizygotic (DZ). Likewise for resemblances between parents and offspring, sibling pairs, and so on. Cousins share only a quarter of those genes that make a difference in a trait, whereas pairs of individuals chosen randomly from the population would not share any. Comparing resemblances among such relatives might, in effect, separate the genetic from the environmental effects. A heritability for mental potential or intelligence could be calculated in that way. Or so Burt and his followers have claimed.

The simplest approach, it was realized, would be to compare MZ twins who had been reared apart. This would seem to control for the effects of the environment on resemblances while allowing the genetic differences to be fully expressed. The average correlation between such pairs of twins could be a direct estimate of heritability.

In an influential set of papers, Burt claimed to have done just that and to have measured the IQs of twins reared apart. He arrived at an estimate of the heritability of IQ of 0.83 (83 percent). This means that 83 percent of the variation in IQ is associated with differences in genes; only 17 percent results from differences in experience.

However, separated identical twins are relatively rare, and Burt seems to have been suspiciously lucky in finding so many. An alternative approach has been to compare average resemblances of pairs of MZ twins with those of DZ twins. The degree to which the resemblances correspond with genetic resemblances is also an index of heritability (or so it has been claimed). This is the "classical twin method."

Both approaches have been enthusiastically pursued since the 1950s, with some refinements, but using essentially the same basic logic and procedures. For many reasons, though, they have been shrouded in doubts and controversies. Many psychologists and other scientists now look askance at Burt's results, because suspicions were raised, after his death, of fabrication of data. This came to light after Leon Kamin's forensic scrutiny of Burt's data.[4] And the classical method comparing MZ and DZ twins is open to environmental confounds, as we shall see.

Before looking at methodological problems, it is important to consider the theoretical flaws of the methods. These mainly revolve around the large number of simplifying assumptions that have to be adopted to produce the expectations with which the twin correlations can be compared. In his 1956 paper, Burt used the words "assume" or "assumption" more than sixty times. The paper ends with the remarkable understatement, "The assumptions adopted are perhaps not entirely free from criticism."[5]

ASSUMPTIONS

The formulas that behavioral geneticists use to calculate genetic effects on variation are not straightforward. Because the measures are aspects of populations, they are statistical, such as averages and variations, rather than numbers of individuals in established categories. Because the measures are from samples of a population, rather than from every individual, many things have to be assumed before the statistical formulas can be applied. In the age of opinion polls, most people are at least vaguely aware of such problems. Basically, the logic involves asking what the results (the twin correlations) should look like if the variation were genetic in origin. The assumptions create a model, and the procedure is called "model fitting." Here is the basic model.

Basic Model

The model assumes that intelligence is an ordinary quantitative trait. That means that it develops and varies like milk yield in cows, back fat in pigs, or height in humans. I correct that grossly false notion in chapter 3 and in later chapters. In addition, the model assumes that all genes associated with a trait (including intelligence) are like positive or negative charges, G+ or G–: like batteries in a flashlight, but in their hundreds or thousands. For the statistical model, it has to be assumed that these charges are randomly distributed among individuals to constitute, in each, the individual's total "genetic charge" (or genotype; figure 2.1). Accordingly, the so-called power of an individual's intelligence, say, lies in the particular permutation of strong or weak alleles. So behavioral geneticists speak of "intelligence-enhancing alleles" and "intelligence-depleting alleles."

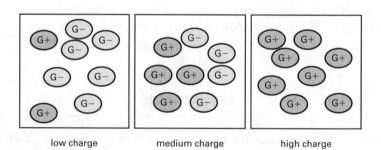

low charge   medium charge   high charge

FIGURE 2.1

Genetic effect for three individuals in the behavioral genetic model.

The trait value is then thought to be an expression of the corresponding genetic charge, more or less attenuated by the environment (of which I have more to say later in the chapter). You do not need to be a statistical wizard to appreciate that in a large population of such random gene combinations, the distribution of trait will be like the *normal curve*: lots of individuals having middling values and diminishing numbers of them toward the extreme values. This is the famous bell-shaped curve. It is shown for human height in figure 2.2. Behavioral geneticists have to assume that intelligence is made up and distributed in exactly that way. It is the only way their statistical models can be applied to data. We know that height is normally distributed. But does that distribution apply to intelligence?

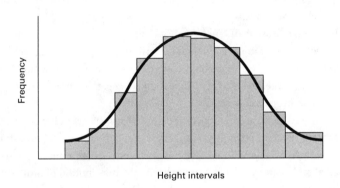

Height intervals

FIGURE 2.2

The normal curve of height measurements—shorter values to the left, taller values to the right.

Bell-Shaped Curves and Unicorns

The problem is that heritability estimates can be seriously flawed if the population distribution is not bell shaped—for example, if it is skewed one way or the other. What the behavioral geneticists fail to mention is that IQ tests have been deliberately constructed so that the scores will exhibit such a distribution. This is done by devising test items and trying them out in advance. Then those items that around 50 percent of testees get right are kept in the test, along with smaller proportions of the items that either very many or very few testees get right: "It is common practise to carry out item analysis in such a way that only items that contribute to normality are selected."[6]

As it happens, few natural biological traits are distributed in that way. Measures on almost all basic physiological processes—visual acuity, resting heart rate, basal metabolic rate, and so on—do not display normal distributions. And of course, myriad human traits—language variation, tastes in dress or food, use of tools and technologies, knowledge of the world, and ways of thinking—vary at a completely different, cultural, level compared with physical traits. In the 1980s, Theodore Micceri amusingly titled a review of such distributions "The Unicorn, the Normal Curve, and Other Improbable Creatures."

Ironically, those traits that are most important to survival are the very ones that *do not* have a normal distribution. Natural selection itself produces more and more phenotypes with values that would have previously been above average. That is, of course, the aim of artificial selection in crops and animals.

We would surely expect most aspects of human potential to come into that category of "important to survival." Indeed, some recent brain studies by György Buzsáki and Kenji Mizuseki concluded that "at many physiological and anatomical levels in the brain, the distribution of numerous parameters is in fact strongly skewed . . . suggesting that skewed . . . distributions are fundamental to structural and functional brain organization. This insight . . . has implications for how we should collect and analyse data."[7]

Buzsáki and Mizuseki go on to review the overwhelming evidence against the bell-shaped distributions of psychological functions. From sensory acuity and reaction times, to memory word usage and sentence lengths, individuals simply do not vary that way.

I return to this question below. In the meantime, it needs to be stressed that if the bell-shaped curve is the myth it seems to be for these traits, then the model used in behavioral genetics is already suspect, the statistics are inappropriate, and estimates for causes of variation may be seriously wrong.

Additive Gene Charges

The second assumption is bound up with the first. Remember the subtitle of Fisher's paper: "On the Supposition of Mendelian Inheritance." In other words, it assumes that the overall genetic effect on individual differences consists of sums of Mendelian genes having independent effects on the phenotype (only now the effects are increments or decrements in a quantitative trait rather than manifesting as distinct categories).

Another way of saying this is that there are no interactions among the individual genes, or that they have no effect on one another. The product of an individual gene always makes the same contribution to the overall result, irrespective of which other alleles and their products are present. As a consequence, the model assumes that individual differences directly reflect underlying genetic differences.

A related assumption is that a given gene will make exactly the same contribution, whatever the current environment happens to be. That is, there are no gene-environment interactions. The statistical methods of behavioral genetics have been honed around the absence of gene-gene or gene-environment interactions. The twin method, to be discussed further below, is predicated on this assumption.

Bizarre Genes

Although computationally convenient, everyone knows that this is a bizarre model of genes, as of biological systems in general. There are probably thousands of genes involved in development and variation in human potential. How can anyone seriously imagine that all these products are used without at least some sort of coordination, integration, or other effects on one another, either directly or indirectly?

Standard books on behavioral genetics of intelligence present details to the point of overkill on the structure of DNA and its transcription into RNA and then proteins. But there is usually little of the really important matter of how those products are utilized in the creation of form and variation.

Yet that is the crucial bit. The simple additive model works when there is a direct association between a genetic variant and phenotypic variant, as for eye color. However, evolved, complex traits do not develop and vary through such simple, independent associations, as we shall see in chapters 4 and 5.

Imagine each of the thousand legs of a millipede contributing to movement independently of what every other leg is doing, or with what is happening in the environment. They would be useless without the evolution of some sort of coordinating function. Yet behavioral geneticists' computations suggest that most of our individual differences in mental functions are due to hundreds or thousands of genes having independent effects in that sense.

Alternatively, of course, there might be something badly wrong with their computations. In their article in the journal *Social Science Research* in 2015, John Daw and colleagues argue that "these conclusions are erroneous due to large violations of the additivity assumption underlying behavioral genetics methods—that sources of genetic and . . . environmental variance are independent and non-interactive."[8]

This should hardly be surprising. Just as coordination guides function, and millipedes would not have evolved without it, so genes do not act alone but have evolved, and are utilized, through elaborate coordination functions. The effect of one or a few wayward legs behaving independently would not simply produce marginal increments or decrements of performance, resulting in the imagined bell-shaped curve. It would disrupt the whole system and result in an uncoordinated shambles.

In chapters 4 and 5, I show that the evolution of complex functions involved the utilization of genes, and much else, in coordinated dynamical systems. When responding to ever-changing environments, there can be no such thing as gene independence for complex traits. The product of any gene automatically becomes the environment of other genes. The wider environment gets changed in the process, to recruit further genes, and so on. Some products of genes even limit their own production. In that way, genes even create their own environments. So variation is created from such interactions. As in static photos of any team or cooperative group, the team members may look independent, but their real-life activity is the result of intense interactions among them.

The problem is that the statistical models employed cannot cope with such interactions. Behavioral geneticists may talk about interactions, but only in the shallowest form (see figure 2.3). Instead, the interactive sources of variance are simply interpreted as additive effects and called "heritability." This is what has been shown in a recent study by Or Zuk and colleagues, who called the result "phantom heritability." They explained how "quantitative geneticists have long known that genetic interactions can affect heritability calculations" but have paid little attention to them. The phantom heritability grows steadily with increasing numbers of interacting inputs, they show. In consequence, "current estimates of missing heritability are not meaningful, because they ignore genetic interactions."[9]

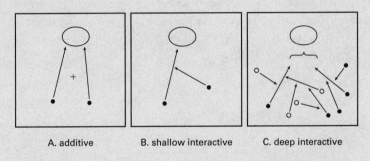

A. additive   B. shallow interactive   C. deep interactive

FIGURE 2.3

Effects of genes or gene products (red spheres) on a phenotype. (a) Independent/additive model; (b) Shallow interaction with one product modulating the effects of another; (c) Deep interactions involving modulations among numerous products, including (open circles) environmental factors (gene-environment interactions).

As we have seen, the results have misled generations of psychologists and students, with serious consequences. For instance, even if a set of data fits a statistical model, that does not make the model correct. And the interactions automatically nullify attempts to predict phenotype from genotype—for example, children's potential from that of parents or from a DNA "sample."

The odd thing is that those who use such models to assess the heritability of human potential tacitly know this. They regularly acknowledge that such interactions are likely, and even talk of gene-environment interplay. At the very least, they may well acknowledge that genes do not act of their own volition, but have to be turned on or off—thereby implying that something *else* is doing the turning on and off.

All this has been suspected for decades. The leading geneticist Sewall Wright explained in the 1930s that interaction among gene products is ubiquitous: "Our knowledge of biochemistry and physiology strongly suggests that interaction among gene products is ubiquitous." In a more recent review, Hongxia Shao and colleagues show how gene products and other cell components "interact in complex ways" that are "pervasive" in the cell.[10]

But the behavioral geneticists of IQ do not ignore interaction effects in their models out of naiveté. They do so out of pragmatism. My worry is that pragmatism further opens science to ideology—that is, obfuscating and omitting details that might spoil a story already derived from what is supposedly socially obvious.

The "Environment"

No one ignores the role of the environment. As with genes, though, the environment is conceived in a special, idealistic, way to render the behavioral genetic model testable. We are asked to assume that, in the creation of individual differences, the environment merely provides resources for, stimulates, obstructs, or otherwise attenuates what is essentially in the genes. It is usually presented in terms of vague "factors" contributing independently of one another, again to result in a bell-shaped curve to make the statistics easy.

This is the idea behind common expressions like "The genes determine potential while the environment determines how much of it is attained." This is what is meant when behavioral geneticists acknowledge the role of gene-environment interactions. Attempted disclaimers like "Most things are down to both nature and nurture," or "Genes and environments work together," are common. But we are left in no doubt over what is really in control. In fact, when comparing twins, the environment that is said to vary is not identified or measured at all; it is treated as the components of variation left over from the correlations between the twins.

Environments and Horticulture

Again this model flies in the face of reality. Hundreds of studies show that environments cannot be broken down into elements with independent effects on form and variation; they are interrelated and depend on each other in nonlinear ways. But then, as mentioned in chapter 1, both sides of the nature-nurture debate are quite vague about the environment. They often complain about not being able to identify it, or at least those parts of it that create variation in human performance. In a 1997 paper, leading twin researcher Thomas Bouchard lamented that "in spite of years of concerted effort by psychologists, there is very little knowledge of the trait-relevant environments that influence IQ."[11] Much the same applies to variation in other aspects of human performance, such as school attainment.

Consequently, the environment in twin studies is defined only in very broad and ambiguous terms. Perhaps it is hardly surprising that there have been numerous reports of the limited influence of the family environment on creating individual differences. I mentioned in chapter 1 how searches for the environment end in disappointment. But there is little attempt to characterize them, anyway. Most investigators, implicitly or explicitly, fall back on the horticultural view of the environment, as mentioned in chapter 1.

In subsequent chapters, I show how the environment has amazing complexity, with deep hidden patterns or structure: indeed, it is structure and not elements that provide the real information for development and variation. And it operates at several different levels in the evolved systems of living things with quite unpredictable consequences, mostly regardless of the specific genes. But by then I will have moved far beyond this crude strength/power model of intelligence, anyway.

Other Assumptions

The above are the most crucial assumptions of the standard behavioral genetic model. There are many other assumptions, all self-evidently false, and in ways that can severely distort heritability estimates. The following are just a few of them.

The statistical models must assume *randomized mating*. That is, parents do not select one another by compatibility of perceived mental traits, or anything else (called *assortative mating*); instead, they select one another randomly, like names out of a hat. But this assumption is most obviously false. In their 2015 article in the journal *Molecular Psychiatry*, Robert Plomin and Ian Deary note that IQ correlations between spouses is around 0.4, and is greater for intelligence (i.e., IQ scores or some approximation of it) than any other behavioral trait.[12]

The statistical model also assumes that there has been no natural selection for human potential. This assumption implies that plenty of variable genes are randomly spread across individuals in the species. In reality, it is a logical consequence of natural selection that traits important to survival have usually been through intense selection. This implies progressive reduction in genetic variation and heritability across generations as deleterious alleles are eliminated (the bearers fail to reproduce). I return to this below.

There is also the assumption that the genes, in passing from parents to offspring, do so as random collections of independent units. In fact this is known to be unlikely, especially for genes that have been subjected to natural selection and that form integrated (i.e., linked) sets or are close to each other on chromosomes.

A NAÏVE MODEL KNOWN TO BE FALSE

All these assumptions are known to be false. As a consequence, heritability estimates based on them are likely to be distorted. The strange thing is that those who make loud claims about the genetics of intelligence also know that the assumptions are false. Occasionally they admit it. This was indicated in the quote from Burt and Howard cited above. In a 2007 paper in the same tradition, Wendy Johnson admitted that one of their model-fitting assumptions (random mating) is "contrary to evidence" and that other assumptions "are generally oversimplifications of the actual situation, and their violation can introduce systematic distortions and estimates."[13]

In sum, the independent/additive model—the basis of nearly all pronouncements (to scientists, governments, or the general public) about the importance of genes—seems to be the solution to another spherical horse problem. This is why critics like Stig Omholt complain of "the abyss that currently exists between quantitative genetics theory and regulatory biology."[14] The only reasonable conclusion is that if the behavioral geneticists' data fit such unlikely models, there must be something seriously wrong with the data, the interpretation of them, or both. So now let us look at those data.

THE "GENETIC" DATA

The logic of behavioral genetics indicates that the ideal sources of inferences are pairs of twins. In the case of IQ, the degree of IQ resemblance among such pairs should, the logic says, be an indication of the importance of genetic resemblances and therefore of genetic differences (assuming that IQ is a measure of human intelligence; see chapter 3). There have been two main types of study.

Twins Reared Apart

The existence of separated identical twins has been seized on as a convenient natural experiment for the direct estimation of heritability in humans. The average degree of resemblance between pairs must reflect their shared genes, because they have not shared environments—or so it is argued.

There have been a number of such studies since the 1930s. I mentioned above how Burt's study is now disregarded because of suspected fabrication of data. The most famous study by far is the Minnesota Study of Twins Reared Apart (MISTRA), conducted since the 1970s by Thomas Bouchard and colleagues. Their first major publication was in 1990 in the prestigious journal *Science*. They began by saying, "Monozygotic and dizygotic twins who were separated early in life and reared apart (MZA and DZA twin pairs) are a fascinating experiment of nature. They also provide the simplest and most powerful method for disentangling the influence of environmental and genetic factors on human characteristics." They conclude that the MZA correlation "directly assesses heritability."[15]

Of course, identical twins who have been reared apart are rather rare. Bouchard and colleagues have accumulated increasing numbers of pairs as the study has progressed; up to the eighty-one pairs in a final total in 2000. They have consistently reported high average correlations between them in IQ and other tests. This proves, they claim, a heritability for IQ of 0.75 (i.e., 75 percent of differences in IQ between individuals are attributable to genetic differences). Papers have been readily accepted in other prestigious journals and books, and popular renderings have grabbed the media's attention on a worldwide scale.

It is difficult to overstate the influence of these results on the nature-nurture debate surrounding human potential. Sandra Scarr, president of the American Psychological Association in 1993, said that "the remarkable studies of MZ twins reared in different families challenges many cherished beliefs in developmental psychology." In 1993, Robert Plomin likewise wrote about "the powerful design of comparing twins reared apart" suggesting "a heritability for g of 80%."[16]

However, regardless of whether the theoretical assumptions are correct, there are many dubious aspects of these data. They have been subjected to forensic dissection by investigators like Leon Kamin and Jay Joseph. I do not review their work in detail here, but quickly note the main problems they have noted.[17]

They Are Not Separated Twins

At the very least, the method demands that twins have developed in truly separate environments. The first problem confounding this requirement, of course, is that the twins share an environment in the womb before birth. We now know that experiences in the womb can have lasting effects on development throughout life (see chapter 10). That could explain much of the reported resemblances in cognitive and other potentials and disqualify the method.

But even after birth, the usual meaning of the word "separated" seems to be stretched beyond credulity. Placement of pairs is, after all, usually down to family convenience and circumstance rather than careful research design. It turns out that many, if not most, of the earlier studies of twins were simply moved into different branches of the same family and brought up by grandparents, aunts, or cousins. They very often lived in the same neighborhood, remained friends, sat next to each other in school, and so on. Some even returned to live with each other in adulthood.

We cannot be sure how well the MISTRA twins were, in fact, separated, because full details have never been published. The investigators do mention in an early report that "circumstances of adoption were sometimes informal" and that "time together prior to separation" was 0–48.7 months—that is, members of at least some of the pairs had spent as much as four years together before being separated. We also know that at least some of these "separated" twins later had reunions and spent considerable amounts of time together.[18]

The consequences for purportedly objective study are obvious. However vague psychologists are about environments important for similarities in IQ, anyone might suspect that something about families, neighborhoods, schooling, and contact time would be among them. But there are other problems.

They Are Not Representative

The twin samples should be typical of the general population in the sense of covering all environmental factors and their possible effects. Lack of true separation already breaches that condition. But so do other peculiarities of the MISTRA. Twins generally tend to be self-selecting in any twin study. They may have responded to advertisements placed by investigators or have been prompted to do so by friends or family, on the grounds that they are alike. Remember, at least some of them knew each other prior to the study. Jay Joseph has suggested that the twins who elected to participate in all twin studies are likely to be more similar to one another than twins who chose not to participate. This makes it difficult to claim that the results would apply to the general population.

Results Are Not Fully Reported

Finally, the experimental results have been partially and selectively reported. For example, of two IQ tests administered in the MISTRA, results have been published for one but not the other. No explanation is given for that omission. Could it be that they produced different results?

Moreover, the MISTRA sample included DZAs, and the investigators have acknowledged that comparison of MZA with DZA correlations would be an important control. But they have seemingly refused to publish such comparisons. In addition, they have restricted access to data by critical investigators like Jay Joseph, which constitutes a breach of a crucial condition of scientific research. Could it be that correlations for DZAs are unexpectedly high—perhaps even as high as those for MZAs— suggesting other imperfections of sampling or that there really is no difference? Indeed, in the partial reports published to date, with incomplete numbers, the latter does appear to be the case.[19]

What is remarkable is that these studies have been published, and their conclusions reiterated, in highly esteemed journals, in full knowledge of the profound message thereby conveyed to psychologists, teachers, parents, a sensationalizing mass media, members of the general public, and policy makers. That the situation has not changed much since then is shown by Jay Joseph's recent analyses.[20] There seems little doubt that studies of "separated" twins are universally—and probably intrinsically—flawed.

Comparing Identical with Nonidentical Twins

The alternative method of study is to compare the average resemblance between MZ and DZ pairs reared together. If the logic of the independent /additive model is correct, then on average the resemblance of DZ twins should be half that of MZ twins. That is, they should correlate around 0.5 compared with 1.0 for MZ twins. The extent to which that difference is found is a measure of heritability of the trait. So the latter is estimated by doubling the difference between the MZ and DZ correlations or by using other model-fitting statistical maneuvers based on the same assumptions.

Most of the conclusions about the heritability of human cognitive potential have come from this "classical twin model." It is this approach that Robert Plomin and colleagues have adopted in their massive Twins Early Development Study started in 1994 and involving more than twelve thousand pairs. However, this approach is also seriously flawed, making the results difficult to interpret.

Equal Environments Assumption

The first of these flaws is fatal and insurmountable. Simply taking the difference between MZ and DZ correlations as an index of genetic effects makes a big assumption. This is that the environments experienced by MZ pairs are no more similar during development than for DZ pairs. This is called the equal environments assumption (EEA). It is unlikely to be valid for a number of reasons.

For one thing, even the environment in the womb before birth is likely to be more similar for MZ twins than for DZ twins. Most MZ twins share a placenta, whereas all DZs have one each. One indication of this derives from an examination of the *epigenetic* markers in different kinds of pairs. These are chemical signals from the mother imprinted on specific genes in embryos before birth. They reflect environmental experiences of the mother or the early fertilized egg and strongly influence the way that the genes are used in subsequent development (see chapter 4). Such signals are found to be much more variable for DZ twins compared with MZs.[21] These environmental effects could reduce the DZ correlations in comparison with the MZ correlations. But they will be mistakenly read in the blind statistical model as "genetic."

More obviously, environmental aspects of the home and other experiences are found to be markedly more similar for MZ twins than for DZ twins. For example, parents treat MZ twins more similarly, and the twins are more likely to dress alike, share bedrooms, friends, activities, and so on. In a review in 2000, David Evans and Nicholas Martin, themselves leading behavioral geneticists, reported that "there is overwhelming evidence that MZ twins are treated more similarly than their DZ counterparts."[22]

In another review, Jay Joseph cites questionnaire studies revealing very large differences between MZ and DZ pairs in experiences such as identity confusion (91 percent versus 10 percent), being brought up as a unit (72 percent versus 19 percent), being inseparable as children (73 percent versus 19 percent), and having an extremely strong level of closeness (65 percent versus 19 percent). It is also known that parents hold more similar expectations for their MZ than DZ twins.[23]

In other words, this is prima facie evidence that it is environmental similarities that, at least in part, make MZ twin pairs more alike in measures like IQ than DZ pairs. In other words, the twin method is unfit for purpose.

So how can twin researchers possibly continue to claim that they have a valid method for separating the effects of genes and environments? How can the Plomin team, for example, continue to argue that "the equal environments assumption has been tested in several ways and appears reasonable for most traits"?[24]

Sarah Norgate and I took quite a hard look at these so-called tests and reported findings in a paper in the *British Journal of Educational Psychology*.[25] These tests are blighted with the make-do empirical culture that pervades this whole area of research. They are not carefully controlled studies based on clear knowledge of relevant environments. They simply attempt to see whether differences in physical appearance, as reported by parents in questionnaires, and *assumed* to reflect how similarly twins will have been treated, correlate with differences in IQ or school test scores. The studies suffer also from selective reporting of results and of inappropriate statistics.

Attempts to justify the EEA have continued with further indirect, often contorted ploys. The latest is that of Dalton Conley and colleagues published in the journal *Behavior Genetics* in 2014.[26] To me the correlations look more like a scatterplot than a coherent pattern. Indeed, the authors acknowledge in various places the weakness of the data (especially the low numbers of participants) for reliable comparisons. But they conclude that the EEA is supported and the twin method is robust anyway. More fundamentally, similarity of environments as such was not—and could not possibly have been—directly measured, and here is why.

We Do Not Know What the Relevant Environments Are

To test the EEA, we need to be clear about the environments specifically relevant for creating differences/resemblances in the trait in question (e.g., IQ, education). Anything else is just guesswork. Of course, there is much guesswork about what is really being measured in IQ or other supposed tests of potential, anyway. But recall Thomas Bouchard's statement, mentioned above, that "there is very little knowledge of the trait-relevant environments that influence IQ." In a review in 1995, Robert Sternberg concluded that "[psychologists] do not have a very good understanding of the factors that affect IQ. Neither does anyone else."[27] That situation has not changed.

In other words, such indirect studies are hopelessly inadequate as tests of the EEA, because they evade the fundamental requirements of a scientific study. Only the following will do. First, identify and characterize the phenotype, so that everyone knows and agrees what we are talking about. Second, prove that the measure used is actually measuring it (i.e., has validity in the same way that a blood pressure or breathalyzer test has). Third, describe in detail the specific environments that *demonstrably*—not merely supposedly—create individual differences in the trait in question; that is, the environments that are trait relevant. Fourth, assess for interactions with each other and with different genotypes (a particularly demanding stipulation). Fifth, and most important, comprehensively assess whether average levels of these trait-relevant environments are indeed equal across twin groups.

Until those conditions have been fulfilled, we have no right to be promulgating definitive conclusions about the EEA and the validity of the classical twin method. Rough and ready approximations simply will not do, especially when there is abundant de facto evidence that, in general, environments *are* much more similar for MZ twins than for DZ twins; and, of course, when the conclusions may have profound consequences for countless children.

Making DZ Twins Different

There are a number of other ways in which simply comparing MZ/DZ twin correlations is not as simple as it seems. One of these is the way that parents may try to reinforce resemblances in MZ twins but do the opposite with DZ twins—that is, they amplify differences. (Ironically, they may do this because of what they have been told by behavioral geneticists about the dominance of genetics). In his 1992 study of twins, Arthur Bryant pointed out that "it is inevitable that parents should constantly compare and contrast . . . and hence look out for differences (which) often leads to the exaggeration of character traits that in fact lie well within the normal range . . . parents and friends soon learn these stereotypes and, perhaps unconsciously, respond accordingly. More serious, however, is the fact that children tend to live up to parental expectations of being the one who is 'good' or 'naughty,' 'quiet' or 'noisy.' "[28]

Related to this is the way that many DZ twins actually strive to make themselves different from one another. This is likely to be an active, not merely a passive, process, in which twins consciously shape different identities. It has been shown that DZs form individual identities earlier than MZs; and that there is less cooperation among DZs than MZs. The development of polarized identities could, of course, shift motivations away from similar learning in literacy, numeracy, academic, and other test-related domains.[29]

These social and cognitive dynamics are rarely considered by twin researchers and could well explain some of the puzzling correlations that crop up in twin research. Such difference-creating environmental effects, greater for DZs than for MZs, would be expected to increase with age. Accordingly, the average correlation differences between MZ and DZ pairs increases with age. Unfortunately, twin researchers keep reporting this increasing gap as evidence that the genetic effects—the heritability—increases with age.

This is rather typical of the tendentious interpretations of murky data. Another example is the way that, instead of accepting that the unequal environments are fatal to the classical twin studies, behavioral geneticists instead argue that it is the more similar genes of MZ twins that produce (or evoke) the more similar environments. There is, of course, no direct evidence for that. But it is not precision science, and all conclusions from twin studies are best taken with a large pinch of salt.

STUDIES OF ADOPTED CHILDREN

There is another way in which behavioral geneticists have tried to show that variation in human potential is largely due to differences in genes. This is to find adopted children and estimate the degree to which their IQs correlate with their adoptive mothers' IQs. This is then compared to the correlation with their biological mothers. A higher correlation with the latter is taken as evidence of the primacy of genes over environment.

Again, though, these are rough and ready observations that bear no resemblance to properly controlled studies. As I pointed out in a review article with Sarah Norgate, there are numerous reasons why adopted children may appear to resemble their biological parents more than their adoptive parents in IQ.[30]

For a start, adoption agencies typically assign children to adoptive families in a nonrandom way. They tend to look for adoptive homes that are "suitable," in the sense of reflecting the social class of the biological mother. In contrast, adoptive parents tend to come from a narrow social stratum. This restriction of range has the effect of reducing the estimate of correlation between adoptees and adoptive parents. Here is a brief list of other reasons adopted children may resemble their biological parents more than their adoptive parents:

(a) The children's average age at testing will have been closer to that of their biological parents than of their adoptive parents (and test scores change with age).

(b) The children will have experienced the influential environment in the womb of their biological mothers, which can have durable effects.

(c) Powerful environmental effects on mothers, such as stress, can alter gene expressions that are inherited by the children. These can mutually affect behavior, such as anxiety in test situations. These effects are described in chapter 4, and are mistaken for genetic effects, even though they are environmental.

(d) Adoptive parents have often been given information about the birth family, which could have biased treatment and their expectations of the child.

(e) Both the biological mothers (by virtue of their unwanted pregnancies) *and* their adopted-out children will have reduced self-esteem and be more vulnerable to challenging situations, like taking IQ tests. In that way they in effect share environments.

(f) Some conscious or unconscious aspects of family treatment policy may prevail to make adopted children different from other adoptive family members. Adoptive parents, from the moment of adoption, worry about the personalities, biological backgrounds, and social histories of the biological parents of their adopted child, and how these factors affect the parent/adopted-child relationship. It has also been reported that adoptive parents hold stronger beliefs about the influence of heredity compared with other parents. This may reduce expectations about the course and targets of adoptees' development. Some reports also suggest that adoptive couples strive to enhance the differences between their adopted children and themselves in an effort to allow the children free development. Later, adolescent adoptees can become highly conscious of their special identity and actually react to adoptive parents' standards and values.

(g) Adopted children also look more like their biological parents than their adoptive parents and may be treated by others accordingly.[31]

All of these factors—and many more—further compromise the adoption study as a method for separating the effects of genes from those of environments. The only adoption study that would avoid such doubts would be one in which adoptees were randomly selected from the newborn population and then randomly assigned to parents, with both groups blind to the treatment (i.e., not knowing whether they were adopted or not)—and all the while controlling prenatal environments as much as possible. All else is little better than guesswork.

DATA STANDARDS

Accounts of this so-called genetic research in the popular media, and even in science magazines, will give readers the impression that the investigators are involved in precision research producing highly reliable results. In fact, the area is suffused with a make-do empirical culture whose results should be treated with caution.

Most outstanding about this culture, of course, is the fact that there is little agreement among psychologists about what is being measured in IQ tests or other supposed tests of potential. In addition, as already noted, many compromises have to be made simply to procure twins and attempt or pretend to randomize/equalize environments. This has meant studying opportunistically rather than by careful, methodological design. In addition, the "field" nature of studies with children and parents has required improvised testing with a vast variety of shortened measures and ad hoc scales. These methods have sacrificed precision for expedience.

Typical of the approach is the large-scale Twins Early Development Study used by Robert Plomin and colleagues to advise us on genes and human potential. It is based on the use of shortened tests (shortened from their standard forms) variously administered by telephone, by post, over the internet, and by parents in the home. Reassurance about the validity of such scales is offered by correlations with more standard tests administered to small subsamples of the children. But the modest level of these, mostly in the range 0.5–0.6, in fact indicates considerable unreliability. These values mean that 75 percent of the time, the test may measure different things.

Indeed, the choice of test has often been based on convenience rather than principles of empirical precision. So disparate are the tests in some studies that it is common practice to estimate a "general factor (intelligence, or g) statistically from pooled results. This practice is known as *meta-analysis* and assumes that, because scores on different tests intercorrelate to some extent, the resulting score is necessarily one of intelligence: another indication of an inexact science proposing exact conclusions.

However, the statistical common factor may well not measure the same "thing" and may not even be cognitive in origin (see chapter 3). As Kevin Murphy says, the assumption that these measures, with disparate properties, distributions, and so on, can be combined as if they describe a single uniform variable can lead to serious problems in the meta-analysis, including "lack of clarity in what population parameter is being estimated." Again, this is not a basis for strong conclusions about anything.[32]

TRIUMPHS OF THE TWIN METHOD: BIZARRE HERITABILITIES

These flaws of twin studies probably explain why nearly all human traits investigated by the twin method have yielded substantial heritability estimates. Here are some individual differences that, according to twin studies, are largely or mainly due to genetic differences:

political party preference
political participation
attitudes toward homosexuality and gay rights
religious denomination chosen in adulthood
experience of suspicion and guilt
gun ownership
loneliness
use of tobacco
happiness
alcoholism
criminality
tendency to take bubble baths
tendency to become a born-again Christian
tendency to vote in elections
preference for tea or coffee
age of losing virginity

This is the message drip-dripped onto a gullible public and student body. It seems that, whatever behavioral differences in humans there are, the twin method seemingly cannot fail to find a "genetic basis" for it. This was nicely summarized in a recent article in the *New York Times* (July 9, 2015). "Over the last 50 years, some 17,000 traits have been studied, according to a meta-analysis. . . . Virtually wherever researchers have looked, they have found that identical twins' test results are more similar than those of fraternal twins. The studies point to the influence of genes on almost every aspect of our being (*a conclusion so sweeping that it indicates, to some scientists, only that the methodology must be fatally flawed*)" (emphasis added).[33]

Many readers may think it quite reasonable to expect such heritabilities. It is surely obvious that some variable genes, as in those studied by Mendel, make a difference to trait values. However, such traits are not typical of complex, highly evolved functions like human intelligence, probably involving thousands of genes and critical to survival.

Fisher himself noted that it is part of the logic of natural selection for important traits to have *low* heritabilities. Indeed, that has been confirmed many times in properly controlled animal experiments and field studies.

TABLE 2.1 Heritabilities of various traits
of the house martin

TRAIT	HERITABILITY
Wing length	0.156
Tarsus length	0.079
Body mass	0.000
Immunoglobulin	0.051
T-cell response	0.007
Leukocyte number	0.059

The heritabilities listed in table 2.1, for example, were estimated, not by fallible twin studies using rough and ready measures. They come from well-controlled breeding studies in the house martin carried out by Philippe Christe and colleagues.[34]

As you can see, these estimates are miniscule compared with those (0.5–0.8) typically reported for human potential derived from twin studies. Little wonder that Peter Schönemann sarcastically warned that heritability estimates for IQ "surpass anything found in the animal kingdom."[35] Any objective observer would surely reach the conclusion that human potential is unlikely to be so unusual. Rather it is the method by which it is estimated that is anomalous.

As I emphasize further in chapter 4, such low heritabilities do not necessarily mean little genetic variation; they simply mean that, for a variety of reasons, there is little association (correlation) between genetic variation and phenotypic variation. Why make intelligence such an exception? We could, after all, apply the same logic in many subjective ways. Perhaps we could establish from twin studies that behavioral geneticists have a surfeit of genes for "belief in heritability." And perhaps critics have a surfeit of genes for "skepticism." In such a gene-determined world, of course, objective science would no longer be possible, because everyone's appraisal of data is hopelessly biased by their genes!

The bottom line is that if we really want to know the causes of variation in human potential, then twin studies and the pursuit of heritability estimates should be abandoned. They are based on false genetic and environmental premises and are fatally flawed methodologically.

AN ALTERNATIVE TO TWIN STUDIES?

As mentioned in chapter 1, gene-sequencing studies have found no reliable associations between gene variants (SNPs) and measures of intelligence. So there is currently widespread dismay at this "missing heritability" problem. Instead of seriously considering the possibility that these are artifacts of the flaws in the twin method, however, behavioral geneticists have sought other ways to rescue their preconceptions.

The main argument is that the genes are not really missing. That is, the heritability must be the result, they suggest, of hundreds or thousands of genes, each making an infinitesimal difference, too small to be detected by the DNA sequencing methods. Their single effects on individual differences in potential are vanishingly small, they argue, but collectively they are still there, just as the twin studies suggest.

So a new and convoluted method has been devised to support that idea. Although involving numerous statistical maneuvers and yet more assumptions, behavioral geneticists have been very excited about it. So convinced are authors like Ian Deary and colleagues of its flawlessness that they claimed it established and unequivocally confirmed previous results.[36] These are terms not widely (or wisely) used in science.

The new method involves sequencing the genomes of thousands of individuals. That identifies which version of each variable SNP (i.e., the component letters in the genetic words) each individual actually has. The degree of genetic correlation between each and every pair of individuals in the sample is then computed. Their IQ correlations are also computed. And the correlation between the two correlations is computed, leading to an estimate of heritability. The method is called genome-wide complex trait analysis (GCTA).

Unfortunately, the method entails a formidable battery of assumptions, data corrections, and statistical maneuvers. One of these is the assumption of genes (and SNPs) as independent units, as just mentioned. Another is that (as usual in this field) all correlations are taken, without further question, to be causal. However, even if we accept those at face value, there are more serious problems. As with the heritabilities arising from twin studies, the results are virtually inevitable because they are hopelessly confounded with environmental influences.

The biggest problem is what is called "population structure." Pairs of seemingly unrelated people may share more genes than average, because they are descended from the same remote ancestors, who have remained (more or less) in the same social strata, even though not from the same immediate family. So genetic resemblances will again be confounded with resemblances in rearing environments. The method makes some effort to correct for such a problem, but only at a superficial level. As Evan Charney noted in an article in *Independent Science News*, there is further cryptic genetic relatedness "omnipresent in all populations and it wreaks havoc with assumptions about 'relatedness' and 'unrelatedness' that cannot be 'corrected for' by the statistical methods [devised]."[37]

This problem arises because of another fatal assumption—that human societies can be treated as random breeding populations with equally random distributions of genetic material in randomly distributed environments. All human populations, however, reflect continuous emigration and immigration. These are usually associated with persecution or the search for jobs. Immigrants with related genetic backgrounds tend not to disperse randomly in the target society but become associated with different classes or strata that are also, for entirely noncausal reasons, associated with different IQs. Examples include the influx of Huguenots to southeastern England to become weavers, the influx of Irish to work in Northumbrian coal mines and Yorkshire and Manchester mills, the influx of Scots to work in Midlands steelworks and car factories, and large numbers of Russian and European Jews fleeing political and racist persecutions to become tailors and retailers.

Again, these genes will have nothing to do with cognitive ability (or mostly anything else of importance). But their coincidence with social class has created an entirely noncausal correlation between social class and genetic background. Moreover, because social class mobility is limited and controlled, correlations among genes, social class, region, and cognitive characteristics will persist across many generations. Finally, because IQ tests have been constructed to correlate with social class (see

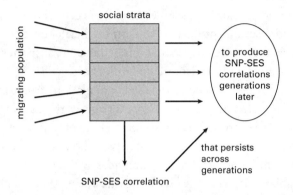

FIGURE 2.4

How nonrandom genetic distributions produce SNP-SES (and therefore SNP-IQ) correlations over generations.

chapter 3), correlations between IQ and SNPs in GCTA studies are self-fulfilling (figure 2.4).

As an illustration, Charney draws attention to the Wellcome Trust's "genetic map of Britain." It shows strikingly different genetic admixtures among residents of different geographic regions of the United Kingdom. Indeed, by analyzing records, Dalton Conley and colleagues found a correlation persisting between genetic relatedness of pairs of "unrelated" individuals and whether their ancestors lived in rural or urban environments, back over as many as twenty-five generations.[38] Likewise, an association between genetic resemblance and a crude measure of social class (parental occupation and education) is what the Twins Early Development Study of UK twins found.[39] (Sadly but predictably, the correlations tends to be interpreted as genetic variation causing socioeconomic status [SES]).

More recent analyses confirm Charney's suspicions. Youngdoe Kim and colleagues tested the model under various assumptions. They concluded that "heritability inflation can be substantial, which indicates heritability estimates should be interpreted with caution."[40] Krishna Kumar and colleagues have also warned that, with so many unlikely assumptions, estimates are "guaranteed to be biased." They conclude that "results obtained using GCTA, and the results' qualitative interpretations, should be interpreted with great caution."[41]

In other words, the method seems to be only good for redescribing the history of the class structure of society. As Charney says, "This state of affairs (not at all unique to the UK) . . . is fertile ground for spurious heritability estimates." In the same paper, he concludes that the GCTA search for thousands of genetic variants of tiny effect "is the last gasp of a failed paradigm."[42]

Yet now, as before, far from reviewing assumptions in their method, investigators call for bigger and more expensive research in the same vein. I explain the deeper conceptual reasons for their problems in the chapters that follow. Let us complete this chapter by clearing up a few simpler misunderstandings.

MISUNDERSTANDING OF HERITABILITY AND MISLEADING TERMINOLOGY

The concept of heritability, as used by behavioral geneticists, is often taken to be the converse of malleability: for example, that individual differences in a trait with higher heritability are quite fixed, and the "genetic" ranking will persist whatever the environmental conditions. This harks back to the fatalism and pessimism about human potential mentioned in chapter 1. For example, Dalton Conley and colleagues expect that reduced heritability estimates might "suggest that many traits might be more socially malleable than previous research . . . would suggest."[43]

This is a widespread misconception, because there is no relation between heritability and malleability in a trait. Heritability says nothing about rankings under different environments. Heritability is about associations between variation (or variance), not identity and causation. A condition like phenylketonuria has a heritability of 1 (there is perfect correlation between genetic and phenotypic variation). But the condition is relieved by a simple environmental treatment (removing the amino acid phenylalanine from the diet). Likewise, a heritability of 0 does not mean that genes are not involved in the development of form/function and phenotypic variation. It simply means there is little correlation between them. Consequently, the heritability of a trait has no implications at all for predicting individual potential—for example, a child's IQ from that of its parents—or the success of intervention.

In fact, genetically minded psychologists will occasionally acknowledge that demonstrable fact—but then rapidly forget it. For example, in their book *G Is for Genes*, Kathryn Asbury and Robert Plomin warn about the limitations of the concept of heritability but then subtitle their book *The Impact of Genetics on Education*—deduced entirely from heritability estimates! Only in a world of pretend genes could we infer such "impact" from heritability estimates. Again, we need to ask, what is really going on in the pursuit of pointless heritability estimates?

We must also note the widespread practice of claiming, from heritability estimates, that genetic variation has caused this or that trait variation. Terms like "genetic influence," "responsible for," or "due to" are routinely used to report what are only statistical associations. For example, in the same book (p. 16), Asbury and Plomin use phrases like "we use twins to estimate how much of the differences between people . . . is *due to* genetic influence." I described in chapter 1 how this is quite misleading wordplay.

Ideology infiltrates science when its underlying concepts are hazy and permeable. In this chapter I have shown how that is the case with the genes said to determine IQ. The model of individual differences based on such genes entails assumptions that are clearly false. Moreover, the latitude built into the model seems to have invited a make-do empirical culture in the methodology intended to test it. Apart from the lack of test validity in IQ tests, generally, scores have often come from shortened forms administered in far from ideal conditions. More seriously, the intrinsic flaws in twin studies—especially the EEA—and of adoption studies, make substantial (pseudo-) heritability inevitable in almost any trait studied. Other statistical attempts to estimate heritability for IQ have their own drawbacks. I have also discussed how the concept of heritability tends to be misunderstood and is pointless in human populations.

Finally, remember that all this effort is expended on the basis of little (if any) agreed-on theory of human intelligence or other aspects of human potential. I begin to rectify that problem in chapter 4. But since so much of the nature-nurture debate hinges on being clear about measures of human potential like intelligence tests, we should examine that subject next.

3

PRETEND INTELLIGENCE

SORTING PEOPLE OUT

Most people in developed countries have probably taken an intelligence test at some time in their lives. They might have been confused by how a few questions and puzzles could really measure their potential as human beings; wondered what decisions about them, personally, it might lead to; yet never challenged the fatal verdict. Such has been the power of the IQ ideology.

I remember taking my first intelligence test, though I did not know what it was at the time, nor that such a thing was supposed to measure my potential for future learning. I remember stalling over what seemed like trick questions and puzzles, the likes of which I had never seen before. But I was in good company. Few children in the mining village where I grew up passed the United Kingdom's 11+ exam. In fact, over the whole period of my six years at a primary school of about eighty pupils, I can only remember two passing.

So we all got shipped off to second-rate secondary school to be readied for mine and factory as soon as we reached the age of fifteen. By then our teachers (occasionally to be overheard referring to "poor stock," or "inbreeding" or asking "what else would you expect?") had done their stuff and convinced us that we had little in the way of brains anyway.

Then, as now, I suspected that this sorting of children in the job and wealth stakes had more to do with social class background than science. The terms "bright" and "dull," used then (as now) to rationalize a class structure, are hardly scientific. But in a society claiming to be democratic, the sorting must somehow be made to seem fair and just.

In meeting that problem, the idea of IQ as a scientific measure of mental power, reflecting biological inequality, has had a crucial part to play. Most of the nature-nurture debate, including the recent hunt for genes for IQ, has revolved around it. So I show in this chapter what a peculiar kind of measure IQ is. I hope you will then begin to see what a slippery notion the contemporary concept of human intelligence is, and how its very vagueness has permitted the kind of ideological infill discussed in chapter 1.

This is a side of psychology of which most people are quite unaware. The general public seems impressed by the huge amounts of money being spent on the search for genes for IQ and by the publicity accompanying it. Funding bodies and even government advisers seem convinced that the measure tells us something about differences in human potential. Yet, amazingly, there is still little agreement about what is being measured and the nature of the potential being predicted.

For example, in a letter to the *Psychologist* (March 2013), the journal of the British Psychological Society, the prominent theorist Mike Anderson complained of the "disgraceful ignorance of lack of theoretical grounding for the validity of the tests." For that reason, he says that "the use of intelligence tests has been a stain on psychology's character," and that "it is tragic that too many of my most intellectually brilliant colleagues have frittered away their talents on largely trivial pursuits instead of focusing on the key question of 'What is intelligence?' "

In the following issue of the journal, John Raven—son of the designer of the famous Raven Matrices test (on which, more below)—argued that "most practical uses of 'intelligence' tests are unethical because they contribute to, and cement, an environmentally destructive hierarchical society." As mentioned earlier, Ian Deary is searching for genes for intelligence, yet tells us that there is no "grown-up" theoretical model of intelligence differences. And in chapter 1, I quoted a number of contributors to the *Cambridge Handbook of Intelligence* (2011) with similar conclusions.[1]

NOT A TRUE MEASURE

In fact, IQ is almost unique in the field of scientific inquiry for the absence of agreement about what is being measured. Defenders will of course huff and puff that this is not true and insist that they are measuring important cognitive differences in children and adults. They point to correlations with school performance, job status, and earnings. But the fact is that intelligence testing is another game of let's pretend, quite different from ordinary scientific measurement.

In scientific method, and in medicine, we do often need to take or infer differences in some observable variable as a measure of an internal function. But we only do that if we can mechanistically (causally) relate differences in one to differences in the other. For example, we can demonstrate why the height of a column of mercury (or digital readout) in a blood pressure machine will genuinely co-vary with the pressure of blood in the arteries. Likewise, we know why a white blood cell count is an index of levels of internal infection, why erythrocyte sedimentation rate is a reasonable measure of tissue inflammation, and why degree of breath alcohol corresponds fairly accurately with level of consumption.

Such measures are said to be valid because they rely on detailed, and widely accepted, theoretical models of the functions in question. In each case, differences in the measure truly correspond with differences in the internal function, and we know why. There is no such theoretical basis for human intelligence nor, therefore, for the true nature of individual differences in it. Differences in IQ cannot be reliably related to such internal functions. Instead, they are calibrated against some other differences of individuals, also assumed to reflect differences in intelligence. That is how IQ testing operates.

So how can we purport to be measuring important cognitive functions with an IQ test? And, if it is not measuring cognitive functions, what *is* it measuring? To answer those questions, we need to look back quite a long way.

IDEOLOGICAL ORIGINS

The pretense started with Sir Francis Galton, cousin of Charles Darwin, in late Victorian Britain. As mentioned in chapter 1, Galton proposed a eugenic, or selective, breeding program for the intellectual improvement of society. And that required a scientific measurement of ability "for the indication of superior strains or races." But, he wondered, how do we measure such ability when we do not know what exactly it is?

The solution he adopted has been the hallmark of intelligence testing ever since. First he devised some simple short tests that he imagined would reveal individuals' mental functioning. These included sensory discrimination, memory, perceptual judgment, reaction time, and so on. He administered them to volunteers and recorded individual differences. But how could he prove that these were really differences in intelligence? "The sets of measures should be compared with an independent estimate of the man's powers," he said. However, the only "independent measure" of cognitive ability he could think of was social reputation or class. "Is reputation a fair test of natural ability?" he asked. "It is the only one I can employ," he concluded, seemingly suspecting that this was second best.[2]

This is, of course, an audacious inversion of logic. It means that we are to take people's positions on the social scale as a true measure of natural ability, and the test score as a measure of it insofar as it reflects such positions. If test scores parallel such status, then the tests are measuring intelligence. As Earl Hunt put it in 1983, "We are presumed to know who is intelligent and to accept a test as a measure of intelligence if it identifies such persons."[3]

In effect, then, Galton's logic was simply that differences in existing social rank and esteem are manifestly ones of intelligence. Any set of tests that reproduces that rank order (or correlates with it) can also pass as a test of intelligence. It did not seem to occur to him that a person's social rank may be a consequence of unequal wealth and privilege rather than of intelligence as such.

However, the self-fulfilling logic is, and remains, the fundamental strategy of the intelligence testing movement. The gloss over test validity has plagued IQ testing and its ancillary works ever since. Of course, the quantitative nature of test scores inspired psychologists searching for scientific respectability. Unfortunately for them, they had to wait a little longer, because scores on Galton's tests did not actually correlate with social class as he had hoped. The development of the strategy into a successful test awaited the invention of different kinds of test items.

ALFRED BINET'S TEST

It needs to be stressed that the motives behind the measurement of intelligence have been predominantly practical, social motives, not theoretical ones. There were many social pressures fueling such motives in the early years of the twentieth century. Among them was the introduction, in many countries of Europe, of systems of compulsory education. These measures brought into schools enormous numbers of children who, for whatever reasons, did not learn the curriculum as quickly as others. There was natural concern about these differences. In 1904, the French minister for public instruction appointed a commission to study how such "retarded" children (as they were then called) could be identified early, so remedial action could be taken. It recommended that no child should be removed to a special school without a "medico-pedagogical" examination to determine his or her ability to profit from teaching in an ordinary school. But how was this to be done?

A member of this commission was the psychologist Alfred Binet. With his assistants, he had been studying mental development, and ways of assessing it, for over a decade. So he was charged by the commission to offer advice on such examinations. Binet's vision was not for a test of a single ability, but of a multifaceted function requiring multiple short tests or "items." In his 1962 book on the history of psychology, George Miller goes on to describe how Binet and assistant Theodore Simon "spent endless hours in the schools with the children, watching, asking, testing, recording. Each proposed test had to be given to a large number of children. If a test did not distinguish the brighter from the duller, or the older from the younger, it was abandoned. . . . The memory tests worked. And the tests of comprehension worked. . . . Binet did not retain the tests on the basis of a theory; he let their behaviour decide which tests were good and which were irrelevant."[4]

Unlike Galton, Binet's test criterion was not social rank but school performance, as judged by teachers, together with age. Again, however, there was no theory to guide test selection, only the yardstick of a collateral measure. Items were selected according to whether individuals' performance on them increased with age and corresponded with their performance in school as assessed by teachers.

On this basis, Binet and Simon produced their first "Metrical Scale of Intelligence" in 1905. It contained thirty items, designed for children aged three to twelve years, arranged in order of difficulty. The items were grouped according to performance at different ages and teacher's judgments, as just mentioned. Here are examples of some of the items:

Imitating gestures and following simple commands
Naming objects in pictures
Repeating spoken digits
Defining common words
Drawing designs from memory
Telling how objects are alike ("similarities")
Comparing two lines of unequal length
Putting three nouns or three verbs into a sentence

They also included some "abstract" (comprehension) questions, for example: (a) "When a person has offended you, and comes to offer his apologies, what should you do?" or (b) Defining abstract words (by describing the difference between such words as "boredom" and "weariness," "esteem" and "friendship").

The tester progressed through the items with each child until the latter could do no more. Performance was then compared with the average for the age group to which the child belonged. If a child could pass half the tests expected of a six year old, say, then the child was said to have a mental age of six. Binet used the difference between the mental age and the chronological age as an index of retardation. He considered two years to be a serious deficiency.

This was how the first modern intelligence test was created. Within a few years, translations were appearing in many parts of the world. In 1912, William Stern proposed the use of the ratio of mental age to chronological age to yield the now familiar intelligence quotient or IQ:

$$IQ = \frac{\text{mental age}}{\text{biological age}} \times 100$$

And so was born the IQ test.

RETARDATION IN WHAT?

From a purely practical viewpoint, Binet's scale appeared to be a brilliant success. It was easily and quickly administered, and it actually identified the children it was supposed to identify. It is doubtful whether it did this any better than, say, teachers could have done on the basis of their experience with the children (after all, this was the only criterion, next to age discrimination, of the acceptability of test items). But it seemed more systematic and afforded immediate comparability across children of the same or different ages, so that degree of retardation seemed to be indicated.

But what it measured retardation *in* is quite a different point. Note that the test result provides no new psychological information; nothing that was not already known about a child. How could it? As George Miller put it, Binet "was not over-concerned with scientific purity; he had a practical problem he urgently wanted to solve, and he did whatever seemed necessary to solve it."[5] And Binet himself insisted that the test merely "classified" children rather than measured some fundamental potential.

THE SUBVERSION OF BINET'S TEST

Binet certainly scoffed at any claims to be measuring innate intelligence. But others were soon to claim otherwise. Scores correlated with school performance, as the tests were selected to do. In addition, though—and unlike Galton's efforts—they were also associated with social class and "racial" background. This is hardly surprising. In so describing children's preparedness for school we are reflecting their social and family background, and, in a sense, the whole social structure of society. But the Anglo-American followers of Galton had other ideas about what the score differences signified: the intelligence and the supposed "superior strains and races" that Galton had hoped to scientifically identify.

The most rapid developments took place in the United States, where so-called feeble-mindedness, especially among the new wave of immigrants in the early 1900s, was seen to be a pressing problem. It had serious implications for education and national social security. Part of the solution was conceived by Henry H. Goddard when he translated Binet's test into English in 1910. Goddard was director of the Vineland Training School for Feebleminded Boys and Girls, in New Jersey. He was convinced that ability and potential were biologically determined and that the test would measure that potential and so suggest the disposal of testees.

In a ghastly tragicomedy, Goddard and associates targeted the waves of immigrants pouring into the United States at that time. Amid distressing scenes at the infamous reception center on Ellis Island, he managed to ensure that all immigrants—men, women, and children of all ages—were given the IQ test as soon as they landed. That was after long and trying journeys, using the tests in English through interpreters. By these means, the country came to be told that 83 percent of Jews, 80 percent of Hungarians, 79 percent of Italians, and 87 percent of Russians were feebleminded. Almost as bad were the Irish, Italians, and Poles and, bottom of the list, the blacks. Only the Scandinavians and Anglo-Saxons escaped such extremes of labeling.

There seems little doubt that such findings contributed to the xenophobia that swept America in the decade of the 1920s and led to the 1924 Immigration Act that introduced an immigration quota system. Moreover, the amount of feeble-mindedness thus enumerated soon had these psychologists pressing ardently for eugenic measures, which eventually became law. Tens of thousands of surgical sterilizations followed. As many historians of science have pointed out, the subsequent growth of the use of IQ testing was that of a blatantly racist tool. Binet was later to protest about the "brutal pessimism" in this subversion of his test, as if the results indicated some fixed quantity of the individual.

Writing in popular journals and magazines, Anglo-American psychologists thus came to present the IQ test as a test of the actual genetic worth of people. Then, as now, enormous benefits to individuals and society were foreseen. In 1916, Lewis Terman, the author of the most famous IQ test, claimed that testing "will ultimately result in curtailing the production of feeblemindedness and in the elimination of an enormous amount of crime, pauperism, and industrial inefficiency." And as a useful side-effect, he said we could also "preserve our state for a class of people worthy to posses it."[6]

So much from a few simple questions and puzzles. These, of course, were chosen because they reflected learning that was more typical in particular social classes. But adding up the results into scores made differences look scientific and biological, as well as socially just and fair. It was the scientific credibility badly needed by a new discipline like psychology. After conversion to pencil-and-paper tests for administration to groups rather than to individuals, the IQ test was soon ready for mass application in school, job selection, and in the army.

THE GROWTH OF IQ TESTING

Armed with such biological "truths," followers of Galton inspired influential campaigns, and Eugenics Societies sprang up on both sides of the Atlantic in the early 1900s. Observing the higher birth rates among the working class, its members were soon warning about the degeneration of the "race" in popular seminars, magazines, and the pages of national newspapers. Recognition of the innateness of intelligence was also good for the individual, they argued. In 1919, famous statistician Karl Pearson—who considered himself a socialist as well as a eugenicist—explained in the *Encyclopaedia Britannica* that "it is cruel to the individual, it serves no social purpose, to drag a man of only moderate intellectual power from the hand-working to the brain-working group."[7]

Under the persuasive touch of Cyril Burt, the logic eventually penetrated educational policy as I described earlier. Abroad, of course—especially in Nazi Germany—essentially the same assumptions were affecting lives in more tragic ways.

CONSTRUCTING THE TESTS TO REVEAL
DIFFERENCES PRESUMED

Terman's test (the Stanford-Binet, as he called it in 1916) subsequently became the model for other tests, and is still, after many revisions, one of the most-used individual tests on both sides of the Atlantic. But it was (and is) constructed using basically the same method that Binet had devised. Terman introduced many more items (ninety in all). And a greater variety of items was used. Some are school-type general knowledge tests ("Can you tell me, who was Genghis Khan?" "What is the boiling point of water?" and so on). But others include memory span for spoken digits, vocabulary, word definition, general knowledge, comprehension, and the like.

Other tests eventually followed and are still being devised. However, although procedures of test construction have been refined, the fundamentals have remained the same. First, large numbers of candidate items are devised by test constructors. These are tried out on samples of people of appropriate ages. The results then dictate which of the items should be selected or discarded to produce the desired patterns of scores.

For example, it was assumed (and still is) that population scores should conform to a normal distribution, as described in chapter 2. It is mimicked in the IQ test by including relatively more items on which an average number of the trial group passed and relatively fewer on which either a substantial majority or minority of them passed. The result gives the test a quasi-biological authenticity.

As also mentioned in chapter 2, few natural (biological) functions conform to such a simple distribution. Yet the possibility is virtually completely ignored in the IQ literature. So widespread is this view, in fact, that nearly all the statistical analyses surrounding IQ testing (including the so-called genetic analyses) have been designed to depend upon it. The falsity of the assumption must therefore raise serious questions about those analyses.

Another assumption adopted in the construction of IQ tests is that, as a supposed biological measure like physical strength, it will steadily increase with age, leveling off after puberty. This property has been duly built into the tests by selecting items on which an increasing proportion of subjects in each age group passes.

Of course, there are many reasons real intelligence—however we define it—may not develop like that. In the case of the IQ test, however, the assumption has produced the undesired, and unrealistic, effect in which scores improve steadily up to the age of around eighteen years and then start to decline.

Again, this is all a matter of prior item selection. The effect is easily reversed by adding items on which older people perform better, and reducing those on which younger people perform better. But large-scale genetic studies on IQ and ageing are conducted on the assumption that the score patterns are real intelligence and not artifacts of test construction and people's experience.

In this way, too, specific patterns of group differences can be engineered in or out of test scores according to prior assumptions. For example, boys tend to do better on some items, girls better on others, whereas overall equality was deemed desirable. It was concluded that any differences were due to "experience and training" rather than "native ability." Accordingly, as Terman and Merrill put it in the 1937 revision of the Stanford-Binet test, items which yielded large sex differences were duly eliminated as probably unfair.

It is in this context that we need to assess claims about social class and race differences in IQ. These could be exaggerated, reduced, or eliminated in exactly the same way. That they are allowed to persist is a matter of social prejudice, not scientific fact. In all these ways, then, we find that the IQ testing movement is not merely describing properties of people—it has largely *created* them.

Some defenders may protest that, with advances in technology and statistical sophistication, IQ tests are far more scientific, and items more objective, than they used to be. Yet they are still based on the basic assumption of knowing in advance who is or is not intelligent and making up and selecting items accordingly. Items are invented by test designers themselves or sent out to other psychologists, educators, or other "experts" to come up with ideas. As described above, the initial batches are then refined using some intuitive guidelines.

A PSEUDO-THEORY OF INTELLIGENCE

Galton had a hunch that intelligence is an all-pervasive natural ability, varying like any physical trait, such as physical strength, but making a difference in everything we do mentally. Beyond that, he did not have much of a clue about its properties. Others have tried to get over this problem statistically, rather than describing actual functions. In 1904, British psychologist Charles Spearman noted that schoolchildren who did well or poorly on exams in one school subject also tended to do well or poorly on others. In other words, the scores were correlated.

TABLE 3.1 Spearman's reported correlations among scholastic
and sensory measures

	1	2	3	4	5
Classics					
French	.83				
English	.78	.67			
Math	.70	.67	.64		
Pitch	.66	.65	.54	.45	
Music	.63	.57	.51	.51	.40

Table 3.1 shows some of Spearman's early results, taken from twenty-two boys in a preparatory school. He assumed that school performance is itself a measure of intelligence and reasoned that performances that correlate significantly like this are probably expressions of the single underlying factor that Galton had proposed. As mentioned in chapter 1, a correlation is really a measure of how two entities co-vary, or vary together, not of its underlying causes. However, Spearman leapt to the conclusion that "there really exists a General Intelligence." He claimed that this was in the nature of an innate energy or power variable, raising or lowering the ability to learn across all subjects, and he called it "g." To the present day psychologists have hailed this finding as perhaps the biggest single discovery in psychology.[8]

That same reasoning—that correlated scores reflect a common ability—is just as popular today, even though the causes of the correlations can be due to many other factors or to none at all. (For example, there is a correlation between the price of coffee and the distance of Haley's comet from the earth). The mistake is in assuming that the correlations reveal the hidden entity generating the observations, a habit of thought referred to as "reification." So g has become another shadowy apparition widely open to ideological infill.

Nevertheless, the habit has been impossible to resist, and further "structures" of intelligence have been inferred from intercorrelations between item subscores. So we get verbal intelligence; mathematical intelligence, and so on. The fallacy is that of assuming that such correlations are describing the real structure of intelligence, when, of course, they merely reflect overlaps in score patterns, themselves derived from the contents of the tests.

What is perhaps most remarkable is that investigators are surprised that performances on such tests are correlated at all. The items are, after all, devised by test designers from a narrow social class and culture; these items are based on intuitions about intelligence and variation in it, and on a technology of item selection that builds in the required degree of convergence of performance. In addition, the items are chosen using fairly uniform criteria.

THE SLIPPERY VALIDITY OF IQ TESTS

We cannot say that IQ tests are valid scientific measures of intelligence. The tests are supposed to measure a kind of cognitive power or strength (g). But we cannot demonstrate that it is differences in such a function that cause differences in test performances, because there is no agreed-on description or model of it.

In consequence, all claims about the validity of IQ tests have been based on the assumption that other criteria, such as social rank or educational or occupational achievement, are also, in effect, measures of intelligence. So tests have been constructed to replicate such ranks, as we have seen. Unfortunately, the logic is then reversed to declare that IQ tests must be measures of intelligence, because they predict school achievement or future occupational level. This is not proper scientific validation so much as a self-fulfilling ordinance.

For example, Robert Plomin and colleagues have justified hunting for genes for IQ because, they say, individual differences in intelligence are strongly associated with many important life outcomes, including educational and occupational attainment. Likewise, Gail Davies, Ian Deary, and colleagues (see chapter 1) say that "individual differences in intelligence are strongly associated with many important life outcomes, including educational and occupational attainments, income, health and lifespan." A special group set up by the American Psychological Association in 2012 to examine IQ, states that "the measurement of intelligence—which has been done primarily by IQ tests—has utilitarian value because it is a reasonably good predictor of grades at school, performance at work, and many other aspects of success in life."[9]

How accurate and meaningful are such associations and correlations?

SELF-FULFILLED VALIDITY

It is undoubtedly true that IQ test scores predict school achievement moderately well, with correlations of around 0.5. The problem lies in the self-fulfillment of this prediction. Some test items are devised to contain knowledge that has been learned in school: "In what continent is Egypt?" "Who wrote Hamlet?" "What is the boiling point of water?" and so on. Or they use a lot of textlike rules and analogies. So it should come as no surprise that performance on them is associated with school performance. As Robert L. Thorndike and Elizabeth P. Hagen explained in their leading textbook, *Educational and Psychological Measurement*, "From the very way in which the tests were assembled [such correlation] could hardly be otherwise."[10]

Rather than independent measures confirming an underlying cognitive power or capacity, IQ tests and school tests are simply different measures of the same learning. This is why correlations between IQ and achievement tests tend to increase with age. This is why parental drive and encouragement with their children's school learning itself improves the children's IQ (as the American Psychological Association group reports).[11]

And that raises another question. If it is genuine predictability we need, why not just rely on teachers' judgments? It has long been demonstrated that teachers can predict with far greater accuracy, and in a fraction of the time, the future achievements of their pupils. One review suggested a correlation of around 0.66, which far surpasses the predictive power of an IQ test.[12] Yet how much more seductive is the mystique of a scientific-looking test of "general intelligence."

Similar doubts arise about the use of occupational level, salary, and so on, as validatory criteria. School achievement largely determines level of entry to the job market. The frequently reported correlation of 0.5

between IQ and occupational level (and therefore, income) is also self-fulfilling. Again, the measures are not independent.

The really critical issue, therefore, surrounds the question of whether IQ scores predict individual differences in the more independent measure of job performance. So let us take a much closer look at that issue.

IQ AND JOB PERFORMANCE

The idea that IQ tests are valid because they also predict job performance is remarkably widespread and durable. Adrian Furnham probably reflects most views when he claims that "there is a large and compelling literature showing that intelligence is a good predictor of both job performance and training proficiency at work." Fritz Drasgow describes the correlation as incontrovertible. John Hunter and Frank Schmidt, in the 1980s, even attached a dollar value to it when they claimed that the U.S. economy would save $80 billion per year if job selection were to be universally based on IQ testing.[13]

The problem is that the "facts" reported turn out to be rather questionable. A large number of studies prior to the 1970s reported that direct correlations between IQ and job performance were very low (around 0.2–0.3). Obviously, this result was very disappointing to IQ testers. Then two statisticians, Frank Schmidt and John Hunter, considered the possibility that those many results were error prone. The errors, they said, arise from three main sources. The small correlations could be due to small samples, just as opinion polls with small samples can bias results. Then the rough-and-ready measures often used could be so unreliable as to give different scores on different occasions, which weakens the true correlations. Finally, the small samples might have yielded reduced ranges of scores (of IQ or job performance or both). Such restriction of range can also reduce correlations.

These arguments are quite reasonable, so Hunter and Schmidt attempted to correct the original correlations using statistical methods and pooling the results (using meta-analysis, as described in chapter 2). This doubled the correlations to around 0.5–0.6. Nearly all studies cited in favor of IQ validity are either drawn from the Schmidt and Hunter meta-analyses or from others using the correction methods developed by them.

The approach has been much admired in some quarters; indeed, it is inventive. However, it has also been controversial and heavily criticized. This is because of the large number of assumptions that enter into the corrections and the poor quality of the original data and reports. Let us have a quick look at some of these, just to indicate how the area has become very murky indeed.

MAKE-DO TESTING (AGAIN)

A major problem with the Schmidt and Hunter validation studies is the haphazard diversity of tests used in the original studies. They have included memory tests, reading tests, scholastic aptitude tests (SATs), university admission tests, specialized employment selection tests, and a vast range of specialist armed forces tests: almost anything, in fact, that yields numbers. Just calling these "general ability tests," as Schmidt and Hunter do, is like reducing a diversity of serum counts to a "general blood test." Moreover, the studies incorporated into the most-cited meta-analyses cover a vast range of dates, some from the 1920s, with the majority of them from before the 1970s.

The point is, of course, that such diverse tests are bound to tap different abilities with various sources of individual differences and with varying distributions. Even todays' attainment tests show little intercorrelation. For example, as reported by the College Board in 2008, the correlation between SAT scores and high school grade point average is only 0.28. And the correlation of either with IQ tests is only around 0.2. Rolling these up together as if they were expressions of the same "general ability" seems little more than guesswork. As Kevin Murphy, himself a meta-analytic expert, reported in a major review, it creates "lack of clarity in what population parameter is being estimated."[14]

JOB PERFORMANCE?

In contrast to the vast diversity of "intelligence" tests, only one measure of job performance has been used in the majority of studies: supervisors' ratings. It turns out that there are a host of problems with such ratings.

The main problem is that it is very difficult for supervisors to agree on what constitutes good or bad performance. In a 1991 article, Linda Gottfredson noted, "One need only ask a group of workers in the same job to suggest specific criterion measures for that job in order to appreciate how difficult it is to reach consensus about what constitutes good performance and how it can be measured fairly."[15]

Consequently, assessments of job performance tend to be subjective, based on inconsistent criteria. They are notoriously biased: age effects and halo effects have been reported; height and facial attractiveness have been shown to sway judgments, and unconscious ethnic bias is also present. Just how poor supervisors' ratings are is revealed by their weak agreement with more objective criteria such as work samples or work output. Correlations near zero are reported in a number of studies.

Another problem is that much of the so-called correction to correlations is based on corrections for measurement error—the way the measure of a function can vary from occasion to occasion (see more on this below). But in the case of job performance, what is called "measurement error" may not be error at all, but normal fluctuation.

After all, no one performs at peak all the time: we all exhibit a difference between maximum and typical performances. In fact individual variation tends to be greater than that across different individuals. Yet correcting these for assumed measurement error boosts the correlations enormously. In describing the difficulties, in his own experience, of seeking objective supervisor ratings across a wide range of jobs, Robert Guion suggested that we should abandon the pretence of objective, true, or hard criteria of job performance.[16]

In other words, far from validating the IQ test and all the other claims that go with it, measures of job performance are themselves unreliable and inaccurate. But let us turn to the corrections themselves, because there are important lessons to be learned from them about the whole exercise.

DUBIOUS CORRECTIONS

Much of the problem of making these corrections is that they require crucial data from the original studies. But those data are often missing. In their report, on which most of this would-be validity is based, Hunter and Schmidt acknowledged that crucial bits of information like test reliability and range restriction, were only sporadically available. Instead, these crucial gaps were filled in by generalizing from those scores actually available. Such guesstimates have obvious dangers.

These problems of missing data are hardly surprising, given that most of the studies in question were done between 1920 and 1970. In 1989, a committee set up by the U.S. National Academy of Sciences commissioned new meta-analyses on more recent studies than those of Schmidt and Hunter. These analyses found much lower correlations than those reported above. As report authors John Hartigan and Alexandra Wignor (themselves leading statisticians) noted in their report, "The most striking finding . . . is a distinct diminution of validities [i.e., IQ-job performance correlations] in the newer, post 1972 set." The corrected correlations came out to be around 0.25, rather than the widely cited 0.50 from the corrected correlations of Schmidt and Hunter.[17]

The committee described the differences as "puzzling and obviously somewhat worrisome." But they noted how the quality of data might explain it. For example, the 264 newer studies have much greater numbers of participants, on average (146 versus 75). It was shown how the larger samples produced much lower sampling error and less range restriction, also requiring less correction (with much less possibility of a false boost to observed correlations). And there was no need to devise estimates to cover for missing data. So, even by 1989, these more recent results are indicative of the unreliability of those usually cited. But it is the earlier test results that are still being cited by IQ testers.

Finally, it seems that even the weak IQ-job performance correlations usually reported in the United States and Europe are not universal. For example, in a study reported in 2010, Eliza Byington and Will Felps found that IQ correlations with job performance are "substantially weaker" in other parts of the world. They include China and the Middle East, where performances in school and work are more attributed to motivation and effort than to innate cognitive ability.[18]

These reflections on correction methods all add to the impression of a large amount of guesswork involved in arriving at corrected correlations between IQ and job performance as well as overzealous claims about test validity.

Note that similar claims have been made about correlations between IQ and training success in various occupations. But they are subject to the same objections as those for job performance: the raw correlations are very low (around 0.2), which are doubled or more in the meta-analyses through estimated corrections. The most quoted results are from training military personnel, while all meta-analyses include dozens of different tests of varying psychometric standards and many very old studies, some dating as far back as the 1920s.

In contrast to such questionable validity studies are a variety of others of real cognition in real working situations. They have shown that IQ-type test scores have little if any correlation with performances on the kinds of tasks individuals regularly encounter in their jobs. Robert Sternberg and colleagues found such null results in studies of managers, salespersons, and university professors. There are many such reports. If there is one area where one would expect a relation between IQ and job performance, it would surely be medicine. But a recent study reported in the journal *BMC Medicine* showed no predictive effect of IQ on clinical performance (e.g., promotion to senior doctor).[19]

This could also explain why members of high-IQ societies like Mensa are not overrepresented in the most demanding jobs. As one member put it (*Guardian Weekend*, April 25, 2015), "I'm almost certainly smarter than you. . . . It's a fact . . . I know what my IQ is (164)." But, she goes on, "I've never held a high-powered job. I don't have a string of qualifications. I don't do terribly clever things . . . I've stayed gainfully employed, but I wouldn't say I've done anything remarkable in any job." And then she concludes, "Mensa . . . is an organisation for the smart-ass rather than the wise!"

So the question of what IQ tests really test remains hotly debated. Let us examine a few alternative possibilities.

Differences in Basic Neural Processes?

As mentioned earlier, Galton viewed human mental ability differences in terms of mental speed. He included reaction time (RT) items in his tests, as did some of his followers. Although they yielded nothing of interest (people from different social classes performed much the same on average), the idea has been recently revived. This time the aim has been to see whether RT correlates with IQ, thus supporting the conclusion that IQ is really a measure of some physiological or neural efficiency. For example, in the 1990s, Arthur Jensen (a well-known supporter of Galton) spoke of "individual differences in speed or efficiency of the various elementary processes," and how "those differences account for the differences in performance on psychometric tests."[20] The hope engendered was that of discovering the mother lode in a psychological gold mine.

Initially, some excitement was created in two ways. First a small correlation (0.2–0.3) was found between IQ and a modified RT test: the appropriate reaction had to be quickly chosen from up to four alternatives (e.g., different buttons for different light signals). Variability of individual performance in this so-called choice-reaction time was also weakly correlated with IQ.

The problem is what to make of it. As always, we must not accept such correlations as causally meaningful without controlled experiments. Such small correlations indicate, anyway, that there are a lot of other things causing differences in performance. And they may not even be cognitive in origin. Douglas Detterman showed how RT involves a lot other than simple response efficiency.[21] Individual differences can stem from misunderstanding instructions, familiarity with the equipment, motivation to do the task, sensory acuity, learned response strategies, time spent on sensory processing and motor action rather than decision time, attention, arousal, task orientation, confidence, and anxiety. Such research appears to be up another cul de sac. But, like frustrated yet hopeful prospectors, IQ devotees keep returning to RT.

Ability for Complex Cognition?

Of course, the favorite claim is that IQ somehow reflects individual differences in the ability for complex cognition (reasoning, thinking, problem solving, etc.). Linda Gottfredson, in an article in 2007, claimed that "the active ingredient in tests of intelligence is the complexity of their items . . . that makes some more difficult than others (more abstract, more distracting information, require inferences, etc.)." She provides examples of such complexity, as seen in test items. For example, completing a 3×3 matrix item like those in the Raven test (see figure 3.1) requires more complex cognition than a 2×2 matrix, even if we are not sure how to describe it. Reproducing a nine-block pattern from one shown in a picture is more complex than reproducing a four-block pattern. Describing the similarity of the words "seed—egg" demands more complex cognition than doing so with "pear—apple."[22]

The trouble with this argument is that IQ test items are remarkably simple in their cognitive demands compared with, say, the cognitive demands of ordinary social life and other activities that the vast majority of children and adults can meet adequately every day.

For example, many tests items demand little more than rote reproduction of factual knowledge most likely acquired from experience at home or by being taught in school. Opportunities and pressures for acquiring such valued pieces of information, from books in the home to parents' interests and educational level, are more likely to be found in middle-class than in working-class homes. So the causes of differences could be differences in opportunities for such learning.

The same thing can be said about other frequently used items, such as "vocabulary" (or word definitions); "similarities" (describing how two things are the same); "comprehension" (explaining common phenomena, such as why doctors need more training). This also helps explain why differences in home background correlate so highly with school performance—a common finding. In effect, such items could simply reflect the specific learning demanded by the items, rather than a more general cognitive strength.

IQ testers might protest, though, that there is more than meets the eye in such simple-looking test items. For example, Linda Gottfredson has argued that even a simple vocabulary test is one of "a highly general capacity for comprehending and manipulating information . . . a process of distinguishing and generalizing concepts."[23] However, that argument must surely apply to all word definitions, not just those devised by an item designer from a specific culture. Remember that items are not selected on the basis of a theoretical model of cognitive complexity, but on how well they produce the desired pattern of scores.

As an example, take the Scots word "canny" (meaning shrewd). Its definition is subtly different from the same word used in northeastern England (meaning amiable, OK). In these different contexts, children are assimilating equally complex conceptual meanings, yet as an item in an IQ test, some would be deemed correct and others wrong. Again the true source of differences is really one of specific learning rather than general cognitive strength.

Of course, IQ test items can look quite plausible: they have, after all, been concocted in the minds of experts precisely because they look plausible, not out of a theoretical model of cognitive functions. Particularly popular, and widely respected, are analogical reasoning items like "DIPLOMAT is to TACT as VIRTUOSO is to . . . ?" (and the word "SKILL" has to be selected from four or five alternatives presented).

Does failure on such items really signify little analogical reasoning ability in general (or low *g*)? In a number of reports since the early 1990s, Usha Goswami argued that failure on these items could simply arise from lack of experience with the specific relations. The solution is, she suggests, to design items based on relations equally familiar to all group(s) being tested. Then we will know that they are truly being tested for complexity of cognition and not culturally related experience. When this is done, indeed, it transpires that even very young children are capable of "complex" analogical reasoning.[24]

The difference is that those items in IQ tests have been selected because they help produce the expected pattern of scores. A mere assertion of complexity about IQ test items is not good enough.

CULTURE-FREE (LEARNING-FREE) TESTS?

One attempted solution to these problems with verbal items has been to devise nonverbal test items that are assumed to be free of prior learning. To that end, the much-used Wechsler test includes five performance subtests. The Stanford-Binet does likewise. But the best example of such a test is deemed to be the Raven's Matrices (or just the "Raven").

The IQ defender Arthur Jensen described the Raven as a test of "pure *g*," dependent only on powers of induction from the information presented, and having nothing to do with prior learning. The example in figure 3.1 illustrates how a correlation or rule has to be induced across the rows but is conditioned by other rules down the rows. The combination

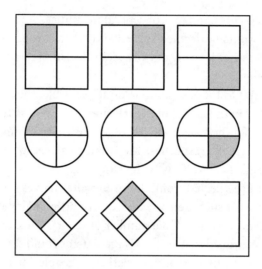

FIGURE 3.1

A 3×3 matrix typical of Raven test items (the matrix is completed by replacing the blank space with an item to be chosen from five given alternatives).

of rules can then be used to identity the missing element from the given options. This is sometimes called *conditional reasoning.* The nature of the rules varies from item to item. And the complexity of the items varies by, for example, having more elements changing with greater depths of conditioning.

Most psychologists seem to be utterly convinced of the validity of the Raven, supposing it to be invulnerable to criticism. Surprisingly, though, as with other IQ test items, there has actually been little in-depth analysis of the cognitive demands of the Raven. The rules—such as adding or subtracting elements along a row—describe the arrangement of elements in the items and not actual cognitive processes.

Indeed, what analysis there has been suggests that the Raven is surprisingly *un*-complex in its cognitive demands. The items may be unfamiliar in appearance, but the cognitive demands turn out to be simple compared with the complex cognition children and adults use in solving their everyday problems in social contexts. Just think of cooking, childcare, driving on a busy road, team work at work or during play, and so on. In perhaps the most thorough analysis to date, Patricia Carpenter and colleagues found little evidence that level of abstraction, as in the complexity of rules, influences item difficulty.[25]

Carpenter and others have suggested that it is differences in working memory that count. This is the sort of memory needed for holding in mind information, such as one particular rule, while more rules are taken in and then combined to solve the problem. Of course, anyone who does regular multitasking (e.g., parents of young children) will be familiar with that kind of problem and be rather adept at it. But attempts to demonstrate that it is the basis of differences in test performance have been inconclusive. Besides, as the American Psychological Association review group concluded, the definition of working memory itself remains vague and disputed (see further in chapter 7), and relations with IQ are still uncertain.[26] So, whatever it is that makes Ravens matrices more difficult for some people compared with others, it does not seem to be item complexity per se. So what is it?

COMPLEX COGNITION IN REAL LIFE

It is easy to look at the puzzles that make up IQ tests and be convinced that they really do test brain power. But then we ignore the brain power that nearly everyone displays in their everyday lives. Some psychologists have noticed how people who stumble over formal tests of cognition can handle highly complex problems in their real lives all the time. As Michael Eysenck put it in his well-known book *Psychology*, "There is an apparent contradiction between our ability to deal effectively with our everyday environment, and our failure to perform well on many laboratory reasoning tasks."[27] We can say the same about IQ tests.

Indeed, abundant cognitive research suggests that everyday, real-life problem solving, carried out by the vast majority of people, especially in social cooperative situations, is a great deal more complex than that required by IQ test items. The environments experienced by all living things, indeed, are highly changeable and complex in a more dynamic sense than static test items. Real-life problems combine many more variables that change over time and interact. It seems that the ability to do pretend problems in pencil-and-paper (or computer) formats, like IQ test items, is itself a learned, if not-so-complex, skill.

John Raven wanted his matrices to test the "eduction of relations and correlates," which Charles Spearman had claimed to be the essence of cognition. But research has shown how such cognition is exhibited in quite simple brains, like those of honeybees and fruit flies. Crows have been shown to be quite good at analogical reasoning. In human social life, complex cognition is demanded by the constant novelty of rapidly changing situations.[28]

This is why highly complex cognition is intrinsic to everyday human social life, and it develops rapidly in children from birth. Research by Alison Gopnik and others has shown, indeed, that complex cognition can be exhibited by infants. Young children are able to draw inferences from higher-order relations and use them to guide their own subsequent actions and bring about a novel outcome.[29]

More direct studies with adults have shown that IQ scores are unrelated to the ability to carry out complex problem solving in social and practical situations. Stephen Ceci and Jeffrey Liker did an entertaining study of betting at a racecourse. The paper was rather frivolously titled "A Day at the Races." But they found the gamblers' predictions of odds to be far from frivolous. Indeed, such predictions involved highly sophisticated cognitive processes requiring integration of values of up to eleven variables, such as race distance, course conditions, weight carried, age of the horse, number of previous wins, and so on. Unlike the spherical horse problem mentioned in chapter 2, these variables were not just added or subtracted from each other as simple sums. Instead, predictions took account of interactions among the variables and nonlinear effects. For example, the effects of weight would increase with distance, but more so for wet compared with dry conditions. Again, this is variation being created by interactions among variables—the kind of deep interaction mentioned in chapter 2. The researchers found individuals' accuracy at such predictions to be unrelated to IQ test scores.[30]

In everyday contexts like the workplace, such cognitive inventiveness is almost always unrecognized by psychologists. In one study, Sylvia Scribner recorded the quantitative cognitive strategies of dairy workers in their daily work: taking and organizing orders, procuring items from stock, inventorying, pricing of deliveries, and so on. As with most established workplaces, the dairy had a set of evolved regulations, shared concepts, work procedures, and instruments. But unlike programmed robots, workers exhibited a vast adaptability and variability in their problem solving. These unsung cognitions improved the speed and efficiency of the business in meeting the wide range of specific orders and delivery targets.[31]

Anyone with experience in workplaces will recognize such creative cognition as the norm. Indeed, most people could probably come up with examples from other contexts, from managing public services to managing a group of children, and their complex ever-changing needs, in the home or classroom.

I recommend quietly watching two or three children playing around a sand and water table in a play group. Dwell on the complexity of interchange and communication and what kinds of brain activity are called for. True cooperation like that demands a huge leap in the complexity of cognition. In even simple tasks like jointly moving an obstacle, each individual must operate while taking into consideration the viewpoints, thoughts, and feelings of others. They must anticipate others' actions and organize their own thoughts and actions to coordinate with them, all rapidly changing in seconds or even milliseconds. Yet the same children— all good enough for the highly complex cognition and learning in most social life—may well have starkly different IQs.

THE FLYNN EFFECT

Another strange revelation flies in the face of the idea that IQ measures fixed potential or complexity of cognition: the steady rise in average IQs over generations. James Flynn, one of the first to report this phenomenon, reminds us how average scores (maximum score 60) on the Ravens test in Britain, for example, gained 27.5 points between 1947 and 2002. The same applies to other popular tests, such as the Wechsler and the Stanford-Binet, in every country where records exist.

The phenomenon remains puzzling to IQ testers, because it could not be due to genetic or other biological changes. That would be equivalent, say, to a population-wide gain in physiological function such as metabolic efficiency of 50 percent over two generations. And as Flynn noted, "IQ gains have not been accompanied by an escalation of real world cognitive skills . . . an evolution from widespread retardation to normalcy, or from normalcy to widespread giftedness."[32]

In other words, intelligence test performance has increased, but intelligence has not. This paradox may be resolved by quite a different view of intelligence. Let us look at some possibilities.

LEARNED COGNITIVE STRUCTURES

Human cognitive functions are not ones that develop from the inside out, as with other skills, like walking or visual acuity. This is a point I will elaborate on a great deal in chapter 9. Instead, they are socially evolved tools of thought that, in individuals, are acquired from the outside in. This is obvious with functions like language, but it is true for our cognitive functions, too. In this perspective, what IQ tests actually assess is not some universal scale of cognitive strength but the presence of skills and knowledge structures more likely to be acquired in some groups than in others.

This point applies to both verbal and nonverbal test items. Since the Raven is widely thought of as a test of pure *g*, I will use it as an example. The test involves deducing rules from symbols in two-dimensional arrays on paper (see figure 3.1). Psychologists have always claimed that these are entirely abstract—meaning that the problems are not dependent in any way on previous experience, but only on inner cognitive strength. But it is not difficult to show that the specific rules are more prevalent in some subcultures than in others (e.g., middle-class compared with working-class families). They are more or less likely to be acquired by children in those families in a way that has nothing to do with their mental potential.

For example, tests like the Raven nearly all require the reading of elements in the matrix from top-left to bottom-right, just as in (Western) text. But they also embed further rules that, far from being experience free, mimic those in record sheets, spreadsheets, timetables, or other tables of rows and columns, with totals and subtotals. The rules involve additions, subtractions, and substitutions of symbols across columns and down rows, and the deduction of new information from them.

These arrangements and rules are far more likely to be part of the culture and mindset in families whose parents have white-collar occupations than in others. Being able to handle them involves acquired mental skills, just as the use of a physical tool develops certain patterns of motion in limbs and muscles. Or, as Richard Nisbett put it in *The Geography of Thought*, "Differences in thought stem from differences in social practices."[33]

It is well known that families and subcultures vary in their exposure to, and usage of, the tools of literacy, numeracy, and associated ways of thinking. Children will vary in these because of accidents of background. Indeed, it is suggested from brain imaging and other studies that background experience with specific cultural tools like literacy and numeracy is reflected in changes in brain networks (see chapters 7 and 9). This explains the importance of social class context to cognitive demands, but it says nothing about individual potential.

In other words, items like those in the Raven contain hidden structure which makes them more, not less, culturally steeped than any other kind of intelligence testing item. This cultural specificity is hardly surprising, because as already mentioned, the items are the products of the cognitions of human beings who themselves have been immersed in a specific cultural milieu.

It has been a great mistake to classify verbal items as knowledge based, and nonverbal items, like the Raven, as somehow not knowledge based, when all are clearly learning dependent. Ironically, such cultural-dependence is sometimes tacitly admitted by test users. For example, when testing children in Kuwait on the Raven in 2006, Ahmed Abdel-Khalek and John Raven transposed the items "to read from right to left following the custom of Arabic writing."[34]

Again, this suggests that IQ tests are not measures of some mythical cognitive strength. They simply reflect the different kinds of learning by children from different (sub)cultures: in other words, a measure of learning, not learning ability, and are merely a redescription of the class structure of society, not its causes. This is further confirmed in particularly striking ways. When children are adopted from lower-class into middle- or upper-class homes, they rapidly gain as much as 12–18 IQ points.

There is another irony in the "cognition from experience" view. Many psychologists, including Linda Gottfredson and Frank Schmidt, argue that intelligence, or g, is basically learning ability. But it will always be quite impossible to measure such ability with an instrument that depends on learning in one particular culture. We may call this the g paradox, or a general measurement paradox. It must haunt some IQ testers.

IQ TESTS ALSO TEST NONCOGNITIVE FACTORS

It is likely that other causes of differences in IQ test performance are not really ones of cognitive ability at all. As mentioned above, the concept of g or general intelligence arose from Galton's intuitions and from Spearman's observation that pupils who do well on one school test tend to do well on others. This has been confirmed with IQ test performances many times. However, as Nicholas Mackintosh explained in his book, *IQ and Human Intelligence*, "We have little idea of the reason(s) for this multiple association."

After all, surveys consistently show that the biggest influence on school performance is parental support. For many reasons (see below), parents vary enormously in many aspects of relations with their children. These include their interest in and support for development, their educational and occupational aspirations, involvement in play, provision of opportunities for learning, and their familiarity with the appropriate cultural tools. Not surprisingly, these aspects are highly correlated with IQ (0.6–0.7). It seems obvious that a child who is being motivated or pushed by parents will tend to put in above-average effort in all school subjects, and conversely for those who are poorly motivated. This factor alone could explain intercorrelations across subjects. Why invent a mysterious mental factor that no one can prove the existence of?

Other noncognitive factors can also explain IQ score differences. David Wechsler acknowledged the role of personality factors like competitiveness or compliance with authority in performance on his IQ test. But levels of self-confidence, stress, motivation and anxiety, as well as general physical and mental vigor, all affect cognitive test performances. These factors tend to affect all tests, so the scores will intercorrelate.

Just being made to feel inferior in a class structure—feeling that others might think you are inferior, or that a test is about to expose you as inferior—has devastating effects on performance. This is known as stereotype threat and is now known to affect lower social class and ethnic groups, women, older people, and other groups. For example, being asked to take an IQ test under evaluative conditions, compared with taking it under nonevaluative conditions—treating it as a game—makes a huge difference in performance.[35] Again, why mystify the situation with an intangible concept like *g*?

On another front, it has been shown that biological effects of stress on parents, as in future response and avoidance tendencies, can alter gene expressions by a kind of molecular tagging or silencing of specific sections of their DNA (see chapters 4 and 5). These tags are then transmitted to children along with parental DNA, affecting the children's physiological stress management processes. The consequence is that offspring will tend to underperform in stressful situations: an effect that appears to be genetic but is really environmental in origin.

SOCIAL *DIS*-INTEGRITY, NOT BODILY INTEGRITY

IQ scores, then, are merely a kind of redescription of the distribution of social power, with its consequences for people's thoughts, feelings, self-confidence and so on. This reproduction in minds of circumstances in societies also explains well-known correlations between IQ and many other consequences of that power structure. These include, not just social class per se, but also health, life expectancy, involvement in crime, and just about any other consequence of social inequality.

IQ testers like Linda Gottfredson, Ian Deary, and colleagues have made much of such correlations. For one thing, they have suggested, without any causal evidence, that they validate IQ tests (i.e., that they really do measure human potential). But they have gone beyond that to suggest that IQ reflects bodily system integrity or "general biological fitness."[36]

Of course, such suggestions reflect a constant strand in the IQ testing movement: the reduction of social class to immutable biological forces.

But the web of correlations merely redescribes the class structure and social history of society and its unfortunate consequences.

IQ IS NOT FIXED

Such fatalistic views presuppose IQ to be a measure of a stable and durable trait—the cognitive strength, or level of g—of individuals. IQ levels are expected to stick to people like their blood group or height. But imagine a measure of a real, stable, bodily function of an individual that is different at different times. You'd probably think what a strange kind of measure. IQ is just such a measure.

Of course, there is a modest average correlation of IQs at one age with IQs at a later age, depending on the gap between measurement times. This may be no more than a measure of the degree to which individuals' circumstances have remained unchanged. Exactly the same may be said about language dialect, or personal antibody profiles, for example. But longitudinal studies have shown just how much IQs can change across generations, as in the Flynn effect, but also for individuals as they age.

Carol Sigelman and Elizabeth Rider reported the IQs of one group of children tested at regular intervals between the ages of two years and seventeen years. The average difference between a child's highest and lowest scores was 28.5 points, with almost one-third showing changes of more than 30 points (mean IQ is 100). This is sufficient to move an individual from the bottom to the top 10 percent or vice versa. In a 2011 report, Sue Ramsden and colleagues showed how individual IQs in the teenage years, in their sample, varied across the mean between minus 18 to plus 21, with 39 percent of the total sample showing statistically significant change. What a strange measure indeed![37]

Consistent with such observations is the frequent demonstration of the trainability of IQ. Test performance improves with practice, even on the so-called tests of pure g, like the Raven, usually expected to be the most experience free and therefore most stable. For example, Susanne Jaeggi and colleagues trained adults on memory tasks, requiring the ability to hold and manipulate information in the mind for a short period of time.

They found a substantial dose-dependent transfer to performance on a matrices test like the Raven.

In a 2011 paper in the journal *Developmental Science*, Allyson Mackey and colleagues showed average improvements of 10 points from training seven to nine year olds. Timothy Salthouse showed that cognitive test experience at any age improves performance at a later age, irrespective of age of participants (from eighteen to eighty years). Sylvain Moreno and colleagues showed that even computerized training in music for pre-school children boosted IQ test performance. Others have shown how factors like physical exercise that improve sense of well-being also improve memory and cognitive test scores. All these results suggest that experience with an appropriate cultural tool and/or a boost to self-confidence or other measure of well-being enhances test-taking ability.[38]

Such results also suggest that we have no right to pin such individual differences on biology without the obvious, but impossible, experiment. That would entail swapping the circumstances of upper-and lower-class newborns—parents' inherited wealth, personalities, stresses of poverty, social self-perceptions, and so on—and following them up, not just over years or decades but also over generations (remembering the effects of maternal stress on children, mentioned above). And it would require unrigged tests based on proper cognitive theory.

EXPLAINING THE FLYNN EFFECT

As mentioned above, the phenomenon of ever-rising average test scores remains puzzling to *g* theorists. Tortuous webs of explanation have been spun in the literature in recent years, but with little agreement.

However, the Flynn effect is readily explained once we accept that IQ tests are measures of social class and cultural affiliation rather than tests of innate ability. These leaps in scores correspond to the demographic swelling of the middle classes over the period in question: the movement of individuals to new levels in the social power structure. On one hand, the increasing class elevation means greater use in families of test-relevant cultural learning, such as conceptual categories and text and number literacy. On the other hand, it means improved sense of place in the power structure together with enhanced self-esteem, self-confidence, and sense of self-efficacy.

That is, the Flynn effect further supports the suggestion that an IQ test score is an index of specific background learning rather than some so-called strength in a mythical *g*. The explanation is supported by several other observations. The rise in scores is steeper in developing countries in which the demographic changes are occurring relatively faster. In contrast, there are reports of the effect leveling off in at least some developed countries in which neoliberal economic policies have reduced social mobility. Finally, as just mentioned, adoption of children from lower-class into middle-class homes results in massive IQ gains.

IQ: THE PHONY MEASURE OF POTENTIAL

Intelligence is viewed as the most important ingredient of human potential. But there is no generally accepted theoretical model of what it is (in the way that we have such models of other organic functions). Instead, psychologists have adopted physical metaphors: mental speed, energy, power, strength, and so on, together with simple genetic models of how it is distributed in society. The IQ test was invented to create scores that correspond with such metaphors, with the distribution—who is more or less intelligent—already presumed.

This circularity in IQ testing must not be forgotten or overlooked. IQ tests do not have what is called "construct" validity, in the way that a breathalyzer is calibrated against a model of the passage of alcohol in the bloodstream. They are constructed on the basis of prior beliefs of who is or is not intelligent. But by creating a numerical surrogate of a social class system, they make that system appear to spring from biological rather than social forces. Such ideas are dangerous, because they demean the real mental abilities and true potential of most people in everyday social situations.

What is clearly missing, above all, is a respectable theoretical model of intelligence. This is what I construct in the next couple of chapters.

4

REAL GENES, REAL INTELLIGENCE

ILLUSIONS OF WHAT GENES ARE FOR

There is a more or less standard narrative about the origins of individual differences in intelligence. The narrative is widely accepted among psychologists and members of the public and has been for a long time. It goes something like this. Each gene produces a protein. The proteins then combine like cog wheels to form the cognitive "machine" in the brain (or produce other proteins that do so). Different individuals have inherited more or less "good" variants of some of the genes. Because of that, the power of the machine (for solving problems encountered in the world) also varies. So we get individual differences in intelligence.

To most psychologists and behavioral geneticists, intelligence—or *g*—is the variation in that power. There is no agreed-on scientific theory about what the power-that-varies actually is; except that it lies in variable genes. This is the genetic model that drives estimates of heritability in human potential, and the genome-wide scans for the variant genes that create it. Of course, everyone tacitly recognizes that a whole lot of developmental processing is going on between the genes and assembly of the cognitive machine, much of it influenced by the environment. But all that is left as a kind of agnostic black box. As we have seen, into that black box is poured an awful lot of hunches, intuition, and unconscious ideology—not to mention the rather crude mechanical model of cognition.

In the past couple of decades, however, that agnosticism has been changing. The box has been slowly prized open by molecular and cell biologists, and many others who study, not statistical models based on dubious assumptions, but the real relationships among genes, the workings of the cell, their environments, and their products. These researchers have shed bright light on the nature of development and the real origins of function and variation. As a result, genes are being redefined and put into context, tearing away the veil shrouding the mystical, animistic blueprints said to determine form and inequalities. Moreover, intelligence—hitherto an anonymous statistical variable with vague definitions—is truly coming to life.

LIFE WITHOUT GENES

Let us begin with one revelation that is usually quite startling to the unreflecting: living forms existed long before genes were available. The widespread idea that life starts, is structured, varies, is controlled by, and ends with genes carries a profound implication. This is that, at its origin, genes must somehow have been there to kickstart life. How else could undirected, unmanaged, physical forces have produced such immensely complex, living states from inanimate substances?

The problem is that the "genes as genesis" theory glosses over a couple of tricky problems. One of these is the mystery of where the genes themselves came from. Francis Crick, the co-discoverer of the structure of DNA, even suggested they must have blown in from outer space. But that only displaces the problem without solving it. Just as problematic is that strings of some original genes could not have just appeared de novo and functioned independently. They would have needed—*in advance*—hugely complex supply chains of ingredients and enzymes (molecular catalysts) to express their proteins and assemble them into a coherent organism. Years of effort in laboratories have failed to find processes through which genes can make themselves, produce components of living systems, and then string them together in just the right order at just the right time.

This standard, rather superstitious, model of the gene has persisted, I suspect, for two reasons. One is that it has been a very useful vehicle of social ideology, as mentioned earlier. It can be made to correspond so easily with our everyday social experience and social structures. The other is the difficulty of seeing how else order and complexity in living things could have originated.

It so happens, however, that another origins story has become apparent in recent years. Biophysicists have been observing how structure and complexity emerge all the time through natural thermodynamic forces. Uneven distributions of energy create disequilibria in physico-chemical systems. The systems are rebalanced by dissipating their energy through the most economical means possible. Doing so usually creates changes in matter itself, often including the emergence of complex structures.

The sun in our solar system is the most prominent source of such disequilibria. The uneven distribution of energy (light, heat, and gravitational) patently drives so many complex, structured activities on earth. But the process has been studied in much simpler systems.

For example, if a layer of liquid between two glass plates is heated from below for a short period, the layer may churn a bit and show signs of random disturbance. But then it quickly returns to its homogeneous, equilibrium condition. If the heat persists, however, the disequilibrium eventually exceeds a critical—far from equilibrium—point at which the liquid forms a new kind of more organized motion. This movement is not random. It can be seen on closer inspection to consist of closely packed and structured convection cells moving in alternate directions.

Such complex formations are known as Bénard cells, after Henri Bénard, who studied them around 1900 (figure 4.1). The tendency to move toward the most efficient means of absorbing and dispersing heat has produced an ordered structure. Structure and complexity have emerged without the intervention of a specific coordinator—that is, the system is *self-organized.*

Researchers have now demonstrated complex, self-organizing structures in so many domains of existence. Spirals, coils, and helical clusters, sometimes arranged as hierarchies, suddenly appear in apparently inert liquids.[1] In the famous Belousov-Zhabotinsky reaction, a mixture of certain acids forms patterns of waves and concentric circles that change over time (figure 4.1b). Many examples of these patterns are displayed on Wikipedia and YouTube. On a larger scale are the turbulence patterns of liquids and gases under shear stress, tornado vortices, and weather changes. As in living things, they suggest some specific form-making agent

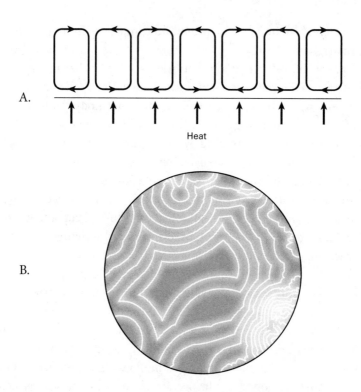

A.

Heat

B.

FIGURE 4.1

(a) Bénard cells forming in a layer of heated water; (b) An early stage in a Belousov-Zhabotinsky reaction. (Photo by Stephen Morris. Diagram from Wikipedia.)

at work, whereas it is only the self-organized dissipation of energy or perturbation. "Order for free," as Stuart Kauffman put it.[2]

The world as we experience it mostly consists of such dynamical systems. The river we may be gazing at is a flow of water molecules dissipating energy. But as the flow encounters obstacles, the molecules self-organize into waves, whirls, eddies, and currents as the most economical way of doing so. The whirlpool you see persists, though the water continues to flow in the same old direction. The whirlpool is called an *attractor* state, because water molecules nearby, caught up in the conflict of flow and obstacle, get pulled into it while the rest of the river flows on.

You may see a leaf caught in an endless circuit, going around and around, under some river rapids without escaping back into the general flow. Again the circuit defines an attractor state. Such states, exhibiting a kind of dynamic equilibrium are also called *limit cycle* attractors. They will persist until factors like volume and flow of the river—perhaps after a storm—change beyond a critical limit. Then the states break down and enter a brief period of reorganization (sometimes referred to as "chaotic") during which some new stable attractor states may emerge.

Note that, in such dynamical attractors, new structure is arising even though the constituent elements are the same. It is happening without any directing "executive," purely as the economic means of dissipating free energy. Finally, the transformation has converted utterly unstructured, random states into ones that are structured and predictable. For living things, predictability means information, possibly for survival (see below).

In sum, changeable environments drive many systems toward critical states. In such states, they can either contain perturbations through equilibrium dynamics or access chaotic dynamics that can rapidly explore novel response possibilities. It is now generally accepted that many aspects of living beings, operating under rapid environmental change, reflect the self-organizing properties of such dynamical systems. I will be referring to them a lot in what follows.

THE ORIGINS OF LIFE

It is now thought that living things originated from such self-organized complexity rather than some gene-centered creationism. That is, systems of molecules evolved from chance beginnings into structured attractor states with energy-dissipating properties.[3] One scenario suggests that organic molecules, as the early components of such systems, could have been synthesized during electrical storms in the atmosphere of the early earth. These molecules rained down on the oceans. Collections of them created chains of chemical reactions, dissipating the uneven energy distributions around them. During those reactions, new components—some of them capable of self-replication—emerged and formed new interactive structures.

Another scenario suggests that such structures originated in the volcanic vents on ocean floors. As Nick Lane (following the work of Michael Russell) puts it, such vents "were the ideal incubators for life, providing a steady supply of hydrogen gas, carbon dioxide, mineral catalysts, and a labyrinth of interconnected micropores (natural compartments similar to cells, with film-like membranes)."[4]

Either way, not only can amino acids so form, they can also readily combine into strings (polymers), the basic structures of peptides and proteins. Many of these strings are also known to have at least weak catalytic properties and so are able to assist in the formation of other molecules, including strings of RNA or DNA from naturally occurring nucleotides. It has been demonstrated in laboratories how chains of such components can create reactive networks, known as "autocatalytic sets." As Steen Rasmussen in *Astrobiology Magazine* (October 21, 2014) puts it, "An autocatalytic network works like a community; each molecule is a citizen who interacts with other citizens and together they help create a society."

So these molecular networks may have constituted primordial "metabolisms": self-sustaining, self-organizing forms. They could have taken in components and energy from outside the system, fueling their chemical reactions, maintaining their structures, and then dissipating wastes back into the wider environment. Different combinations of constituents, with different reaction properties, would have survived in different conditions. And they would have "evolved," as variation (individual differences) arose through more or less random modifications to constituents; for example, peptide structures might change under changing temperature, altering reaction rates and directions and so creating new potentials.

GENES AS RESOURCES, NOT CAUSES

Such discoveries are transforming ideas about the place of genes in the constitution of living things. Clearly, if life originated in such molecular interactive systems, only later did the genetic polymers (RNA and then DNA) arrive on the scene as what we now call "genes." Rather than the structures and order of life being created by pre-existing genes, it seems that the properties of matter that created life might have, eventually, created genes.

And how did genes eventually become established? Probably not at all as the original recipes, designers, and controllers of life. Instead they arose as templates for molecular components used repeatedly in the life of the cell and the organism: a kind of facility for just-in-time production of parts needed on a recurring basis. Over time, of course, the role of these parts themselves evolved to become key players in the metabolism of the cell—but as part of a team, not as the boss.

This has been nicely described by physiologist Denis Noble: "The modern synthesis has got causality in biology wrong. Genes, after all, if they're defined as DNA sequences, are purely passive. DNA on its own does absolutely nothing until activated by the rest of the system through transcription factors, markers of one kind or another, and interactions with the proteins. So on its own, DNA is not a cause in an active sense. I think it is better described as a passive data base which is used by the organism to enable it to make the proteins that it requires."[5]

It is not surprising, then, that we find that variation in form and function has, for most traits, only a tenuous relationship with variation in genes. The rest of this chapter and much of those that follow explain why.

THE REAL ROOTS OF INTELLIGENCE

The role of genes is restricted in the sense just mentioned for a very important reason. They are *not* the creators of "all that we are," because as dumb templates, they can only code for the production of the same thing all the time: the same protein; the same "cog" for the same machine, doing the same thing ad infinitum in a predictable environment.

That traditional view might be fine in stable environments, with recurring, predictable, demands. But during the lifetime of most organisms, most aspects of environments are rapidly changing, making changing demands on the organism. Energy sources may start to fluctuate, a new predator appears, temperature variations become more extreme, and so on, with diminishing predictability. Such conditions cannot be met by the same recurring response from the same system (i.e., the genes). So something different or additional is needed to render the unpredictable predictable.

A dynamical analysis of the environment gives us clues: it explains that a changing environment is not just changing in simple ways, like chemical gradients, or slow temperature changes in water. Rather, at all levels, from the molecular to macro-events, the environment tends to change with complex interrelations. A sugar concentration may change with heat, and heat with light. Darkening skies may portend rain, and a sound over there is followed by the appearance of danger. Things change together.

That means that change in any one variable is reflected in changes in others: levels of one can be predicted from levels of another. A temperature change in the water layer in Bénard cells is correlated with movements of water molecules through space and time. In the movement of any object, location is correlated with time and may well be correlated with movements of other objects. In those inter-dependencies lie crucial prerequisites for living things.

The most important thing to know in changing environments is what is going to happen next? That is one aspect of predictability. Another is what to do about it—or what will be the consequences of this or that action? Think of driving a car on a busy road. You need to predict the order of constantly changing events and predict the consequences of your reactions. Fortunately, you can do that because the unfolding road, signs and signals, and movements of other vehicles are not independent of one another, as if random events. They have structure, or interdependencies, that you, in the course of past experience, with your powerful cognitive system, have condensed into an abstract set of rules: the rules of the road, which form a kind of "grammar."

The environment of all living organisms is providential in that dynamical sense. There is rich information for predictability in the relational patterns of the environment; in the structure of interactions among its components over space and time. As an example at another (macro-event) level, consider the pattern of the day-night cycle created by movements of the earth around its axis. The light intensity correlates with other variables such as temperature and moisture levels. But the relations go deeper than that.

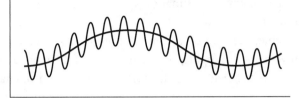

FIGURE 4.2

Schematic illustrating day-night temperature changes changing with (in this case, rather short) seasons.

Figure 4.2 is a mock-up of typical changes in daily temperature over time, dependent upon season (caused by the motion of the earth around the sun). All living things on earth have assimilated such a day-night pattern in their physiological and metabolic systems. But they also have to regularly retune it as the pattern changes with the seasons—a kind of learning. There is a correlation between two variables (temperature and daylight) which living systems have assimilated. But the correlation is conditioned by another variable (time of year, or season). By assimilating that deeper structure, temperature or light intensity can be more accurately predicted. So plants can get their photosynthetic machinery together, with their leaves ready lifted, as they do, just *prior* to the onset of dawn—but not in winter, when the lack of strong enough light intensity makes the effort pointless. In animals, sensory receptors, digestive systems, limb muscles, and so on get primed in advance of dawn. In some animals, over a longer term, coat thickness and coloration may change, and hibernation may occur.

Here is an example at another—more social—level. Imagine some formation dance—a reel, or a set in a square dance. Corresponding movements of members follows a formation changing over space and time but adapted to a background tune (the "environment"). The dancers are following a set of rules (or "grammar") that describes the relations between members over space and time, making the formation totally predictable. Now imagine the background tune changes, from a 4/4 to a 3/4 rhythm, say. A new spatiotemporal formation appears, with a different grammar, adapted to the new environment.

Now imagine, in either case, that a couple joins the set, but their timing and movements are slightly off (perhaps one of them has a limp). Their deviations are nevertheless usually accommodated and the dance holds its shape—varying a little in precision, perhaps slightly changing the rules, but within certain limits. Now imagine they are wildly off. The dance breaks down, the members mill around for a bit, consider alternatives, and then devise a modified formation, with modified rules, to accommodate the wayward couple more comfortably.

We now know that there are such molecular dances taking place in every living cell in every living system. Again, the formation is an attractor or basin of attraction, each shaped according to the flow of energy around it. Different attractors are adapted to their different local environments. The disturbance created by the incursion of an unusual component—a new signal or stimulus, a molecule with an unexpected shape—distorts the attractor. But, like a swinging pendulum knocked out of sync, the signal becomes accommodated, and the attractor eventually recovers its shape (such an attractor is called a "limit cycle attractor" for that reason). Sometimes the disturbance is so way out, however, that a complete reconfiguration is needed. A critical phase ensues; the attractor becomes chaotic for a while and then settles back into a new limit cycle, now accommodating the disturbance or change in a new formation with a revised grammar.

We now know that living systems, from the single cell to brains to complex human societies—all under constant perturbation—spend most of the time in that critical state on the edge of chaos. Such dynamical systems are far more adaptable in changing environments than any mechanical, if-then, input → output system with predetermined "rules." I will be referring to these systems a lot in what follows.

In each case, the information that furnishes the crucial predictability is not signal "elements," but the dynamical structure: "statistical information, based on past observations, about what seems likely to occur in the future," as David Moore puts it in *The Developing Genome* (p. 11). This is the kind of information that all organisms need in circumstances that are, in effect, constantly novel.

But as with the rules of the road or of grammar in speech, it is mostly deep information in the sense of many variables interacting at many different levels. Accordingly, the most important aspect of evolution has *not* been that of genetic adaptations to recurring circumstances. It has been that of intelligent systems able to deal with changing environments by abstracting such information at increasing depths. That involved in driving or human speech is vastly more complex than the intelligence of the cell. But even in cells, it can be difficult to describe. We can just about envisage interaction among three variables; beyond that we require mathematical tools, especially where relationships are nonlinear and changes are not uniform. Moreover, with increasing numbers of variables the system tends to become dynamical, as described above.

INTELLIGENT SYSTEMS

As usually envisaged, a gene is sensitive to a very specific environmental change and responds with a very specific response. An intelligent system is sensitive to how one change is conditioned by other changes, and that sensitivity shapes more adaptable responses.

Living things, then, need to be good at registering those statistical patterns across everyday experience and then use them to shape the best response, including (in the cell) what genes to recruit for desired products. This is what intelligence is, and its origins coincide with the origins of life itself. Indeed, in an important sense, intelligence is life, and life is intelligence.

Accordingly, what is now being discovered is that even molecular networks can "learn" the statistical rules encountered in their environments. The rules are assimilated in reconfigured reaction networks, themselves due to changed reactivities of molecular components. Such abilities are being abundantly revealed in single cells and single-cell organisms. Even bacteria, it turns out, are "dynamic predictors actively oriented toward what comes next."[6] This follows experiments by Ilias Tagkopoulos and colleagues, showing how bacteria can adapt to changing environments by learning statistical associations between variables. The bacterial biochemical networks create internal models of the complex environment.[7]

Hans Westerhoff and colleagues go even further, suggesting that "macromolecular networks in microbes confer intelligent characteristics, such as memory, anticipation, adaptation and reflection." All forms of life, from microbes to humans, "exhibit some or all characteristics consistent with 'intelligence.'"[8] And Frank Bruggerman and co-authors report the discovery of signal integration circuits that "enable the bacterium to 'compute' the optimal physiological response by evaluating its current internal physiological status and the external environmental status." Instead of genes merely producing biochemicals, they say, "Evolution has led to highly sophisticated and quasi-intelligent regulation of that biochemistry."[9]

Today's molecular biologists are increasingly reporting "intelligence" in bacteria, "cognitive resources" in single cells, "bio-information intelligence," "cell intelligence," "metabolic memory," and "cell knowledge"—all terms appearing in recent literature. "Do Cells Think?" is the title of a paper by Sharad Ramanathan and James Broach in the journal *Cellular and Molecular Life Sciences*. And Philip Ball, after a group in Japan had reported a slime mold solving a maze to reach food, informed *Nature* readers that "learning and memory—abilities associated with a brain or, at the very least, neuronal activity—have been observed in protoplasmic slime, a unicellular organism."[10]

These processes, at the roots of intelligent systems, have been called *epigenetic*, meaning above or beyond the genes. It has been tempting to suggest that they simply regulate or modulate the real information (i.e., the true potentials of living things) in the genes. But the truth is, such information never existed there. This is a hard message for people to swallow, even those who are trying to rewrite our understanding of genes. In her book, *The Epigenetics Revolution*, Nessa Carey is worried that "the field is in danger of swinging too far in the opposite direction, with hardline epigeneticists almost minimizing the significance of the DNA code. The truth is, of course, somewhere in between."[11]

To describe further the real truth, let us look more closely at how biological systems manage to be so intelligent. What follows is a little more of the technical detail of intelligent processes in the reaction of a single cell to the environment (either as a single unicellular organism, such as a bacterium, or in a multicellular plant or animal). I want to illustrate how, through the different stages, the intelligent processes of the cell more

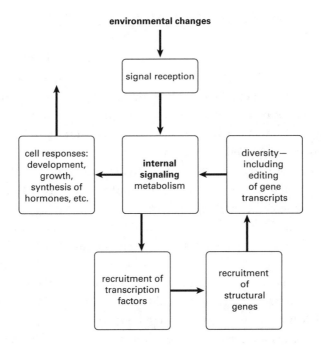

environmental changes

signal reception

cell responses: development, growth, synthesis of hormones, etc.

internal signaling metabolism

diversity— including editing of gene transcripts

recruitment of transcription factors

recruitment of structural genes

FIGURE 4.3

Stages of intelligent responses to environmental changes in cells.

resemble the square dance, with resolutions "democratically" based on the statistical structure of information available, than an executive genome handing up instructions to an assembly line on the basis of something it already "knows."

I refer to such statistical structure in various places as statistical patterns, mutual information, or information grammars. But in all cases, the intelligent processes condense seeming disorder and disharmony into what Denis Noble calls *The Music of Life* (the title of one of his books). And like music, new forms or harmonies can be readily created as demanded from a knowledge of relationships; novel variations that cannot be found in genes.

Readers who prefer to can skim or skip this evidence, at least up to the section "Gene Products Become Further Modified." For those who do, figure 4.3 offers a simplified roadmap.

SIGNALING ENVIRONMENTAL CHANGE

Fixed DNA codes cannot register the changing statistical patterns needed to survive in rapidly changing environments. That can only be done through the intelligent computations among components mentioned by Bruggerman and colleagues. They are not trivial, even in single cells. Making sense of the environment through constantly novel arrays of signals is not easy. Cells are receiving hundreds of thousands of environmental cues every second—a relentless dynamic storm of information. But they are acutely sensitive to changes and to structure within it.

Even single-cell creatures like bacteria pick up tiny changes in their environments. These include concentrations of chemical attractant and repellent substances; physical stimuli, such as light or heat; or mechanical bumps and shocks. For example, the common gut bacterium *Escherichia coli* can detect fewer than ten molecules of nutrient like an amino acid in a volume of fluid about equal to that of its own size—the equivalent of detecting a few drops of foreign liquid in a bathtub of water.

This intense monitoring of the environment, in all single cells, is done through rich signaling receptors on the cell surface (figure 4.4). These receptors are specialized molecules embedded in the cell membrane. They react to physical and chemical changes in their surroundings and pass the information to internal systems. So important is this function that the production of relevant signal-receptor proteins is itself a major activity of the cell involving over a quarter of its genes.

The receptors need to do their work, though, not as independent units but as coordinated teams. They are sensitive to external chemical signals, or *ligands*, but not as isolated cues. A single sugar molecule hitting a cell membrane receptor in *E. coli*, for example, gives no information about direction of source or how to pursue it: its track cannot be visualized. Instead, directions have to be computed from an integration of a series of such signals over time and space on a number of receptors—just as bats use their ears in echolocation; or you use your eyes in fast traffic: in other words, through the detection of statistical patterns in time and space.

Registering the statistical pattern is crucial for predicting where the change is leading and therefore, for the guidance of responses, such as motion, development, and metabolism. It depends on the four-

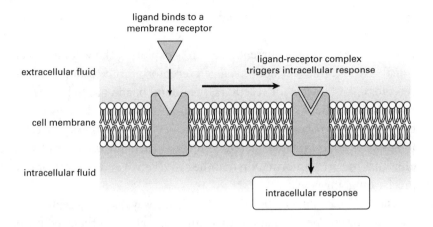

FIGURE 4.4

Cell surface receptors detecting external signals.

dimensional (three spatial dimensions plus time) shape of those cues every bit as much as the bat depends on the shape of its echoes. This is important to cells in multicellular bodies as well as to single-cell organisms.

A good example is the G-protein-coupled-receptors (GPCRs) that copiously stud the cell membranes of every cell in a multicellular body. They mediate most physiological responses to hormones, neurotransmitters, and environmental changes (over a thousand varieties have been identified in different tissues). Like other receptors, they function to initiate or suppress a multitude of biological processes in the cells. Their malfunctioning, in humans, is often manifest in heart disease and blood pressure, inflammatory, psychological, and other disorders. Not surprisingly, then, more than half of all drugs given to patients work by targeting one or another GPCR on the body cells.

None of this GPCR stimulation takes place independently, however, as if a simple stimulus-response trigger. Responses depend on a background of structured activities. For example, many of the GPCRs have two binding sites, one that responds when one factor is present in the environment, the other when another factor is present—but only if a third factor is also present. The order of the appearance of these factors in time and space is also crucial; variations lead to different consequences for the cell and its responses.

Consequently, much current research is exploring the ways in which signals are integrated either on a given receptor or in crosstalk between receptors. In at least some receptors, the bound ligand has to be actively released from the receptor before the signal works. It is done by membrane enzymes (proteases), activation of which depends on specific combinations of environmental factors. "Proteases therefore act as regulatory hubs that integrate information that the cell receives and translate it into precise outgoing signals."[12]

Another kind of interaction at this level is the way that the receptors can signal to one another. In that way they can "negotiate" over the structure of the messages sent on to the intracellular processing. As Christof Niehrs explains, "What emerges is an intricate network of receptors that form higher-order ligand–receptor complexes routing downstream signaling."[13]

In multicellular systems, of course, the cells are not just responding to one another but also collectively to the changing environment outside. That requires an intelligent physiology, as described in chapter 5. However, it is still the statistical structure of the changes that matters and that forms the basis of a living intelligence. Even at this level, closest to the genes, then, the environment is emphatically *not* a loose collection of independent factors to which the cells respond, in stimulus-response fashion, under gene control. This reality makes the additive statistical models of the behavioral geneticist quite unrealistic.

Internal Signaling

Signals already integrated at receptors then initiate a variety of internal responses, depending on the patterns gathered in the signal structure. This is signal transduction. A vast internal signaling network extracts further structure from that passed on from receptors. Once assimilated, that structure will permit prediction of what is likely to follow and will indicate the best response. In their paper in *Science* in 2008, Ilias Tagkopoulos and colleagues showed how even bacteria could "capture the dynamical, multidimensional structure of diverse environments by forming internal representations that allow prediction of environmental change."[14]

We now know that the molecular networks of cells do that through changes in the reactions between molecular components. As with the Bénard cells mentioned above, networks self-organize in response to the disequilibria set up by the signals. It "depends on proteins that are assembled from a toolkit of modular domains, each of which confers a specific activity or function."[15] The structure may then be played out in various ways: cell growth, cell division, cell movement (for a number of purposes), cell differentiation to form this or that tissue, production of a hormone for transfer to other cells, and so on. Response pathways can involve hundreds or thousands of components, calling up genetic resources as required (see below). So crucial are these networks to normal function that approximately 20 percent of the human genome codes for proteins involved in signal transduction.

Some responses, such as those to steroid hormones, involve only two or three steps, from initial ligand reception to genetic transcription and synthesis of product. More typically, though, extensive cascades are conducted through wider networks. These need many intersections as control points where they can "listen" to so many other cell states or events. Box-and-arrow diagrams, like that in figure 4.5, can give a rough impression of the interrelations. But they cannot capture the more abstract nature of the statistical structures. These have only been approximated through the use of computer models.

An example of this coordination of internal and external structure is that of epidermal growth factor (EGF), the subject of thousands of research papers since its (Nobel Prize–winning) discovery in 1962. EGF is a peptide produced in the brain and circulating in body fluids, where it binds to EGF receptors on many cells. This binding initiates, not a single stereotyped response, but one or another of a wide range of possibilities: cell growth, cell division, differentiation, migration, and so on, depending on external and internal contexts.

It is important to note how the pathways—the ripples of chemical reactions—through signaling networks are being continuously reconfigured in response to changing environmental structure. That is, the networks "learn," or alter biases of response, in ways that always makes current responding contingent on past experience. They are actively creating new, adaptive, variations in the process.

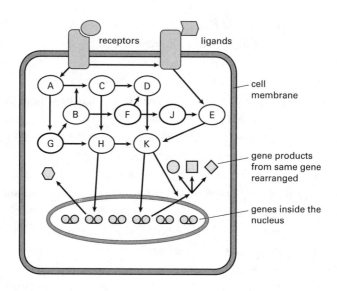

FIGURE 4.5

Simplified illustration of signaling pathways. Circles A–K are various signaling factors (of which there can be hundreds in a typical pathway).

We now know of numerous examples of such plasticity. In the journal *Science Signaling* in 2012, Nancy Gough gives the example of how one particular signaling molecule acts as a coincidence detector, sensitive to the timing of two other signals. The response of the detector, such as "calling up" specific gene products, "only occurs if the stimuli producing the signals are received sufficiently close together."[16]

Writ large, we have to think of multitudes of enzyme-boosted reactions, with modifiable effects on one another and on substrates, the whole sensitive to the structure of change. Of course, some aspects of the environment can be sufficiently recurrent, even across generations, as to permit predictability without such constant retuning. As with gears in a machine, some components may be sized or weighted by natural selection to create biases in signal processing toward particular endpoints (see the discussion of the canalization of development in chapter 5). A monkey will develop a tail whether it grows up in the jungle or a penthouse suite— humans definitely not! Many, or even most, of those components, including signaling proteins, will be gene products. But they will have been selected as harmonious members of a dynamic team rather than as independent players.

Transcription Factors

The endpoints of many of these signaling pathways will be gene transcriptions. That is when important resources are called up from the genes at certain times. The process is commonly spoken of as the genes being "turned on" or "switched off" and then taking over to do their stuff. But what is really in control is the dynamics of the signaling processes. The involvement of genes then takes further self-organizing steps.

The first of these steps consists of merely gaining access to the genes. DNA strands containing the genes are heavily intertwined with other proteins (histones) that spool the strands into space-saving units. These strands first need to be unwound by signaling factors so that other factors can get at them. It constitutes another layer of regulation/control over gene transcription and is part of the epigenetics mentioned earlier.

Even following such access, though, the transcription of genes is no straightforward matter. It involves a variety of further regulatory components: transcription factors, activators, repressors, enhancers, and cofactors. Correspondingly, the genes have evolved with flanking regions on the DNA sensitive to such regulatory factors. Like identity checks to your online banking accounts, they operate in different combinations to further regulate payout (transcription). The combination governs whether transcription will occur at all, and to what degree.

Most prominent in this particular team are the transcription factors (TFs). These are proteins that bind to special "promoter" regions flanking the DNA sequences and help initiate gene transcription. The template is then read and converted into the transcript. As such, TFs are themselves the products of genes and are the first gene products of all the signaling gossip just mentioned. The importance of TFs is seen in that around 95 percent of genes code for TFs, only 5 percent coding for those expressed as the structural proteins used in development and metabolism. Unsurprisingly, therefore, the number of TF genes correlates with the complexity of the organism.

TFs are considered to form the rather deterministic gene regulatory networks mentioned in chapter 1. But they do not function as independent units. As Sara Berthoumieux and colleagues note from their experiments, the transcriptional response of the network is itself controlled by the physiological state of the cell. Accordingly, "the absence of a strong regulatory effect of transcription factors suggests that they are not the main coordinators of gene expression changes during growth transitions." Instead it is appropriate to regard them "as complementing and fine tuning the global control exerted by the physiological state of the cell."[17] This change of perspective, they point out, has important consequences for the interpretation of so-called gene networks. In addition, TFs themselves interact, some serving as co-repressors or co-activators of other TFs, with some TFs regulating other TFs (again the term "orchestration" is sometimes used).

In other words, the whole is a vast self-organizing control network involving feed-forward and feedback loops. TF recruitment depends on the resolutions of signaling networks, which are shaped by input from signaling receptors, that register the structure of changing environmental patterns. Computer models demonstrate that such systems have tremendous powers of control and creativity of response.

Using the Genes

As already mentioned, TFs do not work alone but with a variety of promoters, enhancers, and co-activators. Together with other factors, they recruit RNA polymerase. This is the enzyme that attaches to the DNA and copies the sequence into messenger RNA (mRNA). It is actually the mRNA that serves as the sequence template bringing amino acids together to form peptides and proteins. Imagine how all of these, too, are products of gene transcription under the control of the signaling maze, and you begin to get a picture of how interactive the entire process is.

Already we can see that there is no single command level, no single dominant factor in charge, but a self-organizing global pattern formed among myriad components responding to environmental changes. Together these factors vastly expand the *transcriptome*—that is, ways in which the network responds and genes are utilized, according to context. Genes are very much the followers, not the leaders in the decision making.

Furthermore, trying to account for variation, even at this level, by deciding what is genetic and what is environmental is already quite impossible. Environmental structures regulate gene transcription, and every gene transcript becomes the environment of other genes. The resolution of those in cell responses then changes the external environment with feedback consequences, and so on. It is those dynamic patterns rather than their elements that are important.

The importance of structure over elements is shown in another way. TFs are crucial gene products and gene regulators. But they vary in form only very slightly from flies to mice to humans: "a striking level of conservation, despite dramatic morphological differences resulting from more than 600 million years of evolution."[18] Yet we get tremendously increased phenotypic variation. This is because the form and variation of cells, what they produce, whether to grow, to move, or what kind of cell to become, is under the control of a whole dynamic system, not the genes.

GENE PRODUCTS BECOME FURTHER MODIFIED

What we have just glimpsed is the way that a cell picks up statistical patterns in the storm of signals from its environment and how these are assimilated by internal processes. The implications of the signals are then meshed with internal conditions and needs. The whole ensemble then recruits transcription factors to call up appropriate gene products to create a harmonious response. Already this is more like an orchestra than a knee-jerk reaction, producing a diversity of harmonies as required by changing conditions.

However, what I have discussed so far is by no means the limit of such variation production. It turns out that, under the system dynamics, the transcripts—the gene products—are themselves subject to further extensive modifications. This further editing of gene products is now a vast area of research—one that increasingly rejects the notion of direct correlation between genetic variation and phenotypic variation. But here I can only offer a brief summary of findings.

The sequence of DNA comprising a gene is copied into a corresponding sequence of nucleotides in mRNA. This is the immediate transcriptional product—the transcript—of the gene. It used to be thought that the mRNA template rigidly determines in what order amino acids are strung together to make up one—and only one—protein:

One gene → mRNA → one protein.

The sequence of nucleotides copied into mRNA actually consists of coding sections, called *exons*, interspersed with noncoding sections that are silent and are called *introns*. But the processes that translate these into proteins are less predictable. These processes remove the introns and rearrange the exons in a variety of combinations (called *exon shuffling*; figure 4.6).

One obvious result is that many different proteins can be produced from the same gene, with potentially widely different functions: one gene → many proteins. By 2003, it was known that at least 74 percent of human gene products can be alternatively spliced in this way. We now know that the percentage is much higher than that. In their 2014 paper in *Nature*, James B. Brown and colleagues reported that even in a fly, "a small set of mostly neural-specific genes has the potential to encode thousands of transcripts each through extensive alternative promoter usage and RNA splicing" and that "the fly transcriptome is substantially more complex than previously recognized."[19]

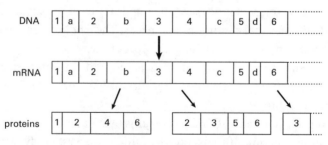

alternatively spliced proteins

FIGURE 4.6

Simplified diagram of exon shuffling. Exons 1–6 are interspersed with introns a–d. The latter are spliced out, while the exons can be combined in many different ways: one gene yields different proteins as dictated by wider cell processes.

Other research suggests that the variety of alternative transcripts is strongly associated with levels of evolutionary complexity. It makes the obvious point: it is difficult to claim that this or that variation in a trait is caused by this or that gene when a gene can be associated with so many different products. But that is by no means the end of such complications.

OTHER PROTEIN MODIFICATIONS

The protein products of the genes are further modified by system dynamics in a variety of other ways, all depending on current context. There are only about twenty-two thousand genes in the human genome, but at least a hundred thousand protein products are known. Such is the plasticity of gene transcription and translation.

A particularly well-studied example is ubiquitination. Ubiquitin is a small regulatory protein that binds to gene products in a number of different ways. The binding can speed up degradation of the product or change its location in the cell, its level of activity, degree of interaction with other proteins, and so on. However, the form and activity of ubiquitin is itself under the regulation of the signaling/transcriptional networks just mentioned. Indeed, such modification to proteins after translation from the mRNA has evolved to become critical for animal, including human, learning. It helps coordinate the changes in connectivity after activity between brain cells.

RNA REGULATORS

In the standard model of development and variation the variable gene codes are transcribed into mRNAs that are translated into correspondingly variable proteins (resulting in correspondingly variable phenotypes). We now know, however, that mRNAs have numerous functions other than serving as protein-coding templates. For example, they can alter the speed of initial transcription in a feedback loop; improve transport of products around the cell; and increase rate of decay of the mRNA product (the peptide or protein) itself, affording more rapid responses to external signals.

In addition, thousands of other small RNA transcripts (other than mRNA) have recently been discovered that are not translated into proteins at all. It is now estimated that although 70 percent of the mammalian genome is actively transcribed, only 1–2 percent of the transcripts are involved in making proteins. Instead, an ever-growing number of regulatory functions is being discovered for them. These functions include activation of hormone receptors, modulation of promoters (and thus transcription rates), silencing (or blocking) specific gene transcriptions, and acting as co-activators for transcription. This "hidden transcriptome" also clearly operates in the context of wider signaling networks. Again, these regulatory networks seem to have become more important in more complex organisms.

OTHER EPIGENETICS

Genes are inherited by offspring, of course. But environments experienced by mothers before or during pregnancy can modify the way those genes are utilized during the offspring's development. These modifications can, in turn, affect development throughout life and even on to subsequent generations. It was illustrated in the Netherlands in 1944 when a Nazi blockade followed by an exceedingly harsh winter led to mass starvation. Babies born at the time grew up small. But the effects persisted in smaller grandchildren, too, even though their parents had been well fed (see also chapter 10).

This is the kind of epigenetics—or epigenetic inheritance—described by Nessa Carey in her book, *The Epigenetics Revolution*. It is a set of processes that challenges the traditional doctrine that the only modifications that can be inherited are those that occur accidentally in the genes. Some mechanisms have now been identified through which the genetic material can be altered through environmental experience and passed on to the next generation. For example, in a study reported in 2014, mice were trained to fear a specific odor before conception. Subsequently conceived second- and third-generation pups also had an increased behavioral sensitivity to that odor but not to other odors. This was correlated with changes in some aspects of neural structure.[20]

In fact, many stressful effects in the parental environment are now suspected to cross generations in that way. Even effects of experiences preceding pregnancy can sometimes seem be passed on to offspring: for example, in humans, a child's future risk of obesity, diseases such as diabetes, poor response to stress, and a general anxiety-prone behavioral tendency.

It needs to be said that these observations are tentative and mechanisms not always clear. At least some are now known to arise from chemical tags placed on the DNA to silence certain genes: a kind of memory of experience then passed on to the next generation through the mother's eggs.[21] Again, the molecular architecture of a whole developmental system, itself a product of evolution, is "instructing" the genes, rather than vice versa. In an article in *Proceedings of the National Academy of Sciences* (2012), John Mattick reported that this maternal epigenetic inheritance is far more widespread and general than previously thought. He said it is "rocking the foundations of molecular genetics." And Denis Noble agrees that "its implications are profound for biological science in general."[22]

Note that this environmental source of variation will appear in the behavioral genetics twin-study statistics as genetic variation: quite probably another way in which heritability estimates are distorted.

REPAIRING AND ALTERING GENE SEQUENCES

In the standard model, the crucial information for development and variation is in the genes. It may be altered accidentally by mutations, or its expression attenuated by favorable or unfavorable environments. But that picture is too simple. Gene sequences are, indeed, occasionally damaged through the commotion and bustle of molecular activity, and mistakes can be made during cell division and replication. However, the signaling networks can sense the damage and its dangers, and relay this information throughout the cell. Metabolic pathways can then be redirected to promote DNA repair. Again, this is not the genes in charge but a global cellular intelligence.

Even more surprisingly, gene sequences, we now know, can be deliberately modified (mutated) during development by the demands of changing environments. As Mae-Wan Ho explains, "Numerous mechanisms for generating mutations are involved that appear to be under the control of the cell or organism as a whole in different environmental contexts."[23] Other ways have come to light of how "environment directly instructs the organism how to vary" and how such variations are inherited.[24]

This is now referred to as natural genetic engineering. In a paper in *Physics of Life Reviews*, James Shapiro says that "the standard model of a 'Read-Only' tape that feeds instructions to the rest of the cell about individual characters" is a "dangerous oversimplification." Now, he says, "we have to reconsider the genome as a 'read–write' (RW) information storage system."[25]

On his blog in *HuffPost Science* (April 30, 2013), Shapiro says, "NGE [natural genetic engineering] is shorthand to summarize all the biochemical mechanisms cells have to cut, splice, copy, polymerize and otherwise manipulate the structure of internal DNA molecules, transport DNA from one cell to another, or acquire DNA from the environment. Totally novel sequences can result from de novo untemplated polymerization or reverse transcription of processed RNA molecules."

One startling implication is that organisms can help direct their own evolution as well as their development. Biologists discuss what they call the "evolvability" of organisms. This refers to the probability of actively generating beneficial genetic diversity for evolution through natural selection. In such a view, it is easy to associate the potential of organisms with their genomes. However, the potential of intelligent systems lies in their ability to develop: actively creating potential, not merely expressing it. As Robert Lickliter explains, "The process of development generates the phenotypic variation on which natural selection can act."[26] Thanks to intelligent systems, the "developmentability" of organisms has greatly enhanced, and tended to eclipse, their evolvability.

An important consequence is that environmental experiences at other levels—physiological, psychological, and/or social—can exert top-down influence on the utilization of genes. In an article for the *Institute of Science in Society*, Mae-Wan Ho also says, "Researchers are identifying hundreds and thousands of genes that are affected by our subjective mental states. Feeling constantly sad and depressed can genuinely turn on genes that make us physically unwell and prone to viral infections and chronic diseases, just as feeling particularly relaxed and peaceful can turn off those genes and activate others that help us heal and fight infections. The emerging field of human social genomics is demonstrating that social conditions, especially our subjective perceptions thereof can radically change our gene expression states."[27]

All this indicates how the model of the cell and the organism as a machine, with form and variation written in the genes, is obsolete. Instead, adaptability in changing environments (as opposed to the static circumstances of the model) has required integration of bottom-up and top-down processes across numerous levels. They are self-organized, dynamical systems with emergent properties, in which genes are servants, not masters. They allow cells to have dramatically different forms and functions through regulations beyond the genes. And they transform development from a passive executor of genetic instructions into creative adaptabilities.

We will be visiting all those different levels in the rest of the book. For the remainder of this chapter, let us look at some immediate manifestations of those dynamics.

Form and Variation Are Unpredictable

Even a brief survey of complex forms and functions reveals unpredictable relationships between variation in genes and variation in the phenotype. In complex, adaptable traits there is no direct mapping from genes to phenotype. And that, in changing environments, is for the best of evolutionary reasons. Instead we get a dynamical system that is far more adaptable to changing environments.

Most Genetic Variation Is Irrelevant

As with choosing which font you type with, most gene mutations and subsequent protein variations are, except in rare circumstances, irrelevant to function. Most traits crucial for survival are buffered against such genetic variations; they are canalized in development, as described in chapter 5, or are subsumed into developmental plasticity. With the exception of categorical disorders, natural selection tends to eliminate deleterious gene variations, resulting in reduced heritabilities, as described in chapter 2.

Many Alternative Pathways

A deficiency in the provision of a metabolite, either from the environment or through a genetic mutation, can usually be overcome by recruitment or creation of an alternative biochemical pathway. So it is with some astonishment that experiments have revealed that large portions of the genome can be deleted without noticeable effects on basic functions. The common yeast *Saccharomyces cerevisiae* has six thousand genes. Experiments have shown that up to 80 percent of them can be deleted without detriment to normal function under optimum conditions. This observation attests to the robustness of biological networks even at that level.[28]

Andreas Wagner and Jeremiah Wright studied fifteen different signal transduction pathways and two large networks regulating transcription and found many alternative pathways between demand, on the one hand, and response in metabolism, on the other. They concluded that "multiple alternative pathways . . . are the rule rather than the exception . . . such pathways can continue to function despite amino acid changes that may impair one intermediate regulator. Our results underscore the importance of systems biology approaches to understand functional and evolutionary constraints on genes and proteins."[29]

Frederik Nijhout and colleagues were surprised to discover that, in humans, many of the genes for enzymes in critical metabolic pathways actually exhibit large degrees of variation. But as they also discovered, although the effects of the gene variations are quite large at the molecular level, the epigenetic processes "greatly reduce their effect at the phenotypic level."[30] They do not matter.

Identical Genes Do Not Restrict Variation

In complex forms and functions, the relationship between genetic variation and phenotypic variation is extremely fuzzy. Thus, individuals with identical sets of genes can be markedly different from one another. Vertebrates contain hundreds of different cell types. They all contain the same (or closely similar) genomes but develop and maintain their separate identities by a combination of genomic and epigenomic regulations (see chapter 5).

Similarly, a population of animals of identical genotypes, reared in almost identical environments, can exhibit the full, normal, range of individual differences in behavior expected for the species as a whole. This has been shown in laboratory animals that are bred to be genetically identical and raised in the same cage. They exhibit normal variation in physiology and behavior, even in immune responses.

The members of a famous seven-hundred-year-old herd of ancient wild cattle (known as the Chillingham herd in Northumberland, United Kingdom) have become so inbred over countless generations as to be genetically identical. Yet they still exhibit the normal range of morphological and behavioral characteristics.[31]

It Is the System That Counts

There is no direct command line between environments and genes or between genes and phenotypes. Predictions and decisions about form and variation are made through a highly evolved dynamical system. That is why ostensibly the same environment, such as a hormonal signal, can initiate a variety of responses like growth, cell division, differentiation, and migration, depending on deeper context. This reflects more than fixed responses from fixed information in genes, something that is fatally overlooked in the nature-nurture debate.

THE INTELLIGENT CELL

In this chapter, I have taken a closer look at the origins of genes and how they enter into the development of form and variation. I have also scrutinized the environment, its true nature, and how it interfaces with genes at the level of the single cell. In the standard picture of behavioral genetics, individual differences arise through this interface as additive sums of the variable genetic and environmental factors. I have described how the real picture of genes and environments is far removed from such simple additive mechanisms.

The main difference is that the interface is itself a dynamical system that makes the real decisions about form and variation in constantly changing environments through vast signaling networks. The system is needed because unpredictable environments can only be made predictable by the informational structure lying deeper within. That cannot be achieved through the linear code in DNA. So the dynamical system of the cell has evolved to abstract the statistical patterns in changing environments and guide responses accordingly. In this network, genes are utilized as resources for a self-organized, intelligent system.

It is now clearer than ever that, as Eric Turkheimer claimed, that systematic causal effects of genetic and environmental differences are lost in the developmental complexity of the network (see chapter 1). In adaptable systems, correlations between genetic and phenotypic differences (except in rare deleterious conditions) are transcended in those networks. All form and variation (and, therefore, potential) is created from the evolved dynamics responding to informational patterns at many different levels.

The great deception of the behavioral geneticists has been in mislabeling such system variation as genetic variation and convincing large swathes of psychology, the general public, and governments, that they are really separating the effects of genes from those of environments. And all without a scientific model or theory of intelligence or other potential.

As we shall see, intelligent systems evolved in more spectacular ways as living things encountered more complex environments. Above all, they are systems for embracing and creating variation to extents far outstripping that allowed by genes. The next few chapters follow that evolution.

5

INTELLIGENT DEVELOPMENT

DEVELOPMENT AND FATE

Development is usually considered to be the means by which potential becomes realized. It tends to be described, more formally, as growth combined with differentiation, or increase in numbers of different components like tissues and organs. Informally, it strikes nearly everyone as a wonderful, but mysterious, transformative process in which an insignificant speck of matter becomes a coherent, functional being so automatically that it appears to be magic. Development seems so self-fulfilling that it is all too easy to imagine the homunculus with all its potential already there, inside that speck, either in material form or in code.

That popular impression has, indeed, been encouraged by leading scientists. "Development," said leading evolutionist Ernst Mayr, consists of "the decoding of the programmed information contained in the DNA code of the fertilized zygote." So it is hardly surprising when newspapers tell their readers: "At the moment that a sperm penetrates an egg, that single-cell zygote . . . is pure potential . . . it has in it the finicky instructional manual that will direct the building of the body's every fiber."[1]

Again that message is the ideological mix of hope with fatalism, orienting parents and childcare givers not to what might be, but to making the best of what fate delivers. Little wonder that parents start to worry about their child's potential almost as soon as conception, especially with regard to the child's future intelligence. Thinking that they might be constrained by a predetermined plan, they are concerned about how to do the best for their children: how to help the plans "unfold" and blossom, provide the right environment, ensure the appropriate nourishment, equip the nursery, get the right baby books and toys, and perhaps even save for private school or some educational coaching.

Sometimes, too, parents think of child rearing as a bit like cooking: they cannot do anything about the basic ingredients, but those ingredients can be enhanced with the right skills in the kitchen/development area. In this chapter, I show how development is not a predetermined unfolding or assembly line, or "cooking" from a DNA recipe. It is a hugely adaptable set of processes that actively constructs potential and variation rather than merely expressing it—and that, in complex changeable environments, it could be no other way.

DEVELOPMENT IN CHANGING ENVIRONMENTS

Scholarly thinking about development has been undergoing something of a slow revolution over the past few decades. It took a major step forward when it was realized in the 1990s that, rather than being predetermined, the dynamics of the developing system itself plays a crucial role in the creation of complex form and variation.[2] At first, these advances were largely conceptual. But research since has increasingly supported them. Discoveries of real developmental processes have raised important questions about the origins and nature of potential, the origins and nature of variation, and the nature of the environment that supports and promotes development.

Development cannot be an assembly line, obeying a fixed plan in the genes, for the reasons outlined in chapter 4. As living things evolved in increasingly changeable environments, the "plan" had to be one that is subject to constant revision during the individual's life. Quite different systems of revision have emerged in the evolution from single cells to human social cognition, as organisms have faced and survived more unpredictable environments.

We have seen in chapter 4 how single cells can learn and adapt. But development as we now think of it really started when cells aggregated into multicellular organisms beginning about 2.2 billion years ago. Multicellularity itself evolved as a way by which single cells could become more adaptable to changing environments. It began as temporary arrangements established during hard times. Even some bacteria and slime molds occasionally form multicellular groups in response to extreme conditions, such as nutrient depletion. In the slime mold *Dictyostelium*, for example, it culminates in striking individual differences in cell form, including functional specializations that could not have been foreseen from reading DNA "instructions" in the single cells. Some cells form a stalk to prop up others supporting a fruiting body, from which still others produce and shed spores that emerge into new individuals when conditions improve.

One quite instructive aspect of this early example of development is how we get so much sudden variation from cells previously quite uniform and having essentially the same genes. There is no overall controller or plan for the process—no recipe for specialization in each cell. It is a dynamical, self-organized process that emerges from interactions and structural information shared among the individual cells. There is no distinct executive or supervisory agent.

It was a foretaste of much more spectacular evolutionary possibilities. The divisions of labor and mutual support in these first pioneers imparted adaptability to changing environments. Permanent multicellularity eventually evolved. Rapid diversification of species followed. It culminated in the Cambrian Period about 550 million years ago, which founded all the major branches (or phyla) of animal life we know today, with the exception of vertebrates (appearing a little later).

That diversification of species itself changed the environment in an ever-increasing spiral. As animals have been forced to inhabit those more changeable parts of the environment, the living world has become increasingly dynamic and multifaceted. With the evolution of increased motility, for example, the environment increasingly became more about interactions among animals themselves. This is a world in a different league of changeability from the flowing concentration gradients experienced by the first primitive organisms. It required development as itself a major instrument of adaptation, often on a lifelong basis.

DEVELOPMENT AS IT APPEARS

From the start—the fusion of egg and sperm—development in the individual is a process of creating spectacular variation from seemingly homogeneous beginnings, including identical genes. The single fertilized egg is the zygote that, over the next few days, divides to form a ball of two, four, eight, and then sixteen cells. Of course, this process would simply produce an ever-growing ball of undifferentiated cells if it were not regulated in some way. So the first problem is how to start the process of differentiation, or true development, into the beginnings of body parts.

Over the first few hours after conception, the start of differentiation can be seen quite distinctly under a microscope. A cavity, the blastocoel, forms and fills with fluid to define formation of the blastula and then the gastrula (figure 5.1). The cells of the blastula have the amazing capacity to become any tissue and cell type in the adult body. They are the *totipotent* stem cells. But they soon change to cells displaying varied sizes, shapes, and types that come to form the gastrula. Now the cells are separated into three layers: the ectoderm, mesoderm, and endoderm. These form the basis of the different cell types and tissues that follow with bewildering variety.

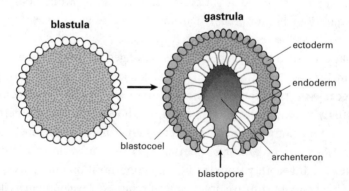

FIGURE 5.1

Early cell division and formation of gastrula. (From Wikipedia.)

Take just one small aspect of the early embryo, the neural crest, a transient pool of migratory cells unique to vertebrates. Under the microscope, it can be seen to form from the folding of a sheet of cells into a kind of tube—the neural tube. Some cells in this tube give rise to the spinal cord. But others creep in different directions in the embryo to generate a further diversity of cell types and tissues. They include neurons of the autonomic nervous system (ultimately regulating body organs); dozens of different nerve cells for the central nervous system (brain); sensory neurons of touch, smell, hearing, and vision; endocrine cells (producing hormones in adrenal and thyroid glands); various cardiac tissue cells; pigment cells of the skin and internal organs; and the blood vessels. Neural crest cells also give rise to the cranial tissues that generate facial bone and cartilage, the cornea of the eye, meninges (membranes around the brain), roots for teeth and eye muscles, and many others.

All the other tissues of the body emerge from the primordial, totipotent stem cells in similar ways. The body axis of vertebrates starts, for example, as serial repetitions of clumps of cells called "segments" or "metameres." The residue of this process is especially conspicuous in the adult's spinal column. From these basic segments, the rest of the general body form—bones, muscles, skin, limbs, and organs—emerges. This is creation of a vast variety of cells, none of which could have been predicted from even the fullest specification of the cells' DNA (figure 5.2).

The proliferation also involves not only changes of cell form but massive migrations over long distances. It inevitably entails growth, or just getting bigger, due to multiplication of cells through binary fission. But this growth and differentiation both happen in a perfectly coordinated manner, preserving the required proportions of size, location, and timing of development among different tissues.

It seems obvious to even the casual observer that such a highly integrated and harmonious process must be under the close supervision of some executive function. But it is not. It all progresses through a remarkable extension and amplification of the intelligent systems already apparent in single cells.

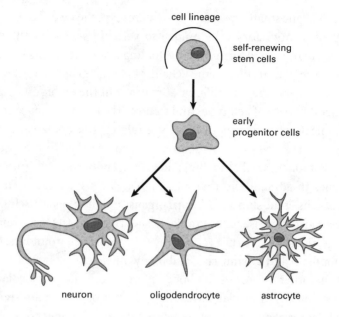

cell lineage

self-renewing
stem cells

early
progenitor cells

neuron   oligodendrocyte   astrocyte

FIGURE 5.2

Cells with the same genes (stem cells) can develop many different forms according to signals assimilated from local environments; just three of the dozens of different nerve cells are depicted here. (Redrawn from the National Institute of Neurological Disorders and Stroke, http://www.ninds.nih.gov/disorders/brain_basics/ninds _neuron.htm.)

HOW IT HAPPENS: DEVELOPMENTAL INTELLIGENCE

Biological Pattern Formation

The question is really about how one cell type becomes a host of different cell types performing so many different functions. After all, each cell in the body has (nearly) exactly the same genome. The answer is an important illustration of adaptive production of variation, at least at this level of biological systems.

Indeed, the answer was partly given in the previous chapter. There I offered a glimpse of how single cells adapt to changing environments by internalizing external structure. The information in that structure is condensed by the cell's signaling maze and then used to activate TFs, followed by patterns of gene transcription. The gene transcripts are edited and passed into the developmental machinery to construct, adapt, and maintain the form of the cell, its organelles, and functions.

Similar processes arise in the cells of multicellular organisms but are much more amplified. As with single cells, the spatiotemporally organized inside must be closely integrated with what is happening outside. But most of the outside now consists of a multitude of neighboring cells and the storms of signals from them. So the activities of each cell need to be integrated and adjusted with what is going on in all the others. The plan—or potential—for the emerging form of each cell cannot be predetermined in the genes, because all cells contain the same genes. Instead it must be created through the signaling traffic going on *among* them—and there is a lot of that.

Signaling between cells occurs through releases of chemical messages, or ligands, into the extracellular spaces. In some cases, it occurs by direct contact between neighboring cells, known as juxtacrine signaling. Paracrine signaling is that which occurs over short distances. But most people are familiar with endocrine signaling occurring over longer distances. The signals have evolved to match with the cell surface receptors of other cells (although a few can enter the cell directly through its membrane). The docking of a ligand with its receptor results in signal transduction and the activation of the second signaling system inside the cell, leading to various physiological responses.

Of course, what any individual stem cell needs to "know" is what kind of cell to change into and where and when to do to it. In general terms the process is not about isolated signals as informational triggers or elements. That could not take the context of other cells into account, and how they are changing. Instead, the signaling is structured in space and time (spatiotemporal structure). In that structure is the deeper information needed by recipient cells to determine their future states in relation to the whole. Let us look at this process a little more closely.

As a single cell, the young egg may look like a homogeneous sphere. But there is already intelligence in the cell. For example, the apparent sphere is already rendered uneven by the structure of its environment. The point of entry of the sperm cell into the egg provides it with polarity—an anterior and posterior (or a front and a rear). Other aspects of the environment of the egg (e.g., the uterine lining, or the surface of the soil) may also help define polarity.

In addition, the mother has deposited specific mRNAs or TFs into the egg cell that are unevenly distributed. When the zygote divides, some of the daughter cells will contain more of those chemicals than others. Their concentrations influence the decisions as to which genes are transcribed. The different proteins so produced alter the structure and function of cells in different ways. Cell fates, that is, are already unevenly distributed in the egg.

In other words, we already see how a uniform ball of cells becomes differentiated according to the structure of its environment—involving genes, but not under instructions from them. The different cells must now signal to one another about their relative positions, so that each further "knows" what it should become.

There is a long and fascinating history of ideas on how this happens. It was long ago realized that some kind of positional information is needed, as in the coordinate systems in today's land maps and GPS systems. In a paper in 1952, Alan Turing suggested that concentration gradients of diffusible chemicals might serve that purpose. These chemicals were duly dubbed "morphogens," and the theory was elaborated by Lewis Wolpert and Francis Crick in the 1960s.

The idea is that morphogens are secreted by cells and diffuse through the tissues of an embryo during early development, setting up concentration gradients. Each cell is given positional information by its place in the gradient. The cell then switches genes on or off to take control and direct the assembly of components that follows. Cells far from the source of the morphogen will receive it in low concentrations. They express only low-threshold target genes and associated products. In contrast, cells close to the source of morphogen will receive it in higher concentrations. They will express both low- and high-threshold target genes and products.

This was the idea superbly represented in Lewis Wolpert's French flag model. In this case, the morphogen targets genes that, metaphorically speaking, produce different colors according to concentration. High con-

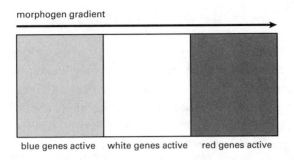

FIGURE 5.3

French flag model of pattern formation in development.

centrations activate a blue gene, lower concentrations activate a white gene, with red serving as the default state in cells below the necessary concentration threshold (figure 5.3).

The first direct evidence for the theory came in 1982 with the discovery of such a morphogen, called "bicoid," in the early embryo of the fruit fly. It turned out to be the product (in fact, mRNA), not of the cell's genes but of a maternal gene. It is produced by the mother and deposited in the egg before laying such as to form a concentration gradient across the egg. Bicoid is translated in the egg as a classic TF that then regulates transcription of genes in cells according to their location in the gradient. Knockout studies—chemically inhibiting bicoid's action—showed that it is critical in the formation of the embryonic head and body axis. The axis in turn creates a crucial spatiotemporal framework for the action of other morphogens in the unfolding of the body plan (organs, limbs, and so on).

To Wolpert, the coordination of the patterning rested with the genes. The information is just a simple cue, and the cell itself an obedient respondent. However, in his 1989 paper "Positional Information Revisited," Wolpert was already warning that "patterning by positional information provides a relatively simple mechanism for making a wide variety of patterns. Alas, compared to 21 years ago, that simplicity now seems more like simple-mindedness. Things seem, at this stage, much more complicated."[3]

Needing More Than Simple Cues

What has been discovered since is that each cell is not simply the passive target of isolated morphogenic cues. It is being literally bombarded with such morphogens in cross-cutting gradients in different sequences, in different places, at different times. Like the fading in and out of notes in music, it is the precise spatiotemporal intersection of these gradients of morphogens that matters for harmonious messages.

So the simple bicoid model for *Drosophila* was an important start. But a vast variety of other factors have since come to light. For example, other maternal gene products are likewise unevenly distributed (see figure 5.4) and involved in head, thorax, and tail formation. These, too, are TFs for genes that produce proteins required for segment formation; for differentiation of head, thorax, and tail; and also for dorsal-ventral (up-down) differentiation. Overall, it is the cross-cutting, spatiotemporal pattern of these factors that determines which genes are transcribed in the egg and where.[4]

Numerous morphogens involved in very early development are now known. They are given exotic names like Hox, Hedgehog, Notch, and Wnt (a reduction of "wingless" and "integrated") and BMP (bone morpho-

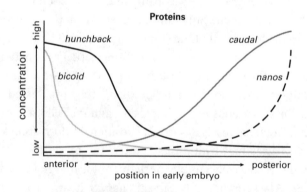

FIGURE 5.4

Distribution of four maternal gene products in the oocyte determining early cell differentiation. (From Wikipedia.)

genic proteins). Most of them are derived from knockout studies and describe consequences of chemically inhibiting a signal. They have pressed home the need for a new perspective on the causes of early differentiation and development.

Dynamical Rather Than Deterministic Processes

Wolpert's gene-centered model was already being challenged by Brian Goodwin in the early 1980s. Goodwin spoke of a more "self-organized entity" born out of the "relational order" among active entities. He suggested that the genes play only a secondary role in establishing the patterns of development. That has been confirmed in recent molecular biological studies. For example, enumerating all the genes turned on in a particular location does not predict the developmental outcomes. The latter are determined by the global physiological state of the cell in the morphogen gradients, all dynamically self-organizing in attractor landscapes, as described in chapter 4.

Take, for example, Wnt signaling proteins. These are a large family of nineteen proteins helping coordinate a daunting complexity of signaling regulation and function in development: cell fate, cell motility, body polarity and axis formation, stem cell renewal, organ formation, and others. But Wnt proteins are themselves tightly regulated in feedforward and feedback loops. As Yuko Komiya and Raymond Habas noted in a review, Wnt proteins and their antagonists "are exquisitely restricted both temporally and spatially during development."[5] They are heavily modified prior to transport and release into the extracellular milieu. Then their activity levels, their shape, and degree of binding to target cell membranes are regulated by a number of co-factors, including other morphogens.

Wnt signaling is just one of a host of morphogenic pathways. The profusion of them explains why many embryologists and developmental biologists are now resorting to mathematical modeling of developmental systems. As described in chapter 4, the form and drift of interactions among large collections of signaling networks, TFs, RNAs, and so on, is best described in terms of attractor states.[6] It may be remembered from chapter 4 that one of the characteristics of such systems is that they are

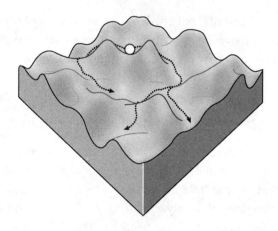

FIGURE 5.5

Developing cell (ball) is drawn into one or other final attractor states (cell types) in an attractor landscape. (From B.D. MacArthur, A. Ma'ayan, and I. R. Lemischka, "Systems Biology of Stem Cell Fate and Cellular Reprogramming," *Nature Reviews Cell Biology* 10 (October 2010): 672–681. Reprinted with permission.)

both very responsive and suitably adaptable under changeable conditions. This contrasts with the rigidity and lack of adaptability in predetermined states and processes.

A diagrammatic representation is shown in figure 5.5. Here a previously undifferentiated stem cell is drawn by morphogenic fields into one or another attractor states (cell types). The topographical figure shows only three dimensions. In reality, there will be many more variables in the multidimensional attractor space.

In this way, we can see that there is nothing radically new about the intelligence of development. The operational logic among components in single cells has been extended to regulate interactions among cells. That is how they become coordinated, responding cooperatively to an even more changeable outside. As before, potential—its form and (now more extreme) variation—is not encoded in genes but emerges in self-organized systems. With these general principles in mind, let us continue the story.

THE BUILDING OF BODY STRUCTURES

In a matter of hours, the cells of the gastrula have started to differentiate into hundreds of different cell types. The essential body structures, body axis and segments, and organs soon begin to form under dynamical regulation.

The brain is, of course, the most complex organ of the body. The first step in its formation is the transformation of part of the ectoderm (see figure 5.1) into a neuro-ectoderm. It involves interactions among a number of morphogens forming an underlying "organizer." This was shown in experiments in the 1920s in which transplanting part of this organizer to another part of the ectoderm induced the development of an almost complete second nervous system, as well as the supporting body axis.

Embryologists seem to have had some fun showing how potential originates in morphogen interactions rather than in codes in genes. As with the brain, teeth in mammals originate as interactions between the mesenchyme and ectoderm. The general impression is that birds (*Aves*) lost their teeth when evolving from reptiles and adopting different diets. But then mouse dental mesenchyme was experimentally grafted onto chick epithelium. It resulted in the development of a hen with a variety of dental structures, including perfectly formed crowns! As the experimenters explained, "The results suggest that the loss of teeth in *Aves* did not result from a loss of genetic coding . . . but from an alteration in the tissue interactions requisite for odontogenesis."[7]

In a more recent review (2008), Irma Thesleff and Mark Tummers (in *Stem Book*, http://www.stembook.org/node/551) explain how even something as seemingly simple as tooth development involves a host of morphogens. They are used reiteratively during advancing tooth development at a number of hierarchical levels and are coordinated by other factors, including a BMP. BMP is actually a family of morphogens that regulates several developmental processes, including patterning and differentiation of limbs as well as teeth. Ultimately, at least two hundred genes are recruited in the regulation of dental organogenesis.

In sum, changes in the cells of developing tissues appear to depend on a seeming storm of signals from other tissues, ebbing and flowing over time, with criss-crossing gradients in many directions. The storm is more like a symphony than a discordant racket, operating with nonlinear dynamics rather than a linear command sequence. Development, that is, constitutes a responsive, intelligent system, with many compensatory mechanisms and creative options in an ever-changing environment. This process has many other striking consequences for understanding the construction of form and individual differences.

UNIFORMITY AND VARIATION IN DEVELOPMENT

Canalization

Of course, we do know that human individuals, like all animals, do vary genetically. But not randomly so. Remember that it is a basic principle of natural selection that this genetic variation will be reduced for forms and functions important to adaptation. Genes that do not supply the goods needed for survival are simply eliminated as their bearers fail to reproduce. This is why any two humans are genetically more than 99 percent identical.

However, even 1 percent of genes is still a lot of genes to vary, especially when we remember that each gene consists of thousands of single nucleotide polymorphisms that themselves can vary. What are the implications of that for development?

In fact, the evolved intelligent system of development usually copes with such genetic variation perfectly well by using alternative resources and/or different routes of development. Biologists have noted that the development of important forms and functions is remarkably durable, or "robust." And, as mentioned in chapter 4, many genes can be eliminated without affecting function. So the physical bumps and shocks of the environment that might be expected to disrupt development seem to have little effect in most cases. Development continues regardless.

In experiments in the 1940s, C. H. Waddington discovered that for characteristics crucial for survival, development is strongly buffered. That is, fairly standard forms and functions appear in all individuals, regardless (within wide ranges) of variation in the individuals' genes, or in the environment during development. More broadly, Waddington realized that there must be some layer(s) of regulation between genotype and phenotype that keep development on a uniform course. He introduced the concept of *epigenetics*, meaning above or beyond the genes.

Such "canalization" of development, as Waddington called it, is found in all basic aspects of the body: eyes, limbs, internal organs, and so on. At the level of behavior, canalized development accounts for what has often been described as instincts, also often (mis)attributed to genes. When people speak of a trait as being genetic or innate, it is just such a complex of developmental regulations that is actually being referred to: it will require gene products, but it is not in the genes.

Canalization suggests intelligent interactive systems that can adapt by modifying pathways of processing. By helping to reveal it, Waddington was instrumental in contrasting preformationist theories of development (potential present in the fertilized egg) with the theory of epigenesis (potential emergent from interactions among simpler components). This history is well described in the article by Paul Griffiths and James Tabery.[8]

Until recently, the details of how those interactions work were still only theoretical. Although Waddington devised the term "epigenetic," it still connotes cellular processes subordinate to quasi-executive genes, whereas it now seems more likely to be the other way around: the genes are used as a resource by the intelligent system of the cell. Gilbert Gottlieb has noted this persistence of "preformationism" in biology beneath a superficial layer of epigenetic clothing (as also noted by Griffiths and Tabery).

Today, that relationship and the processes supporting it are a little clearer. It is thought that changes in protein folding, enzyme activity, TF recruitment, and so on can create alternative metabolic pathways that compensate for proteins that are absent or have been modified by gene mutations. In dynamical terms, we would say that alternative routes to the attractor state have been created.[9]

There is much research to support that view. For example, Orkun Soyer and Thomas Pfeiffer showed how biodynamics "leads to evolution of metabolic networks that display high robustness against gene loss . . . underlined by an increased number of multifunctional enzymes and independent paths leading from initial metabolites to [full development]."[10] In other words, there is self-organized patterning as with the Bénard cells described in chapter 4. Only in cases of rare deleterious gene mutations will the system be unable to cope with the loss of resource and a disease state will ensue.

Developmental Plasticity

Canalization of development is obviously advantageous when a single form or function fits a predictable environment—one that recurs from parents to offspring across many generations. Having initially appeared accidentally, such a single, well adapted, form would have been hugely favorable to survival. It and any gene variants supporting it would have been targets of strong natural selection.

However, offspring can sometimes experience significant aspects of environments not experienced by immediate parents. This demands a different kind of development: an adaptability to changed circumstance in a way that could not have been foreseen in the information in genes. The phenomenon came to be called *developmental plasticity*.

A popular example is that of the water flea, *Daphnia*. If the juveniles develop in the same body of water as a predatory midge larvae (*Chaoborus*), they develop a protective neck spine or helmet and extended tail spine. These defensive structures allow the *Daphnia* to escape from their predators more effectively. This structure is completely absent in their parents, who have developed in predator-free water.[11]

Numerous other cases of intelligent, predator-induced, plasticity have been studied. Tadpoles of the wood frog (*Rana sylvatica*) that grow in water previously containing predatory dragonfly larvae—and, presumably, some chemical produced by them—develop bigger tails that allow faster swimming and turning. A species of barnacle reacts to the presence of predatory snails in its environment by developing a bent shell form that is more resistant to predation compared with the more typical flat form.

The logic of such processes is fairly obvious. It would be wasteful of developmental resources if the defensive structures were produced when they are not needed. It makes sense to leave the direction of growth and reproduction until after the adult habitat has been reached. So they are induced on the spot by a chemical substance released by the predator.

Sometimes gross changes in morphology are involved. Some of the most striking examples are the different castes in bees and ants. Here, developmental plasticity radically alters behavior and physiology as well as anatomy and is unrelated to genetic variation. Locusts also develop physiologically and behaviorally distinct morphs in response to current population densities. Likewise, the sex ratio in certain reptiles is known to be developmentally plastic. Each embryo develops into a male or a female depending on local conditions, such as temperature, at the time (investigators incubating eggs in the laboratory were initially astonished to find that offspring were either all male or all female). Classic metamorphosis in frogs and other amphibians involves the remodeling of almost every organ in the body, and radical changes in behavior from filter feeder to predator and in locomotion and visual systems.

Not surprisingly, in some cases of developmental plasticity, the differences between the morphs have been so stark as to lead them to be classified as different species and presumed to have different genes.

Of course, the most celebrated example of such life-long developmental plasticity is that in the brain. Canalized development seems to ensure that neurons in different layers in the cerebral cortex—the most recently evolved aspect of the brain—form the requisite variety of processing types. These will then be "good enough" for the demanding tasks ahead. But the wiring up of these into specialized areas, with their specific response properties, seems to depend on context and experience.

Such plasticity of development has been demonstrated spectacularly in many ways. Many years ago Mriganka Sur and his colleagues surgically rerouted visual nerve connections from the eye in newborn ferrets away from their usual destination in the visual cortex of the brain. They directed them, instead, to what usually develops as the auditory cortex (i.e., processing auditory information). That auditory cortex subsequently came to process visual information like an ordinary visual cortex.

Somewhat similarly, it has been shown that plugs of cortex transplanted from, say, visual to somatosensory areas (responding to sensations of touch), develop connections characteristic of their new location rather than those of their origins. Finally, it has been shown how the functions of one area of the brain, surgically removed, can be taken over by another.

Developmental plasticity obviously reflects potential and variation being created by intelligent systems, rather than from the blind codes in genes. It will be favored when there is environmental heterogeneity in space and time. However, the cases considered so far are limited in that they involve a one-shot developmental trajectory. Development has a fairly definite endpoint, when the adult structure/function has been constructed. Some further maturation may take place, and some environmental change can be coped with, but only that which varies within a permanent set of rules and range of variation. That model of development and function has also been prominent in much cognitive theory (see chapter 7).

Living things that have been forced into more complex environments, however, will continue to experience change on macro- and microscales throughout life. One-shot developmental plasticities cannot cope with such environments. What is required for survival in such conditions are living structures and functions that can adapt to such changes throughout the life of the individual.

Lifelong Plasticity

In the early twentieth century, James Baldwin was the first to suggest developmental plasticity as a strategy for dealing with environmental novelty throughout life. Simple examples abound. They include tanning in the sun; growing protective skin calluses at points of friction; increasing muscle bulk and strength with exercise; and, in animals, changing hair/fur color and thickening with the seasons. General physiological fluctuations in response to rapidly changing environments have also been described.

Lifelong plasticity seems to have emerged quite early in evolution, as an extension to developmental plasticity. Certain plankton species really need to be transparent for protection against fish predators, yet they also need protection against UV light. Accordingly, they can reversibly develop body pigmentation as and when needed. Whereas the skin coloration of most reptiles is fixed at the end of development, that of the chameleon retains a lifelong changeability. Birdsong control systems exhibit seasonal plasticity in many species. These include dramatic volume changes of entire brain regions in response to photoperiod (length of day) and its impact on circulating levels of sex steroids.

Sometimes, lifelong plasticity is seen in spectacular body changes, as in certain coral reef fish. As described by Gilbert Gottlieb, "These fish live in spatially well-defined social groups in which there are many females and few males. When a male dies or is otherwise removed from the group, one of the females initiates a sex reversal over a period of about two days, in which she develops the coloration, behavior, and gonadal physiology and anatomy of a fully functioning male."[12]

A wide variety of such inducible phenotypes, responding to environmental signals across the lifespan, are known. In some social insects, soldier castes may be induced by collateral changes in their prey, such as the appearance of more defensive phenotypes in aphids. These changes are akin to rabbits changing to porcupines because wolves have been smelled in the woods—and then calling on more of their peers to do the same.

By far the most spectacular example of lifelong plasticity, of course, is that in the brain and in learning and behavior. Such plasticity has already been mentioned, and chapter 6 is largely devoted to it.

Much developmental plasticity depends on other epigenetic processes that alter maternal genes. You may remember from chapter 4 that these are chemical tags placed on the DNA and/or histones (proteins surrounding the DNA and controlling access to genes) during egg formation. It is known, for example, that the DNA in the egg is heavily tagged in this way. Conversely, there is a massive, but highly selective, detagging in the very early embryo. This is another phenomenon that shapes developmental futures in the light of current environmental conditions.

I have shown how the principles of development in multicellular organisms are extrapolations of those evolved in single cells. They permit even greater flexibility in the face of more changeable environments. Waddington drew diagrams like that in figure 5.6 to illustrate development and differentiation of the hypothetical organism along an epigenetic landscape—but here depicted as emergence of attractor states, through perturbations, in an attractor landscape.

In tracking lifelong, rapid environmental changes, however, these principles have been hugely augmented by adaptable physiology, behavior, and the brain and cognitive systems. Before turning to them, it is worth considering some evolutionary implications.

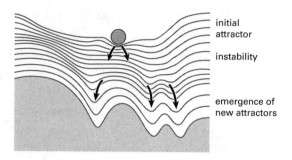

initial
attractor

instability

emergence of
new attractors

FIGURE 5.6

A developmental landscape depicted as progression of an initial state (the ball) with the emergence of new attractors (differentiated tissues, organs, individuals, etc.) over time. (Redrawn from E. Thelen and L. B. Smith, *A Dynamic Systems Approach to the Development of Cognition and Action* [Cambridge, Mass.: MIT Press, 1994].)

DEVELOPMENT AND EVOLUTION

It is natural to think of these intelligent developmental processes as direct products of evolution:

evolution → development.

However, we now know that it can work the other way:

development → evolution.

There are many possibilities. For example, by buffering development, canalization allows genetic variation from mutations to accumulate. Normally, such genetic variation in important functions would be reduced or eliminated by natural selection. Under canalization, however, it may be sheltered and consequently accumulate in the genome. (This is another way in which genes that appear to be independent are actually brought into highly interactive processes). However, this cryptic genetic diversity can be subsequently exposed by extreme changes in the environment that canalization cannot cope with. It might then become available to natural selection and contribute to further evolution.

A classic example involves "heat shock" proteins in fruit flies. As their name suggests, this is a collection of proteins that seem to offer protection against sudden temperature changes during development. Temperature changes might affect variation in traits, recruiting variable genes in the process. But the absence of such variation in most critical traits in flies, through canalization, gives the impression of little underlying genetic variation. However, when transcription of heat shock proteins is chemically inhibited, huge variations in nearly every structure of the body suddenly appear.

The effect is rather like a population of humans suddenly producing a generation of offspring exhibiting a tenfold increase in variation in things like height, facial appearance, numbers of legs, as well as developmental defects. It indicates how much of the genetic variation in organisms is normally buffered in development by regulatory interactions. Only under extreme circumstances does the formerly hidden variation become evident. By being correlated with trait variation, such genetic variation is now available for natural selection, sometimes resulting in altered evolutionary trajectories. If stability of the environment is subsequently restored, then canalization of new traits may once again evolve.

From at least the early 1920s, it was suspected that evolution and development may be intertwined, and the suspicion has grown in recent times.[13] It is now thought that dynamical developmental processes, by innovating new responses to changed environments, can account for evolution of new species, often in rapid bursts, rather than gradual progress. This echoes Robert Lickliter's point that such aspects of development "generate the phenotypic variation on which natural selection can act."[14] Many biologists now believe that development has been a major cause of evolutionary innovation. In reviewing the evidence, Mae-Wan Ho concludes that, in such ways, living things become active participants in shaping their own development and future evolution.[15]

IS POTENTIAL IN THE EMBRYO?

Intelligent developmental systems ensure that all complex forms and functions are created in the course of development, often in surprising ways, and with little correlation between genetic and phenotypic variation. Nevertheless, the general public, as well as many scientists, still hold on to the idea that differences in potential lie in differences already coded in genes in the embryo. That view implies that differences in individuals in important functions are largely due to differences in genes. As we have seen, though, things are far from being so simple.

In the case of the individual, the original cells are totipotent; in spite of the same genes they have the same potential to become any kind of differentiated cell for a particular organism. In the French flag model, every cell has the potential to develop as white, blue, or red. Indeed, recent advances in the laboratory prove that it is possible to change virtually any cell type into another cell type—to recover their potential for diversity. This is epigenetic reprogramming, in which a developed specific potential is turned back to become a totipotent stem cell.[16]

But what about different individuals? When people think of original potential, what they really mean is how it differs from individual to individual, with the implication that some will have more potential than others for traits like intelligence. And those differences, they have been told, reside largely in differences in the genes.

However, it is very difficult to describe potential in such simple terms, as demonstrated by in vitro fertilization clinics. The clinics talk to anxious would-be parents about "egg quality." But the concept turns out to be rather vague. In rare cases, worrying variation can be spotted as chromosomal or other defects, usually visible under the microscope in the first two or three days. Otherwise, quality criteria consist of eye-balled (under the microscope) physical features, such as cell number in the three-day embryo, cell regularity and uniformity, or degree of any fragmentation.

Indeed, most labs admit that any generalizations about quality made from grading embryos are rather inaccurate. As one site puts it: "We see some cycles fail after transferring 3 perfect looking embryos, and we also see beautiful babies born after transferring only one low grade embryo." The best test of egg quality in fact seems to be female age. Much the same applies to sperm quality.

What parents tend to worry about, of course, is "genetic potential"— usually for critical functions like brain and cognition. Thanks to the kind of hype and publicity mentioned in chapter 1, these parents are mostly convinced that the potential for desirable qualities like intelligence, or special talent of some sort, resides and varies in the genes. And they think of it as already residing in the egg, in that homunculus state mentioned earlier.

As we have already seen, though, what is assumed to be genetic is really the manifestation of a whole developmental system. So, with the exception of rare disorders, the anxiety is probably unwarranted. And there is certainly no test for the genetic potential of eggs or sperm from different individuals, however much the idea inspires the pipe dreams of some behavioral geneticists.

INTERIM SUMMARY

So far this chapter has dealt with development in multicellular organisms. The general message is that development itself is an intelligent process, not a fixed program. It generates almost incredible variation of form and function on the basis of structural information gathered from internal and external environments. Developmental processes among cells also incorporate and regulate the intelligent systems within them, adding a new level to a regulatory hierarchy. It all ensures that functional potentials are not merely expressed; they are made.

Some developmental systems provide plasticity throughout life. But they remain limited in their speeds of response to change. Other intelligent systems—physiology, brains, behavior and cognition—are needed to deal with more rapid environmental changes. The emergence of such intelligent systems has been the most significant and exciting manifestation of evolution, though it is often overlooked as such. The rest of this chapter deals with the first of these systems.

PHYSIOLOGY

Intelligent developmental systems ensure that the myriad different cells differentiate and find their proper places, at the right times, in a whole assembly. But their activities then need to be coordinated so that the whole functions harmoniously in the ever-changing environment. As with development, that coordination is achieved through signaling systems among cells that are also sensitive to environmental changes outside the body. This is what is called "physiology." The nature of physiology, and how that coordination occurs, tells us more about the nature of potential and its variation.

Physiology is based on communication, a crucial obligation of cells and tissues in multicellular organisms. One cell separated from the rest in a culture dish soon dies without the usual storm of signals from others, even when provided with all the nutrients and other conditions it needs. Very early in evolution, then, a wide range of chemical messenger systems (including pheromones, prostaglandins, interferons, and hormones) became the crucial aspect of the physiology of organisms.

The most prominent and best known of these forms the endocrine system: a collection of glands that produce hormones that influence many functions of the body. Released into the circulating body fluids (or bloodstream in more complex species), they find their way to specific target cells and inform them about what to do. More than fifty human hormones have been identified, all acting by binding to target cell receptors, then firing up the internal signaling systems as described in chapter 4.

It used to be thought that physiology performs an essential equilibrium or homeostatic function, reflecting idealistic views of the environment as essentially stable or cyclically recurring. In that view, each physiological function independently maintains, as far as possible, some constancy of part of the internal milieu—blood sugar, temperature, salt balance, or whatever—in the face of disturbances from inside or outside. As Steven Rose pointed out in his book, *Lifelines*, "No modern textbook account of physiological or psychological mechanisms fails to locate itself within this homeostatic metaphor."[17]

A good example would be body temperature regulation. To be sure, regulation includes a thermostat-like center located in the hypothalamus at the base of the brain. Nerve receptors in the skin and spinal cord provide feedback about changes to outside and inside temperatures. The body is then driven to either conserve heat (piloerection, or hairs standing on end), produce heat by shivering, stimulate adrenaline production to activate muscles, or increase heat loss through sweating or panting.

However, the unconditioned thermostat metaphor is too simple. Changes in the heat regulating process are constantly undergoing modifications that go unnoticed. The set point is constantly being revised according to circumstances. It is not a static process but an ever-moving one that is important to normal body function.

For example, invading bacteria and other bodily breakdown processes (e.g., from extreme exercise) produce pyrogens. These are confronted by the immune system as toxins. In response, signaling molecules (cytokines) are produced and released from immune cells. The cytokines reach the hypothalamus, and the thermostat gets jacked up. The elevation of body temperature produces the familiar symptoms of fever. But it also suppresses bacterial metabolism and stimulates the movement, activity, and multiplication of white blood cells, as well as the production of antibodies. When the new temperature has been reached, the thermostat is turned down again.

It sounds like a simple cue-response mechanism with feedback. But it is much more than that. The cytokines have to operate through a number of intermediaries in the brain in different brain areas. It is the interactions among these that ultimately orchestrate the body-heat-changing mechanisms via the autonomic nervous system (causing shivering and vasoconstriction). The reason for all of this is so account can be taken of what else is occurring in the rest of the body and in the wider environmental context. Instead of a crude categorical change, the degree of reset can be calibrated more exactly to a wider structure of needs.

That circumspection is important, because many other systems, such as circulation and respiration, are affected by body temperature. Regulation is also coordinated with the animal (including human) circadian rhythm. This includes the rise and fall of temperature responsible for night-time (and afternoon) drowsiness and later recovery. The process is a highly dynamic orchestration, rather than a simple cue-response switch. This general principle has become increasingly obvious in studies of hormone functions.

Hormones Interact

That hormones interact and do not usually act as independent switches has been known for a long time. All major functions depend on their integration and coordination: digestion, metabolism, respiration, sensation, sleep, excretion, lactation, response to stress, growth and development, heart function and blood circulation, reproduction, motivation, mood, and so on. One example is thyroxine from the thyroid which helps regulate temperature and general metabolism. Another is the wide range of steroid hormones, produced in the adrenal glands or in the gonads. They are involved in an even wider range of anti-inflammatory, anti-stress, and sexual functions, as well as integrating general cell/tissue metabolism. Some of them regulate sodium and potassium levels and the state of body hydration (and thereby—e.g., through thirst—influence behavior).

As with other intercellular signaling, the integration of hormones is mediated through receptors on cell membranes, to which the hormones bind before activating internal transduction pathways. But they meet with interactive rather than independent effects. The effects depend on wider bodily states and events in the outside world.

Most common are synergistic effects in which two or more hormones produce effects greater than the sum of their separate effects. One hormone may enhance the target organ's response to another hormone, even one secreted at a later time. Conversely, there are antagonistic effects in which one hormone inhibits the effects of another. For example, insulin lowers blood glucose level, while glucagon, produced by the liver, raises it. Ordinarily these will be well orchestrated, but the effects of hormone imbalances are well known in medicine and can be difficult to diagnose and treat.

The neuroendocrine stress axis—or, to give it its full name, the hypothalamic-pituitary-adrenal axis—is a key physiological system. It regulates responses to stress, either from internal or external sources. As such it affects, and is affected by, many body functions, such as digestion, the immune system, energy metabolism, and emotional aspects of psychology. Part of the classic stress response consists of secretion of corticotropin-releasing hormone from the hypothalamus in the midbrain. It passes quickly to the nearby pituitary gland, which actually releases the hormone corticotropin. This then enters the bloodstream to reach the adrenal glands above the kidneys. The steroid hormone cortisol is released into the bloodstream, followed by aspects of the classic stress response, such as feelings of alarm, muscle tension, and increased heart rate.

Again, it sounds rather like a straightforward cue-response (or reflex) cycle. But there are many other players in the stress-response system. One is noradrenalin, produced in the *locus coeruleus*, a small nucleus in the hindbrain. This center is connected with all the primary senses, and also with cognitive and social activities. The noradrenaline is released via nerve fibers into numerous parts of the brain in response to perceived challenge. It promotes a state of excitement and awareness (as well as the release of corticotropin-releasing hormone from the hypothalamus, as just mentioned).

The other player is adrenalin, which is released from the adrenal cortex following stimulation from nerve branches terminating there, themselves being stimulated by fibers from the *locus coeruleus*. Together the hormone interactions produce the classic preparation for "fight or flight," including increased heart and respiratory rates, dilation of arteries to muscles, constriction of peripheral blood flow, release of blood sugar for energy, increase in blood pressure (to get blood to the muscles), and suppression of the immune system.

Although superficially plausible, therefore, the general stress-response concept has turned out to be too simplistic. Responses now appear to be more variable, depending on individual histories and current contexts, with diverse psychological and neurological consequences in both animals and humans. These include emotional dysregulation, panic attacks, post-traumatic stress disorder, and many other states. They are further evidence of an intelligent system attempting to learn from the past to prepare for the future.

In contrast, more congenial conditions, permitting the systems to coordinate as they were evolved to do, fosters great benefits for the body as a whole. As an article in *Nature Reviews: Neuroscience* explains, integration "results in the stress instruments producing an orchestrated 'symphony' that enables fine-tuned responses to diverse challenges."[18]

Research is making it clear that, as for cell metabolism and development, there is deeper structure in physiology. As with processes in the cell and in development, function and variation in physiology are best modeled by nonlinear dynamics rather than cue-response reactions. Indeed, nonlinear dynamical models are being increasingly used as analytical and conceptual tools in studies of physiology.[19]

Remember that a major advantage of nonlinear dynamical systems over deterministic ones is the rapidity and creativity of response to perturbations. So the confluence of numerous variables is constantly moving physiology from one attractor state to another. It suggests the (lifelong) emergence of multiple metastable states for any physiological function rather than a single steady state. That is, these systems hold, in their developed networks, numerous possible states (attractors) in readiness for changing conditions.

From this perspective, disease or other malfunctioning reflects a breakdown of the deeper structure of the system. As Bruce West says, it involves a new understanding of physiology and life processes. It suggests that disease and aging are associated with the loss of complexity, or loss of interactions among component networks, and not with the loss of simple homeostatic regulation as such.[20]

The view is supported by, for example, nonlinear dynamical studies of the physiology of cardiac functions. Variation in heart rate reflects the totality of our physical, mental, and emotional state as we interact with changes around us. Under such real-life conditions, "mode locking" on a single steady-state would restrict the functional responsiveness of the organism. Ary Goldberger and colleagues agree that "a defining feature of healthy function is adaptability, the capacity to respond to unpredictable stimuli and stresses." Simple cue-response mechanisms "would greatly narrow functional responsiveness."[21] In other words, it is the breakdown in such deeper, integrative responsiveness to changing conditions that produces disease states.

Physiology is another evolved intelligent system: so smart, in fact, as to suggest almost brainlike processes at work. This is no doubt why Charles Darwin attributed brainlike activity to the developing tips of plant roots: "It is hardly an exaggeration to say that the tip of the radicle thus endowed, and having the power of directing the movements of the adjoining parts, acts like the brain of one of the lower animals; the brain being seated within the anterior end of the body, receiving impressions from the sense organs, and directing the several movements."[22]

Richard Karban says something similar about physiology in his book, *Plant Sensing and Communication*: "Plants communicate, signaling within [themselves], eavesdropping on neighboring individuals, and exchanging information with other organisms." They have adaptable responses, he says, such that, if they happened at speeds humans understand, would reveal them to be "brilliant at solving problems related to their existence." Or, as an article in *New Scientist* (May 30, 2015) puts it: "They are subtle, aware, strategic beings whose lives involve an environmental sensitivity very distant from the simple flower and seed factories of popular imagination."[23]

Of course, we now know that physiological functions incorporate the intelligent processes of the cell, described in chapter 4, but integrated at a higher level. As in the cells, that involves using genes as resources, not as commands. As a result, physiology amplifies and extends developmental functions and becomes a new intelligent system—a new level of intelligence. It achieves greater adaptability of living things by creating wider variation in responses and response tendencies than the information in genes alone could ever create. On a lifelong basis, physiological processes are constantly responding to numerous variables in the internal and external environments. Thus they create the major life transitions (e.g., metamorphosis in reptiles and insects, or puberty in mammals). And they constantly recalibrate the system as a whole to environmental change.

Individual Differences in Physiology

Multicellular organisms evolved from single cells, but they needed physiology to coordinate their cells. As we have seen, though, this was only possible with an intelligent physiological system based on dynamical principles. So what does this tell us about the causes of individual differences in physiology? Physiology is often assumed to be a good model for a "biological" view of psychological intelligence and individual differences in it. The founder of the IQ testing movement, Francis Galton, was convinced that natural ability varies as if it were a physiological trait. For that reason, he and his followers based their first intelligence tests on sensory motor tests, like visual discrimination and reaction times.

Somewhat similarly, Ian Deary and colleagues, pursuing the notion of a general intelligence factor, *g*, have suggested the existence of a general biological fitness influencing the growth and maintenance of all bodily systems. That is what IQ tests measure, they say. And the same notion is reflected in attempts to relate IQ scores to physiological measures, like speed of processing or using fMRI scans to relate IQ with cerebral activity levels (on which, more in chapter 6).

The main aim, of course, has been to imply something genuinely fundamental in IQ testing—and the concept of *g*—when there have always been doubts about their validity. The logic is that individual differences in IQ are expressions of the same so-called biological forces that are found in physiology, with some of the variation traceable to genes and some to environment. So it should be interesting to look at individual differences in physiology to see whether there really is a parallel.

No doubt individuals vary on a wide range of physiological measures. Take, for example, a simple measure of physiological state, such as basal metabolic rate. This is the minimal rate of energy expenditure per unit time by individuals at rest. In humans there are big individual differences in basal metabolic rate. One study of one hundred and fifty adults in Scotland reported basal metabolic rates between 1,027 and 2,499 kilocalories per day. The researchers calculated that 62.3 percent of this variation was explained by differences in fat-free body mass (i.e., bigger bodies use more calories). Variation in fat mass accounted for 6.7 percent and age for 1.7 percent. The rest of the variation (26.7 percent) was unexplained, but it was not due to sex differences or the size of different energy-demanding organs, like the brain.[24]

Estimates of how much of the variation in such physiological functions is associated with variable genes (i.e., the heritability) have, in humans, relied on the twin method. As described in chapter 2, such estimates are unreliable, because they include unknown amounts of environmental and interactive sources of variance (mislabeled as "genetic"). Estimates of the heritability of basal metabolic rate in animals tend to be small, although subject to measurement and other difficulties.[25] However, this should not be surprising: genetic variance for important traits will tend to be much reduced by natural selection or by epigenetics.

Blood pressure in humans and animals also shows great individual differences. Heritability studies have produced widely divergent estimates of heritability, again based on the unreliable twin method. Already we are seeing that trying to draw a parallel between IQ and physiology is not so easy.[26]

There are measures of many other aspects of physiology. These include heart rate, skin resistance, skin temperature, muscle tension, neuroelectrical activity, and dozens more. However, one of the most striking things about physiological functions is the extremely wide range of variation within which normal, adequate function seems to operate. The following list presents the normal ranges of markers of physiological functions in a standard full blood count. These levels are all critical indicators, or biomarkers, of complex physiological functions. Only deviations beyond these wide limits suggest abnormality:

Red blood cells: 4.5–6.5 trillion cells per liter

White cells: 4.0–11.0 billion cells per liter

Platelets: 140–400 billion per liter

Neutrophils: 2–7.5 billion per liter

Lymphocytes: 1.5–4.0 billion per liter

Monocytes: 0.2–0.8 billion per liter

Vitamin B12: 150–1,000 nanograms per liter

Serum ferritin: 12.0–250 micrograms per liter

Serum folate: 2.0–18.8 micrograms per liter

Vitamin D: 50–75 nanomoles per liter

Serum urea: 2.5–7.8 micromoles per liter

Serum creatinine: 75–140 micromoles per liter

Serum albumin: 35–50 grams per liter

Alkaline phosphatase: 30–130 microns per liter

Serum globulin: 20–35 grams per liter

Immunoglobulin A: 0.5–4.0 grams per liter

Immunoglobulin G: 5.5–16.5 grams per liter

What these figures suggest is that, within these very wide limits of variation, the system functions well enough. In a dynamic, interactive system, so long as a basic threshold is reached, the exact levels do not matter. Indeed, as described earlier, it is part of the function of physiology to create adaptable variation in changing environments. And such functioning seems, in the vast majority of individuals, to apply quite adequately. It is also worth mentioning that few of the distributions assume a normal curve, a basic assumption of IQ test construction and of statistical attempts to measure heritability in human intelligence.

That appears to be the case for other aspects of physiology. Take immunocompetence, which is defined as the ability of the body to produce a normal immune response following exposure to foreign proteins, pathogens, or other toxins. It evolved early in vertebrates and in humans; it rapidly develops in the first month or so of embryonic life. Various specific measures of level of functioning can signify disease states. But functioning is affected by many experiential factors, including physical and mental stress, nutrition, and age.

In other words, the immune system interacts intensively with many other aspects of physiology, resulting in wide variations in indices at particular times. Again, there are categorical disease conditions, some of which can be associated with rare genes. But, so far as I know, no one dreams of ranking individual differences on a general scale of immunocompetence. Most of the time in most people it functions well enough and only draws attention in the relatively few cases when something is seriously wrong.

In sum, no physiologist would dream of suggesting the following:

(a) that within the normal ranges of physiological differences, a higher level is better than any other (as is supposed in the construction of IQ tests);

(b) that there is a general index or "quotient" (à la IQ) that could meaningfully describe levels of physiological sufficiency or ability and individual differences in it;

(c) that such "normal" variation is associated with genetic variation (except in rare deleterious conditions); and

(d) that genetic causation of such variation can be meaningfully separated from the environmental causes of variation.

All this further suggests that the popular conceptual model of IQ and its heritability have far more to do with social ideology than with physiology. Like the intelligence of the single cell, described in chapter 4, physiology is an intelligent system that can buffer large ranges of deficiencies and, in the vast majority of cases, can create compensatory pathways to adequate function.

A preoccupation with ranking variations, assuming normal distributions, and estimating their heritabilities simply does not figure in the field of physiology in the way that it does in the field of human intelligence. This is in stark contrast with the intensity of the nature-nurture debate in the human cognitive domain. But perhaps ideology has not infiltrated the subject of physiology as much as it has that of human intelligence.

6

HOW THE BRAIN MAKES POTENTIAL

POTENTIAL IN THE GENE-BRAIN

To most people, it seems, potential originates in the genes but develops in the brain, and there it gives rise to intelligence; that is, the potential in the genes makes the brain. This is also scientific orthodoxy. As summarized by Toga and Thompson (see chapter 1), alleged individual differences in potential are "partly mediated by brain structure that is likewise under strong genetic control. Other factors, such as the environment, obviously play a role, but the predominant determinant appears to be genetic." The United Kingdom's Royal Society says something similar in its *Brain Waves* initiative.

In the same groove, behavioral geneticists often say that their objective is to describe the pathways from genes, via brains, to intelligence. The fact that no such pathways exist (at least in that genetically deterministic sense) probably explains the difficulty experienced in finding them. Unfortunately such gene-centered thinking has also obscured what the brain really does in relation to intelligence. As Robert Plomin and colleagues say in their book *Behavioral Genetics*, "It has been difficult to connect the dots between genes, brain, and behavior."[1]

This chapter aims to explain that difficulty. Yet again, behavioral geneticists have cause and effect the wrong way around. We have already seen how the intelligent systems of the cell create potential; physiological and developmental systems do likewise under more changeable conditions. Just as Charles Darwin attributed brainlike activity to the physiology of plants I will show how brains further evolved from physiology as even more intelligent systems in more changeable environments. In this chapter, I show how (very dynamic) brains are even better at making potential.

MORE METAPHOR, MORE IDEOLOGY

The root of the problem in connecting the dots is that the brain, like the concepts of potential and intelligence themselves, has also become a vehicle for ideological expression. We pour into it intuitions derived from our social structures and then appeal to brain sciences—as we do with genes—to justify those social structures.

The first hint of that process is found in the widespread resort, not to actual brain functions that might constitute intelligence and individual differences in it, but to metaphors. Most of those are crudely mechanical, like a calculator or computer or some other machine. "High cognitive ability," says James Flynn in his 2013 book on IQ, "begins with genetic potential for a better-engineered brain." Elsewhere, he claims that "a good analytic brain is like a high-performance sports car."[2]

Accordingly, individual differences in general intelligence (or g) are said to reflect differences in the speed, efficiency, or power of the thinking brain; while the brain itself is reduced to an engine with different capacities, with more or fewer interconnections among the parts. The underlying assumption is that, like a machine or computer, there is a fixed logic of operation, reflecting built-in (innate) rules or processes, with individual differences in efficiency.

Other metaphors have more roots in social structures and labor organization, as when the brain and its functions are likened to a factory with its divisions of workers and hierarchy of departments, all under the control of a "central executive." For example, executive functions are often ascribed to the frontal lobes of the brain (partly because they are bigger in humans than in other species). As Elkhonon Goldberg puts it in her book *The Executive Brain*, "The frontal lobes are to the brain what a conductor is to an orchestra, a general to the army, the chief executive officer to a corporation."[3]

A second hint of ideological influence is that the metaphors are as vague as the popular notions of potential and intelligence: there is still little agreement among brain scientists about what the brain is really for. Obviously, those who are in the business of ranking brains should do so on the basis of a clear understanding of what brains are actually supposed to do—what they are supposed to be good at in the first place. Believe it or not, this still is not very clear.

Almost heroic research efforts have been made, with huge advances around very specific aspects of brain functions. They have produced mountains of particular "findings" about nerves and brains. But relatively little progress has been made on integrating them into a global synthetic view of how the brain works.

Thus investigators like Raquel del Moral and colleagues have complained about "the absence of a consistent central theory in the neurosciences." Jonathon Roiser explained in the *Psychologist* (April 2015) how "we lack a generally accepted neuroscientific explanation of how brains make minds." In his book, *The Lives of the Brain*, John S. Allen notes that, "while it would be going too far to say that it is a mystery or an enigma—we have collected an extraordinary amount of information about the brain over the years—an accurate rendering of the big picture, and lots of the little pictures as well, eludes us."[4]

Here I argue that the main reason why the "big picture" eludes us is the same as the reason there is still no agreed-on theory or model of human intelligence and potential for it. The search, at least with regard to intelligence, is for individual differences in ideological grounded metaphors, with little indication of where, in biological evolution, that intelligence came from, or what the differences are really differences in.

In this chapter, I present the remedy to that deficit. To do so, I need to go deeper into the information processing of a real brain in a real world. I argue that brain functions are extensions of the already-evolved intelligent systems in physiology and its predecessors. This also involves a closer look at the dynamics of changing environments. In doing so, I emphasize that individual differences in brain functions are vastly more interesting than ones of speed or power. Above all, I argue that a proper look reveals functional complexity that flies in the face of those who want to, in effect, declare a huge portion of humanity to be saddled with inferior brains.

A QUICK RECAP AND WHAT THE BRAIN
MIGHT BE FOR

Chapter 4 spelled out just what a real-world environment means. The upshot was that complex organisms evolved in highly changeable worlds, but ones full of spatiotemporal patterns. The patterns contain information deeper than simple associations and are assimilated, even in molecular networks, as statistical (relational) parameters. These are a kind of knowledge and help predict the consequences of responses. So responses can be optimized even in novel situations.

This is the intelligence of the cell. Such information cannot be found in genes per se, laid down by the experiences of past generations. It has to be abstracted de novo from contemporary experience.

Chapter 5 showed how the intelligence of the cell has been extended in development and physiology in multicellular organisms. These functions evolved to provide increased adaptability in even more changeable environments. A major feature, even at those levels, is the presence of compensatory (alternative) pathways to form, function, and their development. With so many idiosyncratic expressions, it is impossible (as well as pointless) to arrange individual differences along a single functional axis valid for all circumstances. For the vast majority of cells and organisms, their functions are good enough for the challenges faced.

This chapter extends that theme. I show how the evolution of nervous systems and brains was the continuation of a trend in the evolution of intelligent systems. This implies that the essential currency of the brain, and the cause of its development and its variation, is also the statistical structure of the environment. But now that structure is on a stupendous scale.

Instead of broadcast hormones, or local cell junctions, communication in the brain is through electrochemical signals channeled down long fibers (axons) culminating at specialized connections (synapses) on shorter fibers (dendrites) of receiving cells. A neuron can form synapses on potentially hundreds or thousands of other cells. At each of them, it sends and releases a neurotransmitter to activate receptors on the synapses.

Not only is this fast and to the point, but synapses are also specialized for integrating the statistical (structural) information from large numbers of other neurons. They can modify connectivity in the process, enhancing or inhibiting signals and so permitting enormous plasticity of the nerve networks in the brain. Advanced systems have millions of cells connecting with thousands of others through billions of connections. A staggering depth of statistical structure can now be captured for predictability and behavior.

All this creates much to discuss, but for illustration, in this chapter, I concentrate on vision as in a fairly general vertebrate brain. Most of what it reveals can be extended to other sense modes and to more general brain functions.

VISUAL GENIUSES

The visual system, in nearly all visually unimpaired individuals, presents a good example of the information sent to the brain and what the brain is interested in. Clearly some hard work has to be done there because, like any intelligent system, much has to be deduced from rapidly changing, incomplete information. The information deduced in the "mind's eye" is much richer (orders of magnitude richer) than that actually immediately presented to the physical eye. In vision, in particular, sensory inputs are in constant flux, complex, noisy, and full of ambiguity. So what exactly is the work, and how is it achieved?

Again a simple mechanical metaphor—this time the camera—has sometimes passed as the popular model of vision. In that model, the light-sensitive cells in the retina, in large numbers at the back of the eye, each picks up a tiny part of that image as tiny light spots, like pixels. After further processing, the whole snapshot is passed on, as an electrochemical discharge, down the optic nerve to the brain (this is *sensation*, or sensory reception). The brain reassociates these features, through a series of linear convergences into an image of the original object (this is *perception*). This image can than be handed over to "higher" centers in the brain for recognizing, classifying, memorizing, thinking, action planning, constructing a motor program, and so on (this is *cognition*).

And that is how we make sense of reality: a process of simply rebuilding the sensory image.

In the 1970s, this view was encouraged by experiments in which electrodes were inserted (in anesthetized animals) near cells in the retina or parts of the "visual" brain. The cells seemed to show sensitivity to one of a range of simple visual features, such as light spots, lines, or bars, shone into the eye. These activating features seemed to get more complex the "higher up" in the visual system the electrodes took recordings from. For example, some cells seemed more sensitive to angles, movement in a specific direction, or even whole objects. The tacit assumption is that each level in this cascade comprises some processing rules built into the circuitry for building the simple features into complex ones. An internal model of the external world is thus created from converging features.

It sounds simple, but there are many reasons such feature detection alone cannot create the functional image required. One of these is that the world is populated with objects that are three dimensional. These become collapsed as two-dimensional images on the retina. And the same two-dimensional image can be created by a variety of different objects/features. So the model implies a great loss of visual discrimination (the opposite of what actually happens).

Another problem is that the two-dimensional image would be grossly distorted on the back of the eye, because the eye is hemispheric (imagine watching photos projected onto a bowl-shaped screen) and the image is upside down. Moreover, the receptor surface is not a continuous sheet (like camera film) but an aggregate in the form of millions of light-sensitive cells. So the visual field is really detected in the form of a cloud of light spots.

Yet another problem is that the image of any single object is never still. We will be moving toward and/or around it; it moves to different places at different distances (with changes in apparent size); it may rotate, move behind other objects, become partially hidden, and so on. The movements of the image also have to be distinguished from the apparent motion created by ordinary eye movements. This is further exacerbated by the fact that the eye is constantly moving, from side to side in its socket up to ten times a second. There is simply no stable image to record.

In other words, the image of every object is under constant spatiotemporal transformation; the cloud of light points is in constant motion. And there is another problem. It takes thirty to one hundred milliseconds for sense receptors to respond and pass the signal on. So by the time we have perceived an object in motion, it is no longer exactly where we now see it. This makes actions like catching a ball more difficult than you think— not to mention catching prey or avoiding predators, like our ancestors did: actions need to be ahead of vision, as it were.

Finally, a single scene may contain dozens of objects. This is why William James, in 1892, famously referred to the "blooming, buzzing confusion" of sensory input. It has been described more recently as *"an onslaught of spatio-temporal change."*[5]

For all these reasons, what we see (our percepts) cannot be reduced to the sensory input. One argument is that some kind of inferential construction must be involved. Irving Rock called it perceptual "problem-solving," or "intelligent perception." Like the other intelligent systems we have discussed so far, this means going beyond the information given.

So investigators have been astonished as the extent of the work done by the visual system has become more obvious. In a 2012 article in the journal *Neuron*, Richard Masland noted that "although the retina forms a sheet of tissue only 200 μm thick, its neural networks carry out feats of image processing that were unimagined even a few years ago." They support Donald Hoffman's view that those of nomal vision are, in fact, visual geniuses naively unaware of their rich talents.[6] But how are such feats achieved?

BEYOND THE INFORMATION GIVEN

The traditional mechanical model is based on a linear convergence of recorded visual features or elements. And it demands some built-in rules of processing of some sort. However, the "inferencing" just mentioned seems to be based on something other and more than that. Again, the demand is for predictability (e.g., that a fuzzy input is really from a specific object with all its implications). However, as with cell metabolism, and the intelligence of development and physiology, it is not so much the surface presentation of inputs that are informative. Rather it seems to be the structure beneath the surface that yields the predictability.

With a little reflection this is apparent in other ways. For example, when we move a book around on a table, something invariant about its "bookness" persists in spite of radical changes in appearance. Likewise, the end-on appearance of a car is quite different from the view from the side. Yet we somehow retain a coherent image of a single object and not diverse snapshots—so long as it can be integrated with our own movements.

So, what a sensory system particularly needs is the structure that comes from motion, that is, the deeper statistical information it provides. Indeed, the importance of visual structure has been suspected—and neglected—for decades. In 1953, Hans Wallach showed that if a stationary three-dimensional figure (e.g., a wire-form cube) is illuminated from behind so that its shadow falls on a translucent screen, an observer in front of the screen will see a two-dimensional pattern of lines. But if the same object is rotated, the observer will immediately identify it as a three-dimensional cube, even though only two-dimensional information is presented. Wallach called it the *kinetic depth effect*. (Many demonstrations are available on the internet.)

Another popular demonstration indicates how whole figures can be recognized from a few light points moving with appropriate coordination but not in static presentation. One of the best illustrations is the point-light walker (figure 6.1). There are many other examples. So crucial is the structure in motion that most animals fail to respond to a perfectly motionless object, as if they cannot see it. If a dead fly on a string is dangled perfectly still in front of a starving frog, the frog literally cannot sense this meal to save its life. In humans it has been shown that perception of a retinal image held perfectly still on the eye (by pasting the image to a kind of contact lens) quickly fades and disappears. Again, it seems to be the structure, not the particulars, that is crucial to the evolved visual system.

So what kind of structure can be so informative? A widely accepted idea is that the structure exists in the form of statistical associations between the variables in motion. For example, the way that light spots, or other aspects of a viewed object, move together in space and time, or appear and disappear together. As Horace Barlow put it, "The visual messages aroused by the world around us are full of very complicated associative structure." He demonstrated that single cells in retina and brain are sensitive to them, and he suggests that the brain may use "higher-order" associations.[7]

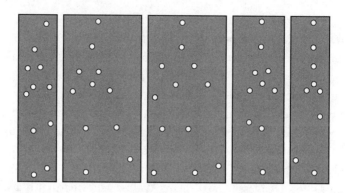

FIGURE 6.1

A sequence of stills from a point-light walker. Volunteers find the images unrecogniz-able when presented individually on a computer screen. But presentation as a se-quence at normal speed evokes almost immediate recognition of a person walking.

Joseph Lappin and Warren Craft also suggest that structure has to mean spatial relations among points emanating from a light source. They showed that it is the change of these relations over time—in the course of motion—that reveals the structure within and between features.[8] A simple illustration is the way that integration of slightly different images from our two eyes depends on abstracting the spatiotemporal correla-tions between them.

The associations are informative because they permit one changing aspect or variable to be predicted from one or more others, as when we predict the presence of a whole dog merely from a tail sticking out from behind a sofa. Deeper, or "higher-order," structure is present when such associations extend to three or more variables—the strength of associa-tion between any two variables is dependent on the values of others. The variables, that is, interact.

These interactions are analogous with those occurring between signal-ing proteins, TFs, noncoding RNAs, and so on, as explained in chapter 4. Likewise, the interactions among morphogens have been shown to guide cell differentiation and motility in development. Those interactions can furnish information about future states and how to respond now. Only now, in the brain, they extend across many more variables—in this case, light points—at much greater statistical depth.

Take almost any pair of points in figure 6.1 and follow them across the five frames. You may soon be aware that their movements vary together, or covary, over space and time. (More formal measures confirm that this is the case.) The association may be rather weak, but we can say that this is a primary association. And the association—call it parameter 1— is information for predicting the position of one from the other in its absence (e.g., when it is obscured behind another object).

The predictability can be strengthened, however, if the association depends on (i.e., interacts with) levels of another variable. We say this is a second-level association (or first-level interaction, or dependency). This is further information for predictability. Call it parameter 2: an association being informed by the value of another variable.

Indeed, we can carry this analysis through to many other levels or depths. The interaction just mentioned can itself be conditioned by one or more other variables—and so on, across the whole picture of moving dots to reveal still further accumulations of useful information. So the whole experience exhibits parameters 3, 4, and so on, embedded in one another at increasing levels or depths. Those parameters are a rich source of predictability for a system that can register them.

Indeed, estimation of such parameters in the point-light walker (or point-light representations of other objects) shows them to be statistically significant up to the fourth and even higher levels of dependency. And this is only in the nine spots of light, compared with the thousands or more that would radiate from a whole object.[9] The upshot is that an object, presenting whole "clouds" of moving points, will project abundant information for creation of a recognizable image. This is important for its implications about the nature of brain processes and cognition.

So the information captured in the nerve networks is not in the form of photographs or movies. What is stored are those parameters coding statistical relations among variables. It is recordings of those parameters (by means to be described below) that form a visual attractor to which visual images bearing them will tend to be pulled. A fleeting pattern in the corner of your eye makes you think you have seen a familiar object; a faint pattern across the street looks like a familiar face; a smudged word you can nevertheless read: all examples of structure pulled into corresponding attractors for verification and further processing.

That, of course, depends on the system (from eye to brain) having assimilated and stored those structural parameters. This is the important point. The system cannot rely on fixed processing rules as if working with recurring stimuli. That would not work with the usual fuzzy, ambiguous inputs like those from the point-light walker. Instead, the visual system must abstract the relational parameters from previous experience with similar objects (like real people walking). The parameters in the attractor basin serve as a kind of grammar from which to generate a reliable prediction from the unreliable inputs. As with speech, it is through the grammar that novel, or imperfect, inputs can be interpreted and from which novel responses can be created.

This is, in principle, the same kind of grammar used in cell functions and in development and physiology. But in the nervous system, there are vast numbers of cells with vast numbers of modifiable interconnections. Each neuron may receive inputs from thousands of others, and also send signals to thousands of others through its axon branches, some nearby, others at possibly great distance. Statistical patterns of great depth can be abstracted (learned) and permanently updated to provide predictability in changeable environments.

THE BRAIN PROCESSES STRUCTURE, NOT ELEMENTS

That the brain is most interested in such statistical structure rather than trigger features has been suspected for some time. In the nineteenth century, physicist and physician Hermann von Helmholtz described the visual experiences of patients who had gained sight for the first time after operations for congenital cataracts. One patient was surprised that a coin, which is round, should so drastically change its shape when it is rotated (becoming elliptical in projection). Another found it seemingly impossible that a picture of his father's face, the size of which he knew from touch, could fit into a small locket.

That vision in particular, and the brain in general, are not interested in stable features, images, or other elements is suggested in another, rather startling, way. As explained earlier, stable images are extremely rare or nonexistent in normal experience. Even when we are standing perfectly still, small but rapid oscillations of the eyeball are constantly shifting the image on the retina. Experimenters have overcome this by attaching an image to a kind of contact lens. The result for the participant is not a perfectly formed copy in the perceptual and/or cognitive system. Instead, the very opposite happens—the image disappears.

As Donald MacKay explained in reviewing such phenomena, "Stabilization, even if it does not abolish all retinal signals, eliminates all covariation. If no correlated changes take place, there is nothing for analyzers of covariation to analyze. If, then, seeing depends on the results of covariation analysis, there will be no seeing."[10]

Mackay also explained how, by abstracting correlations from experience, rich predictability becomes available even from meager data. For example, a blind person can discern the pattern on a manhole cover by probing it with a cane. Every time we park a car, we can sense the position of wheels and bumpers, even though we cannot actually see them. In fact we take this covariation structure so much for granted that we scarcely notice it. It only tends to become apparent in systems that have been damaged in some way.

All this has important implications for understanding potential and individual differences. The standard view would claim that the creation of stable images from sensory stimuli in the brain is via built-in rules, genetically determined. And, according to that view, brains vary in efficiency (or some other power metaphor) from person to person because of good or not-so-good genes. In reality we find it is informational structure that is important. In environments with ever-changing structure, the rules themselves have to be created through experience. The potential (and intelligence) is an emergent property of the self-organized system.

THE VISUAL BRAIN

When we turn to the brain in general, we find similar principles at work. The brain operates by abstracting deep correlational grammars, created by experience, and not by built-in rules and linear processing. This is shown in many ways.

We can now be fairly sure, for example, that the messages provided by sensory systems are self-organized into more compressed form. The systems do not simply pass on independent signals. Knowing a grammar is far more efficient than having to store volumes of individual items. The fact that the retina contains more than one hundred million photoreceptors feeding into one million ganglion cells already indicates that the network must perform significant data compression. We do not know exactly how to describe the message sent from the ganglion cells, except that it must be some aspect of that deeper structure.

In the brain itself, it is now clear that neurons function as cooperative ensembles that learn from experience. At the network level, this is attested to by the many studies revealing the interdependence among different centers (e.g., visual, hearing, olfactory) and among neurons in centers. At the synapses, dendrites integrate multiple signals as complex correlations, not as independent inputs. Across the network, "the majority of interactions in highly interconnected systems . . . are multiplicative and synergistic rather than additive."[11] This is sometimes called "context-dependent coding" and makes it clear that the language of the brain uses the structure of multidimensional, spatiotemporal batteries of signals, not discrete cues.

The early studies of neuron responses used single recording electrodes inserted into single neurons (in anesthetized animals). And they used simple artificial stimuli, such as light spots or lines flashed on the eye. It is now possible, however, to record simultaneously from hundreds of neurons in the same or different brain centers. And more realistic, natural stimuli have been used. The results have been striking.

Tai-Sing Lee and colleagues first developed movies of three-dimensional natural scenes and analyzed the statistical structures in the data (including its fractal, or deeper correlational, structure). Then they implanted microelectrode arrays in experimental animals so as to record simultaneously from hundreds of neurons in the visual cortex. By then exposing the animals to the movies, the researchers could assess whether the neurons actually use that structure.

They showed, first, that the neurons do indeed respond preferentially to the statistical structure previously measured in the movies. Then they

showed how such responses in the cortex tuned responses of other neurons and altered neuron connections. For example, reception of specific
structure in the cortex feeds back to "enforce statistical constraints," as
the researchers put it, on incoming stimuli. These constraints help further
extract the higher-order informational structure, eliminate alternative
possibilities, and resolve ambiguity during perceptual and cognitive
inference.[12]

In another study, Jarmo Hurri and Aapo Hyvärinen analyzed videos
of natural scenes by breaking them down into frames (or small patches
of the total frame) with tiny time delays of 40–960 milliseconds. They
produced evidence that responses in cells, even to simple features like
lines, are dependent on higher-order spatiotemporal correlations. As they
went on to say, a more generative model is needed in which hidden variables can be used to interpret the underlying real world.[13]

Using a similar approach, Selim Onat and co-researchers showed that
the continuity of the movie stimulus, rather than a mixture of bits from
different movies, was crucial for forming a coherent image "in which the
inherent spatio-temporal structure of natural movies were preserved."
The results suggest "that natural sensory input triggers cooperative
mechanisms that are imprinted into . . . primary visual cortex."[14]

All this is consistent with what has been revealed about the anatomy
of brain connections. Communication is not one way, from the outside
in, or the bottom up. There are abundant reciprocal or top-down connections from cortex to sensory waystations in the brain. For example, the
retinae send signals to the lateral geniculate nuclei in the midbrain, which
send them on to the visual cortex. But, as with other sensory waystations,
the lateral geniculate nuclei receive far more signals from the cortex than
they send to it.

More generally, studies of rat, cat, and monkey brains show that the
average input/output connection ratio between different regions of the
cortex is close to one. This suggests the strong role of reentrant (feedback)
mechanisms, in which processing of inputs to cortical centers are rapidly
used as feedback to firm up information from the sensory inputs.

This densely recurrent wiring organization has sometimes puzzled investigators. Susan Blackmore, for example, referred to the "currently mysterious profusion of descending fibres in the visual system."[15] But that is

what is needed for definition of informational structures, as opposed to mere assembly of features or other unitary elements.

The highly plastic brain, in other words, does not operate by fixed rules and would not have evolved in the first place if it did. The standard model has genetically determined operations, with individual differences arising from gene differences. What we really have is plastic networks abstracting informational grammars from good-enough circuits present in nearly everyone. They ensure far more diverse, and adaptable, individual differences.

COMPUTER SIMULATIONS

Of course, actually analyzing the brain's use of deep structure is not easy, given the complexity of naturalistic stimuli and the numbers of cells and synapses involved. So, as with the molecular biology of the cell, researchers increasingly resort to computer modeling. The rate of advance has been enormous. It has been shown how artificial neural networks set up on a computer with modifiable "synapses" can easily abstract the statistical structure in the inputs. The structure becomes assimilated in the connection weights between the artificial neurons, reflecting the relational parameters in experience.

In describing the assimilations in their own studies, Vincent Michalski and colleagues have described networks that use "multiplicative interactions to extract transformation . . . so that higher layers capture higher-order transformations (that is, transformations between transformations)." Moreover, because the networks "encode transformations, not content of their inputs, they capture only structural dependencies and we refer to them as 'grammar cells.'"[16] The transformations, of course, are statistical parameters abstracted from sensory inputs and not built-in (or innate) rules.

As we saw in chapter 4, such structural grammars have been found in the intelligent systems of single cells. They have been further crucial for the integration of cells and tissues in the origins of multicellular organisms. They figure in the remarkable pattern formations of bodily development and in physiology. Here they are coordinating the nervous system to deal with more changeable environments.

I continue to use the notion of "grammars" from time to time in what follows. For the moment, note again that, induced from experience, they will vary enormously across individuals and groups. They will also be the bases of vast individual differences in perception, cognition (see chapter 7), and behavior. This puts into perspective the idea that, by testing for highly specific expressions of these (as in most intelligence tests), psychologists are somehow describing individual differences in brain power. We might as well claim that a test of standard English can describe individual differences in language ability anywhere around the world.

THE BRAIN GENERALLY

Although I have concentrated so far on vision, similar principles seem to apply with other sense modes. For example, it is known that receptors in the cochlea of the ear transform sound into spatiotemporal response patterns sent along the auditory nerve to the brain. Tactile senses, too, are able to abstract the deeper statistical structure from the sequences of vibrations produced by a roving finger tip. Studies with bats and owls show how they use the spatiotemporal correlations in acoustic patterns to compute the location and distance of prey. Shaowen Bao has reviewed a number of recent findings on how the "statistical structures of natural animal vocalizations shape auditory cortical acoustic representations."[17]

In fact, direct recordings from auditory neurons confirm that they are highly sensitive to correlational structure in inputs. As one report explained, "Natural sounds are complex and highly structured stimuli. A number of studies carried out on the auditory systems of insects, lower vertebrates and mammals have yielded evidence for evolutionary adaptations which exploit statistical properties of the natural acoustic environment in order to achieve efficient neural representations."[18] As with vision, the full structure emerges progressively as auditory information travels through the auditory tract to primary auditory cortex.

Again, however, further clues come from the use of naturalistic auditory inputs. Neuronal responses in auditory cortex are highly context dependent and "constantly adjust neuronal response properties to the statistics of the auditory scene."[19] Somewhat similarly, the study of rhythm, as in the psychology of music, is increasingly drawing attention to the deep correlational structure in it, or "structure in time." Again, we seem to appreciate such structure because of its tacit qualities of predictability, which are also the essence of harmony and the source of the pleasure in music.

Each of these senses, like vision, plays a crucial part in behavior. But, of course, they can only do so in conjunction with inputs from other senses. In the way that the world is usually experienced, there will be considerable correlation between the statistical structures sensed in different modes: touch with vision, vision with sound, all three at once, and so on. This is particularly the case as any movement, such as action on objects, will also change experience in several sense modes simultaneously. Consider eating a meal, or just walking along the road; vision, sound, smell, and even the stimuli through the soles of the feet will vary together, or correlate.

There are also signals from internal senses. Every time an individual acts on objects, the brain receives a barrage of signals from receptors in the body itself. These include stretch receptors in muscles and tendons, various kinds of touch receptors in skin, pressure receptors in joints and muscles and from the muscle spindles (providing information about body posture). Changes in all of these will covary with one another and with those occurring simultaneously in visual and auditory experience of the event. As well as enhancing predictability in experience as a whole, integration of senses permits expectations from one mode to another. On hearing a sound, for example, even young babies turn their eyes and heads toward it.[20]

Such integration takes place at various levels in the subcortex and cortex. Multisensory cells, sensitive to more than one mode, have long been identified in the cerebral cortex. Recordings of receptive fields commonly find neurons in the visual cortex that are also tuned to *acoustic* frequencies. Cells in auditory cortex in humans are strongly activated by watching silent videos of speechlike facial movements. Seeing something at the same time as hearing it enhances reaction times and discriminability in laboratory tasks. It has also been noted how the strength of coupling of signals from different modalities depends on the statistical co-occurrence between them.

In the 1980s, Donald MacKay argued that much intentional behavior (e.g., hearing while looking, moving eyes and head to sample a range of views, tactile exploration while seeing) has the very purpose of collecting sample correlations. This is because of the predictability they afford. Even very simple exploratory probing of the environment by a fingertip reveals spatiotemporal correlations, including between those emanating from the surface and the exploratory motor action. These will be complex correlations as explained earlier: spatiotemporal, like music and dance, and conditioned by other variables and correlations at deeper levels. They will be more or less matched to the structural grammars (or attractor basins) developed from previous experience. Then whole mental pictures or scenes can be created from a few relational samples. These, in turn, become the fodder of cognition, as explained in chapter 7.

BRAIN DEVELOPMENT: STRUCTURED PLASTICITY

As mentioned in chapter 5, brain cells and their elaborate networks appear almost magically during a brief period of embryonic development. The complexity of the process is truly remarkable. Brain cells (neurons) are produced in the embryo/fetus at a rate of up to 250,000 a minute. These progenitor cells not only differentiate into the many types of immature neurons. They also migrate over extremely long distances to find their correct positions, often in precise layers, at the correct time, in the three-dimensional structure of the brain. Then they have to produce the thousands of processes through which they interconnect. Axons emanating from a cell of around a thousandth of a millimeter in diameter may travel up to a meter to reach other cells in precise locations, often on distinct layers. They may be doing this alongside dozens or hundreds of other axon terminals from other neurons (figure 6.2).

In mammals, especially primates, much of this occurs in the uterus before precise sensory experience. But it is also known that spontaneous signaling activity between neurons is needed for connections to begin to form. By being structured, they also help structure basic connectivity. For example, very early spontaneous signals from the retina are correlated in time and space, as if setting up the brain for the deeper structures to follow in real experience. Injection of drugs that inhibit such firing much reduces the formation of network connections.

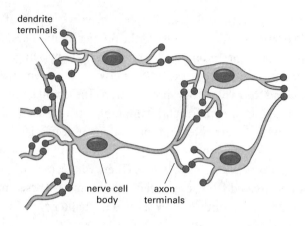

FIGURE 6.2

Getting connected. Axon terminals, guided by molecular signals, find target cells. There, development of dendrite terminals is promoted by structured spike firing along the axons.

This interpretation is supported in other ways. It has been known for decades that actual light experience is important for development of functional connectivity in the visual system. But recent studies show that it is patterned (i.e., spatiotemporally structured) light that is crucial. Confining visual experience to "white light" (i.e., presenting all light frequencies without the patterned quality) actually retards development of connectivity in the visual cortex. This prerequisite also seems to apply in the auditory system: exposure to white (uniform) sound during early development seems to severely disrupt connectivity formation.[21] Although the brain is sometimes referred to as "data hungry," in reality it is "pattern hungry." Far from being a blank slate, therefore, it is geared up for the abstraction of informational structure from experience; that structure is not predetermined by genes.

Multisensory integration is crucial for maximizing information from the environment, as noted above. But it, too, develops according to experience. This was shown in studies by Barry Stein and colleagues. Their studies showed that "neurons in a newborn's brain are not capable of multisensory integration . . . the development of this process is not predetermined. Rather, its emergence and maturation critically depend on cross-modal experiences that alter the underlying neural circuit in such a way that optimizes multisensory integrative capabilities for the environment in which the animal will function."[22]

Nervous systems, in other words, greatly extend and amplify developmental plasticity in other bodily systems. In the process, the plasticity has become indefinite, open to the abstraction of experienced structure. But it is a relentless process. The structure affects behavior, which, in turn, changes the environment in ever-increasing ways. This creates still further challenges for intelligent systems, in spirals of learning and change. The intelligent system of the brain served as a crucial springboard to the evolution of further complexity. I describe how cognition (or cognitive intelligence) emerges from such complexity in chapter 7.

WHAT HAS REALLY EVOLVED?

The account so far provides a coherent evolutionary story. Life itself originated in the spatiotemporal information structures in "molecular soups." Such structures became the means for energy dissipation. The processes evolved as intelligent systems in cells. They assimilate spatiotemporal structures hitting receptors and tune internal signaling dynamics. These grammars are then used to predict optimum responses and utilize genes accordingly.

Such tuning in cells then evolved between cells to regulate intercellular signaling. That all became conspicuous in the development and physiology of multicellular organisms. These already impart, at this level, high degrees of adaptability in changeable environments. But in the upward spiral of changing environments, it was not enough. Behavioral inventiveness, and the brains to furnish and regulate it, became crucial. That required the ability to detect and reflect informational structure that is even more abstract. It emerged in a new form of lifelong plasticity in the modifiable connections among nerve cells. It soon became an evolutionary innovation that eclipsed all others.

In that way a new "law" came into play: the more extensive the neural network, the deeper the informational structure that can be registered and assimilated. So the intricate connectivity of brain networks has increased enormously in the evolution of mammals. It has been achieved largely by increased folding of cerebral cortex to obtain greater surface area. The theme seems to be one of expanding processing capacity for abstraction of ever-deeper spatiotemporal structure.

It is the evolution on a theme of emerging intelligent systems, but now on a much higher plane. As with earlier systems, its very evolutionary function is the creation of adaptable variation. Only now, at the level of the brain, we get effervescences of variation that far outstrip anything that could be generated by gene "recipes." And it becomes adaptable through the internalization of environmental structure.

This should be borne in mind when trying to describe the origins of individual differences in "brain power." What is on the outside, or even measured, is no guide to the abilities on the inside.

THE FEELING BRAIN

Multisensory images are created by conjunctions of sense modes, as we have just seen. They are crucial for computing impressions of the outside world: they go far beyond the information given in sensory inputs, and they guide reactions with predictable consequences. But there is another sensory interface that must be included in those functions.

Since the nineteenth century, the nervous systems of vertebrates have been considered as dual entities. One has been described as a somatic nervous system, mediating sensory-motor responses to the external environment. The other is a visceral nervous system monitoring the internal environment. We now know that the two are much more integrated and reciprocal than the distinction suggests. The interactions between the two create subjective feelings and have a crucial part to play in how the brain interprets inputs and creates responses. This is often forgotten in describing the origins of individual differences in cognition and behavior.

It happens in even the smallest functional brains. In their small brains, even flies and bees have neuronal ensembles monitoring the internal milieu and activating neuro-secretory cells to modulate it. In vertebrates, a small center, the hypothalamus, receives copies of signals from all sensory systems. It also has rich beds of chemoreceptors monitoring the internal state—the homeo-dynamics—of the body. They respond to environmental perturbations (as signaled from the senses) by stimulating endocrine glands (as well as creating many other aspects of feelings). Much of this is done through the neighboring pituitary gland, with which the hypothalamus is densely connected. That, in turn, secretes a wide range of hormones into the circulation to influence functions elsewhere in the body, like the release of adrenaline in readiness for muscle action.

In addition, all the sensory systems share rich reciprocal connections with a number of other subcortical networks (e.g., the amygdala), also concerned with monitoring and modulating feelings. Finally, there is intense interplay between the hippocampus and amygdala as well as other centers constituting the limbic system. The result is to interrelate external and internal sensations, creating emotional states and degrees of motivation for response.[23]

The outputs from these interactions provide the brain with an *affective* image, or "feeling," including motivational state. This promotes, inhibits, or helps shape responses in brain networks, affecting numerous aspects of behavior. Stimuli, and constructs created in the networks, may be quickly identified as potentially significant or otherwise by these affective mechanisms. The latter, in turn, amplify the processing, bring additional attention mechanisms to bear, and so help shape perception in a continuous cycle.

Of course, "hot" feelings or emotions used to be considered to be the enemy of a "cool brain." However, the degree of collaboration among sensory, affective, and other cortical centers now suggests otherwise. All cognitive activities have an emotional value, reflecting inner states. It is such embodiment that contextualizes what goes on in the brain, enhancing it as an intelligent system. Such is often the basis of creative imagination and innovation. This is how we can distinguish cognition from the cold computations of a mechanical robot. Many problems arise from the disrupted integration of cognitive-emotional, self-regulatory processes.[24]

In that way, experienced objects, events, and relations come to have emotional values attached to them. In human social relations—especially in class-structured societies of symbolism, privilege, and deprivation—such values can be very strong. Their meanings (or what they predict) deeply affect perceptions, cognitions, and behaviors, as well as physiology and epigenetics. Those emotional values can in turn be the source of large individual differences in response to challenging situations, such as cognitive and educational tests (of which more in chapter 10). Just feeling inferior, or thinking that others might think of you as inferior, can have devastating effects on cognitive performance. Again, this is rarely if ever considered in the simple linear determinism (genes → brains → intelligence differences) that dominates the literature.

INDIVIDUAL DIFFERENCES

So now let us briefly consider the implications of all this for the description and measurement of individual differences in human brains and the possibility of correlating those with differences in intelligence. (The account here is brief, because I elaborate much more on this issue in chapters 10 and 11).

Unfortunately, studies in the area have been founded on the old view of ranked brains and mechanical metaphors: that individual differences in ability stem from differences in constitutional aspects of the brain-machine. This has meant using metaphors of "brain power"—capacity, size, speed, efficiency, or whatever—as criteria of ability, and always on the assumption they fall into simple ranks in normal distributions determined by the chance permutations of genes.

I mentioned earlier James Flynn's reference to "genetic potential for a better-engineered brain" with likeness to a "high-performance sports car." But such allusions are quite widespread. An article by Kenia Martinez and colleagues in the journal *Brain Mapping* (2015) says that some individuals are more cognitively efficient than others. Ian Deary sees brain functions as having more or less "biological fitness" (reflecting a kind of Darwinian survival-of-the-fittest race to be first).

Almost all these attempts have taken IQ-like measures and some quantitative measure of (assumed) brain structure or function, and looked for correlations between them. Efforts to correlate intelligence with brain size, for example, were started in the nineteenth century. But the methods were acknowledged to be crude and inaccurate. In the past decade or so, the field has been boosted by the use of MRI scans in what is called "brain scanning" or "brain mapping."

These scans are basically of two types. In structural MRI, the scan machine creates a varying magnetic field in the target tissue, thus agitating protons in the atoms within it. When the magnetic field is turned off, the protons gradually return to normal, giving off an electrical signal that can be measured. This happens at different rates, depending on the density of the tissues. So the scan can distinguish, for example, between brain and bone (skull); or between gray matter (such as densely packed neurons) and white matter (more dispersed axons sheathed in fatty glial tissue).

This method has been mainly used for measuring brain size. A number of studies using it have suggested correlations of 0.2–0.4 between IQ and brain size. These values have been widely accepted.

Other research has used "functional" (fMRI). It has focused on more specific aspects of brain, like cerebral connectivity, cortical surface area or cortical thickness in attempts to relate them to cognitive performance (e.g., taking an IQ test). In a review in 2012, Ian Deary claimed that such studies are "exemplifying key empirical advances in the association between brain structure and functions."[25]

Most reports are, like this one, quite bullish, adopting the promissory posture of the "genes for IQ" researchers. Richard Haier, for example, hosts an expansive website making confident claims. One of these is that the density of gray and white matter in regions of the brain is related to differences in how people score on IQ tests. In a glowing note, not untypical of the field, he says that "researchers are now on their way to a detailed scientific explanation of what defines intelligence, where it comes from, and how it operates in the brain."[26]

EMPIRICAL PROBLEMS

There are many problems with such claims, and some make easy targets. One of these is the use of IQ-type tests as measures of brain functions. It is a major error to relate performances on such tests as if they directly reflect a ladder of brain function. As described in chapter 3, they test for the learned factual knowledge and cognitive habits more prominent in some social classes and cultures than in others. That is, IQ scores are measures of specific learning, as well as self-confidence and so on, not general intelligence. In chapter 3, I mentioned the paradox of attempting to measure a supposed culture-free variable with an instrument couched in the terms of a particular culture.

Sensible discussion on the matter could really stop there, in my opinion. But even if IQ scores were genuine scientific measures, there is a much deeper logical flaw to consider. Investigations, interpretations and conclusions are based almost entirely on correlation coefficients. As mentioned in chapter 1, correlations are an investigator's honey-trap and the cheapest weapon of the ideologue. It is inappropriate to infer causation from them without further, properly controlled study.

For example, IQ is a measure of social class that correlates with a host of factors that can also affect physical aspects of brain development (though without necessarily affecting cognitive ability). These factors include stress in pregnancy or early childhood, malnutrition, exposure to toxins, inheritance of cross-generational epigenetic changes, and many others to be discussed in chapter 10. For example, much evidence suggests that moderate to severe stress, which is related to social class and to IQ test performance, affects growth in the hippocampus and prefrontal cortex.[27] And different occupations and cultural behaviors can themselves change the sizes and connectivities in the brain (see below).

In other words, any IQ-brain correlations are probably due to joint association with other factors. Or causation may be in the opposite direction from that imagined. In fact, it seems astonishing that such possibilities are not taken into account when expansive claims about the "neuroscience of intelligence" are made.

There are also acknowledged difficulties with MRI scans, in spite of the basic brilliance of their invention. Remember that the fMRI estimates activity in blocks of brain tissue by measuring changes in the levels of oxygen in the blood passing through it: the greater the oxygen level is, the higher the activity will be. But the scans are known to be subject to a variety of corruptions, such as noise (measurement fluctuations), visual artifacts, and inadequate sampling.

As mentioned in chapter 1, the experience of having a scan, lying in a claustrophobic cylindrical enclosure, is far from natural conditions. A survey in 2007 indicated that 43 percent of participants found the experience upsetting, with 33 percent reporting side effects like headaches. Children in particular are likely to be restless. As Michael Rutter and Andrew Pickles warn, "motion artifacts . . . can lead to quite misleading conclusions about the interconnectivity across brain regions."[28]

Moreover, when the participant is confined in the cylinder, it is difficult to present him or her with realistic cognitive tasks and evoke meaningful responses. For example, speech, which involves muscle movements, distorts the readings. In other words, fMRIs can be quite accurate as indices of categorical disease or trauma states. But they need to be applied more carefully for describing normal variation.

As a consequence, it is quite likely that what are read as cognitive differences are actually affective/emotional in origin. As Hadas Okon-Singer and colleagues warned in a review, the distinction between the emotional and the cognitive brain is fuzzy and context dependent. There is compelling evidence that brain regions commonly associated with cognition, such as the dorsolateral prefrontal cortex, also play a central role in emotion. Furthermore, they go on—and as I mentioned above—"putatively emotional and cognitive regions influence one another via a complex web of connections in ways that jointly contribute to adaptive and maladaptive behavior. This work demonstrates that emotion and cognition are deeply interwoven in the fabric of the brain."[29]

There are other doubts arising from the fuzziness of scans. A study in 2014, for example, compared a variety of MRI scans of axonal projections in monkey brains with accurate maps drawn from standard anatomical methods. It was concluded that none of the scan methods demonstrated high anatomical accuracy.[30]

Finally, as might be suspected in the light of comments mentioned in chapter 1, there is poor replicability and evidence of reporting bias in which only positive results are published. So Rutter and Pickles also warn that, while "brain imaging constitutes a valuable tool . . . so far, its achievements do not live up to the claims and its promise."[31]

In consequence, results have been inconsistent. Some have shown positive correlations between cortical thickness and IQ, others have been negative. As an analysis in 2015 concluded, "These studies have produced inconsistent results, which might be partly attributed to methodological variations." Using three different MRI approaches for measuring cortical thickness and comparing results across two matched samples, the investigators noted "that estimation of CT [cortical thickness] was not consistent across methods," and "there was considerable variation in the spatial pattern of CT-cognition relationships." And finally, "results did not replicate in matched subsamples."[32]

Moreover, functional significance is rarely clear, even in simpler brains. As Cornelia Bargmann and Eve Marder say, describing the connectivity between all 302 neurons of the roundworm *Caenorhabditis elegans* was a stunning success. But it "hides a surprising failure . . . although we know what most of the neurons do, we do not know what most of the connections do . . . and we cannot easily predict which connections will be important from the wiring diagram . . . [and] early guesses about how information might flow through the wiring diagram were largely incorrect."[33]

If we have such difficulties with 302 neurons, imagine the problems arising from billions of them. Bargmann and Marder suggest that such unexpected findings, have created doubts about "big science" efforts to model connectivity without a better understanding of function.

This might explain the results of Hugo Schnack and colleagues, who found that sometimes (at some ages) thicker cortices are correlated with IQ. At other times, it is thinner cortices. As they say in their 2015 paper in *Cerebral Cortex*, "That intellectual functioning is both associated with cortical thinning and cortical thickening is puzzling."[34]

Such flakiness of method could also explain why, in relation to the roots of intelligence, the results have so far been more hype than reality. Although Richard Haier has talked up statistical associations between such brain measures and IQ, as mentioned above, the reality is more complex. Following disappointing results in 2009, Haier and colleagues reported that neural "correlates of the *g*-factor remain elusive . . . the situation with functional imaging studies is no better . . . suggesting that identifying a 'neuro-*g*' will be difficult."[35] They also add, somewhat tellingly, that using a "non-theoretically defined test may produce confusing results"—a point I also made earlier in relation to the fruitless gene-association studies.

Of course, a major assumption driving this research is that brain functions will vary quantitatively across individuals and will be distributed in the population according to a bell-shaped curve. That image, implicitly or explicitly, motivates the research and is also a crucial assumption in the statistics by which results are tested. The fact is that descriptions of real functions in the central nervous system, as with physiological and behavioral functions, flatly dispel the image. György Buzsáki and Kenji Mizuseki explain in a wide-ranging review: "Most anatomical and physiological features of the brain are characterized by strongly skewed distributions with heavy tails and asymmetric variations that cannot be compressed into a single arithmetic mean or a typical example. . . . This is perhaps not surprising, as biological mechanisms possess emergent and collective properties as a result of many interactive processes, and multiplication of a large number of variables."[36]

Unsurprisingly, then, worries about the empirical problems in this field have been increasing. As mentioned previously, a special issue of the journal *Cognitive and Affective Behavioral Neuroscience* reported that "researchers in many areas of psychology and neuroscience have grown concerned with what has been referred to as a 'crisis' of replication and reliability in the field." The authors pointed to functional neuroimaging as "one of the major concerns in the field."[37]

But all these shortcomings are only the empirical problems.

THEORETICAL PROBLEMS

The more fundamental problems in the brain-IQ research, in my view, are its conceptual underpinnings (and their unconscious ideological roots). First, simple associations between region size and specific function are unlikely. The brain is not a machine with built-in functions that vary merely in speed, power, capacity, and so on, determined by gene differences. Function arises not in isolated regions but in the rich interconnectivity among them. As Bargmann and Marder say, "The entire nervous system is connected, but reductionist neuroscientists invariably focus on pieces of nervous systems. The value of these simplified systems should not let us forget that behavior emerges from the nervous system as a whole."[38]

Likewise, in an extensive review, Mike Hawrylycz and associates remind us that "much recent evidence indicates that the vast interconnected network of the human brain is responsible for our advanced cognitive capabilities, rather than a simple expansion of specialized regions of the brain such as the prefrontal cortex."[39] The simple equation, graded-genes → graded-brains → graded-intelligence, looking for "pathways" or "dots to connect," ignores what is now known about what the brain is really for and how it actually works.

In this chapter, I have described the chief functions of the brain as portrayed in recent research. They include computing statistical patterns from the various sensory inputs, assimilating their relational parameters in the networks, relating these across sensory modes, relating them to inner needs, and constructing motor outputs according to the perceived flow of information. Even in single cells, adaptable functions are based not on cue-response routines and linear processing, but on assimilation of environmental structure. That affords greater ability to predict future states, including the effects of response. In development and physiology, such assimilation has evolved to be much deeper.

The operation of the brain, as described here, is based on the evolutionary extension of that strategy of adaptation. As Steven Quartz and Terence Sejnowski point out, cortical development in the more evolved animals is "more extensive and protracted . . . suggesting that cortex has evolved so as to maximize the capacity of environmental structure to shape its structure and function through constructive learning."[40]

As they also point out, the evidence now suggests that cortical connectivity is largely induced by the nature of the problem domains confronting it rather than by predetermined architectures (or, as I would add, predetermined differences). In sum, different individuals of a species have the same complement of initial connections, but the patterns and densities of connections that develop can vary substantially between individual brains as a result of experience. Once more, the vast majority will be good enough for the unknown problem domains to be encountered.

The evidence for this outside-in development is now quite abundant. Brain researchers who attribute differences in brain networks as causing individual differences in cognition sadly overlook the way that learned (often cultural) differences in cognitive and behavioral practice may be the causes of the brain differences. In humans, it was revealed in an oft-cited study of London taxi drivers. These workers are obliged to internalize a detailed map of London streets. It becomes reflected in an increased size (presumably network expansion) of part of the brain, the hippocampus, involved in memory formation. Violin players have been found to have enlarged parts of the brain to do with fine motor coordination on one side but not on the other side. In circus jugglers, three months of training was reflected in gray matter macrostructure in cortical areas. Increases in brain network size even seem to correlate with the numbers of words we learn, at least in adolescence: again differences from the outside in, rather than inside out.[41]

There are many other examples. In a wide-ranging review, Arne May described the phenomenon as follows: "Contrary to assumptions that changes in brain networks are possible only during crucial periods of development, research in the past decade has supported the idea of a permanently plastic brain. Novel experience . . . environmental changes and learning new skills are now recognized as modulators of brain function and underlying neuroanatomic circuitry."[42] The implication is that, whereas the brain-IQ correlations reported from MRI scans have been invariably interpreted as causal, in one direction, the causal effect, if any, may really be in the opposite direction.

May also points to "findings in experiments with animals and the recent discovery of increases in gray and white matter in the adult human brain as a result of learning" and concludes that "understanding normative changes in brain structure that occur as a result of environmental changes and demands is pivotal to understanding the characteristic ability of the brain to adapt."[43]

It has also been demonstrated in monkeys, as well as in simulations using artificial networks, that the existing state of a network can constrain new learning if it is based on an unfamiliar kind of structure. In other words, the current state of brain development will have some influence on ease of learning from new experiences until some catch-up has taken place.

Attempting to describe brain differences in terms of a single overarching speed, or other mechanical function, therefore, is pointless. Like physiology and epigenetics, it is more accurate to describe what the brain does as a cooperative interaction among numerous subfunctions. Nor should we forget how nervous systems are closely interrelated with other intelligent systems: physiology, development, epigenetics, and so on. They function with close reciprocal effects such that, for example, just as brain functions are strongly affected by metabolism, so they can also alter metabolism.

Of course, the evolutionary basis of intelligent systems like brains is that they are capable of creating (adaptive) individual differences on a scale outstripping that possible through previously evolved systems, or through gene variation. Intelligent systems are characterized by dynamic, creative, processes through several layers of regulation. Genes are utilized as resources in such processes but not as brain architects. Except for rare conditions, the functional potential of the brain cannot be predicted, even from a description of the full genome.

This was indicated in research by Julia Freund and colleagues. They collected longitudinal activity data on forty genetically identical mice living in one large enriched environment. In spite of the genetic uniformity, the researchers observed the emergence of notable individual differences over time. These differences correlated positively with individual differences in brain structure. "Our results show," they concluded, "that factors unfolding or emerging during development contribute to individual differences in structural brain plasticity and behavior."[44] Searching for individual—or even cohorts—of genes in the hope of predicting individual differences in potential is pointless.

In sum, huge efforts are being made to clarify what brains really do. These efforts need to be cheered. My complaint in this chapter has been about those that seem to be aimed at dumbing down the brain and ranking brains by using crude criteria. What has evolved, and develops, is a huge complex of dynamical functions: hierarchically nested but with no one at the top, no executor, no boss, no gene controller, no chief hormone, no leading brain center.

If we focus on specific axes of development at particular times, we will find that individuals vary enormously. But that does not predict potential for future development. For the vast majority of individuals, most of the time, development results in functions that are "good enough"—good enough for the emergence of cognitive processes to be fashioned by the experience that follows. Let us now turn to considering those processes.

7

A CREATIVE COGNITION

WHAT IS COGNITION (AGAIN)?

O
ne of the problems of allowing ideology to cloud our science—of grading everyone according to social perceptions of brain power—is that it has obscured relations between brain and cognition. As I mentioned in chapter 1, psychology is currently unclear about where to put itself in relation to neuroscience, even entertaining the suggestion that brain studies will make psychology redundant altogether. Cognitive intelligence is considered to be the most important aspect of human potential, yet the confusion has put a brake on theoretical progress, so that its true nature remains obscure. Those who speak most strongly about cognitive differences among individuals seem to evade this fact.

For example, in their widely used book, *Behavioral Genetics*, Robert Plomin and colleagues obviously skirt the issue. They tell us that "a general test of intelligence is a composite of diverse tests of cognitive ability," that it is "a quantitative dimension," which means that it is "continuously distributed in the familiar bell-shaped curve, with most people in the middle and fewer people toward the extremes." They tell us that the correlations between the diverse tests suggest that cognitive differences can be attributed to a general ability called g and remind us that "g is widely accepted as a valuable concept by experts." But, they go on to say, "it is less clear what g is."[1]

That might incline you to question the experts a little more. It should be clear what we are valuing, especially if we are going to draw strong conclusions from it about people's genes, brains, and potential. But when challenged to be specific, psychologists and behavioral geneticists usually come up with mechanical or social metaphors, together with words like "smart" or "bright," and thus quietly steeped in ideology. IQ test constructors, in particular, have connections with cognitive theory that are rather loose and selective, to say the least.

The fundamental reason for such vagueness is that psychologists in general are not sure what cognition really is. There is little agreement about how or why it originated in evolution; what form or forms it takes in different animals; how it develops; why it has become so much more complex, especially in humans; or the real nature of individual differences in it. A sure sign of such uncertainty is the way that cognition is usually described in the literature by enumerating ends rather than the means.

For example, according to Sara Shettleworth, "Cognition, broadly defined, includes all ways in which animals take in information through the senses, process, retain and decide to act on it." But, she goes on, this definition "is somewhat soft around the edges."[2] In their textbook, *Cognitive Psychology* (2015 edition), the fodder of thousands of psychology students, Michael Eysenck and Mark Keane tell us that cognitive processes include "attention, perception, learning, memory, language, problem solving, reasoning and thinking." They do acknowledge that these are interdependent, without being clear how or with what consequences. But they also mention that cognitive psychology "sometimes lacks ecological validity . . . and possesses theoretical vagueness."[3] While honestly describing challenges in the area, these do not seem firm grounds on which others should be drawing very strong conclusions about people's relative levels of intelligence.

The impression from the literature is of a continuing failure to understand what cognition really consists of—apart, that is, from its various manifestations in perception, thinking, learning, and so on. Instead, the subject exists as an unintegrated collection of findings, models, and approaches yearning for clear theoretical foundations. This is pretty much the view of Martin Giurfa, for example, who says, in his book, *Animal Cognition*, that "despite this diversity and increasing interest, a general definition for the term 'cognition' remains elusive, probably because the approaches that characterize cognitive studies are diverse and still looking for a synthesis."[4]

Sometimes the difficulty is expressed with some despair. In the April 2015 issue of the journal *Trends in Cognitive Sciences*, Ralph Adolphs has a paper titled "The Unsolved Problems of Neuroscience." Some of these, he suggests, are "Problems we may never solve." Number 19 is: "How can cognition be so flexible and generative?" This is the fundamental question put by William James in his *Principles of Psychology* in 1890: "Can we state more distinctly still the manner in which the mental life seems to intervene between impressions made from without upon the body, and reactions of the body upon the outer world again?"[5] We still seem to be struggling with it.

Probably one reason for such pessimism is that investigators are still not clear about why and how cognitive systems have evolved and become vastly more complex, from invertebrates though mammals and apes to humans. This might also explain why cognitive psychologists have turned to raking over the brain (and genes) for clues about "what it really is."

One major difficulty, of course, is that cognitive functions seem to exist somehow separate from, or "above," the web of neurons and synapses in brains and the information they process. What is needed is a model that can describe the necessary powers of abstraction, yet remain firmly grounded in the structure of experience and consistent with the nature of brain functions. That model should provide some clarity both about how cognition works and how it varies.

The purpose of this chapter, then, is to show what cognition really is and how it forms and varies as an intelligent system. I confine the account in this chapter to a very basic system in a generalized living thing. I first describe how the most basic cognitive processes of recognition and classification emerge from neural interactions. I then turn to what are usually considered to be the "higher" cognitive processes of learning, thinking, decision making, motor action, and so on. Finally, I discuss the origins of individual differences in cognitive systems, generally. Chapters 8 and 9 expand on the further evolution of such an idealized system into the spectacular forms we find in humans today.

As with all of science, however, models and theories are based on presuppositions about the entity in question, and much of scientific progress depends on identifying and criticizing them.

BASIC PRESUPPOSITIONS

Theories of cognition today, are, basically, elaborations of one or the other (or some combination) of three sets of presuppositions. These have been around for at least the past two hundred years but have roots stretching as far back as ancient Greece (circa 2,500 years ago). I summarize these assumptions as briefly as possible.

Nativism

So-called nativists have always argued that there must be some very clever data-handling processes built into the cognitive system. This is because— as I mentioned regarding vision in chapter 6—the raw stimuli alone are too fuzzy and unreliable to specify percepts or concepts. This is called the "poverty of the stimulus" argument. So nativists argue that some prestructured operations must be genetically specified and more or less present at birth or early development. Recent exponents of this view include Noam Chomsky, Jerome Fodor, and Steven Pinker.

The view has also been popular with so-called evolutionary psychologists. They argue that cognitive functions consist of specific modules, or dedicated brain circuits, as specific biological adaptations. Through natural selection, such modules have evolved to process the data characteristic of specific aspects of our ancestral environments. These modules automatically deal with the natural, day-to-day variation in experience, because they possess processing rules, like computer programs. They operate by setting up internal representations of current inputs and matching them to ones in memory. Suitable responses, such as categorizations or motor action, can then follow according to these built-in processing rules.

Note that this idea has involved a nature-nurture debate different from the one about individual differences. It concerns the degree to which those so-called rules of cognitive function are shaped by genes in evolutionary time or instead by experience during the course of development. In his book, *The Blank Slate*, Steven Pinker argues that anyone who denies the genetic view of native processes is basically denying human nature.

But the issue is poorly framed, either way. The main problem—as discussed in chapter 4—is that any form shaped over evolutionary time, by genetic selection, will be committed to a specific, recurring environment. It will be unable to cope with rapidly changing environments in current time, such as the need for continuous development of novel rules. That is why more intelligent systems had to evolve. But in the domain of cognition, there are other serious problems, anyway.

A major problem is specification of what is deemed to be built in. It arises because the environmental pressures said to shape such cognitive modules, perhaps a million years ago, have only been imagined in superficial terms. In fact, it has mostly been assumed that they have something to do with broad factors, like climate change and general lifestyle, without making the connection very clear. In his book, *How the Mind Works*, Steven Pinker argues that a "module's basic logic is specified by our genetic program. Their operation was shaped by natural selection to solve the problems of the hunting and gathering life led by our ancestors in most of our evolutionary history."[6] He does not say what, exactly, those problems are.

I argued in chapter 4 that, on the contrary, it was the increasing changeability of environments, on micro and macro scales, within, as well as across, generations, that demanded and produced intelligent systems in general and cognitive functions in particular. Such environments demand frequently changing "programs" that have to be created de novo in the lifetime of individuals. So the nativist approach bears a logical error.

Associationism

In contrast, associationists, on the other hand, have described the basis of cognition as a kind of continual learning of a particular sort. They simplify experience into variations in stimulus components such as sensory qualities (color, shading) and features (lines or other geometric shapes). Those that regularly co-occur in space or time are associated in the mind and form the basic representations—symbols or concepts—of objects and events.

For example, a regularly co-occurring set of features like fur, four legs, ears, and so on, may form the concept of "dog." Concepts with overlapping sets of features form higher-level concepts, just as cats and dogs and other species may form the concept of "animal." Associations over time or sequences can also form if-then or stimulus-response connections in cognition. For example, sheep flock toward the sound of a tractor, because they have learned to associate it with impending delivery of food. The associations—and associations of associations—are kept in memory as knowledge stores. They may be activated by stimuli in current experience, when the learned associations are played out in responses. For example, sight of a tail wagging from behind a sofa retrieves the image of a whole dog.

Associationist models became very popular in the 1970s, when they could be simulated (and further modeled) on computers. Of course this development, as with the original theory, assumed some sort of built-in processing just to register the associations. So investigators started to introduce other forms of ready-made processes in ready-made sequences. As a consequence, basic associationism merged with some basic nativist assumptions to form the computational view of cognition.

The computational view holds that cognitive functions can be described as sets of built-in programs (or apps) in specialized neural networks (modules). Together they register, store, and utilize associations under the supervision of a central executive, much as computer programs work with a central processor, subroutines, and data files to process information and reach decisions or actions.

Since the 1960s, that combined computational model has been virtually synonymous with cognitive psychology. As Tony Stone and Martin Davies explain, "A fundamental principle of contemporary cognitive psychology is that cognition involves the storage and processing of information and that this information processing is achieved by *the processing or transformation of structured mental representations*."[7]

This is the basic model now implicitly or explicitly adopted in applied contexts such as health or education. For example, Roderick Wallace and Deborah Wallace say that "the essence of cognitive function involves comparison of a perceived signal with an internal, learned or inherited picture of the world, and then, upon that comparison, choice of one response from a much larger repertoire of possible responses."[8] (They do, however, go on to discuss cognition as "language.")

Of course, it has been pointed out many times that the elemental inputs of associationism hardly correspond at all with the raw, dynamic flux of experience. They are ready-processed, meaningful, parcels of information, contrived for research convenience. And what the central executive is, exactly, and how it works, is another problem usually side stepped. It is like another little brain put within the brain, rather like the one that is implicitly put in a gene. As mentioned in chapter 6, there is no evidence for such a creature in real brains.

However, these liberties over the nature of inputs, predetermined connections, and central executive have been confronted since the 1980s by a rival version of associationism. This approach sets up a layer of processing units in a computer. The units are programmed to be sensitive to experienced inputs. They then send outputs to a further layer of units, usually in a one-to-all or at least one-to-many correspondence, to integrate the signals (figure 7.1). The trick is that these connections can then be strengthened or weighted by the programmer according to the frequency with which signals co-occur.

The result is that patterns of co-occurring inputs can then be registered in the interconnections among the units. This is considered to reflect what happens in real brains (so they have been called "artificial neural networks"). Moreover, the process captures not only the simple pairwise associations. The deeper statistical structure of the inputs to

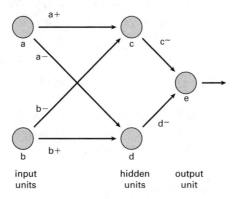

FIGURE 7.1

Fragment of an artificial neural network. Stimuli (usually some well-defined features, as conceived by the experimenter) are input at the keyboard and received by the input units. These are passed to "hidden" (association) units. After integration of a series of signals, the outputs of those should reflect statistical associations among them. +/− signifies features present/absent and c~/d~ represents their integration (or probability of co-occurrence).

multiple units can also be represented. The connection weights thus correspond to the structural parameters (and collectively the grammars) captured by brain networks, as described in chapter 6.

All of this, of course, requires suitable training of the network by the (very human and separately conscious) programmer. It usually involves inputting examples—digitally specified elements such as "eyes," "fur," "legs," or whatever—from the keyboard and reading whatever output is created at the next layer. This is then corrected to what it should be—for example, "dog"—by adjusting the connection weights and then inputting the next examples. Through successive approximations of input and feedback, the network behavior expected by the modeller can be obtained.

One of the striking indications of successful training is the predictability the network then furnishes even on partial or novel inputs, just like real cognition. For example, a partial input, like a few doggy features, will elicit a correct recognition, as if a whole image is being constructed in the network. Experimenters have thus been excited by the way the networks can approximate complex, abstract rules, as in categorization and very simple language learning.

In other words, the networks exhibit emergent properties like those typically found in cognitive systems. Bold claims have been made that connectionist networks explain the construction of knowledge, the epigenesis of cognitive development, emergent cognitive abilities, a decisive refutation of pure nativism, or the like. All this is furnished without a detached central executive, prestructured rules, discrete representations, and so on.

Is this the answer to William James's question? Unfortunately, such connectionist models have many problems. Although pretending to be brainlike, they are not biologically very realistic. The elementism at the roots of the approach is not entirely persuasive. For example, inputs to the system are keyboarded in as predefined, static features or other elements. This is most unlike the natural, dynamic flow of real experience.

Moreover, getting such networks to learn with more complex structures, like segments of human speech, has required elaborate data preparation and built-in (i.e., quasi-nativist) processing propensities working in slow, progressive, stages. Finally, there is obvious artificiality in the way the network has to be trained by the computer operator, according to numerous prior assumptions about the structure of experience. Accordingly, confusion arises about what exactly is being learned or acquired in such networks during training.

All this is reflected in the fact that connectionist models so far apply only to artificial, stable worlds, suitable for computers but not for the natural world of continuous, interactive change. It is important to note, though, that work continues to overcome these artificialities. Self-supervised networks, which automatically acquire structure and generate feedback are now in common use. And there has been clarification of what is being learned and how. I refer to some of these findings below.

Constructivism

These developments with artificial neural networks reflect in some ways the main principles of a third line of cognitive theory. Constructivism proposes that, instead of an executor of predetermined programs, or a mere copier of shallow associations, a cognitive system is itself a constructor of programs. It suggests that nothing is built into the cognitive system except very basic processes of data sorting. From these processes, concepts, or "schemas," corresponding with domains of experience, are built up.

So the Gestalt psychologists of the early twentieth century claimed that cognitive processes went beyond the information given to impose order and organization on sense data. Later, theorists like Wolfgang Köhler argued that there is structure in experience to which activities in the brain must have some correspondence. But he remained vague about the nature of that structure.

Without doubt the most prominent constructivist psychologist has been Jean Piaget, who made a distinctive contribution. As we have seen, the other theories of the cognitive system envisage elements as the fodder and currency of the system. For Piaget, elements may stimulate sensory receptors, but beyond that the cognitive system is only interested in how they are coordinated (or co-vary) in space and time. Moreover, as they become assimilated from experience, coordinations in domains become increasingly coordinated with those in others: features are coordinated in objects, objects are coordinated in events, events in wider events in and across domains, and so on. It is the dynamic structure of these "coordinations between coordinations" at various levels that form the logic of perception, knowledge, thought, and action, according to Piaget.

Coordinations must have precedence over elements, according to Piaget, because in both the world of experience and behavior, there is a continual construction of novelty. From birth, coordinations in experience are revealed by the infant's normal activity in the world. At first, they are registered as a result of random movements, but increasingly activities are guided by the growing knowledge of those coordinations themselves.

Take, for example, a ball of clay rolled out into successive shapes. According to Piaget, the actions of the subject on the object reveal not only isolated properties of the clay but also the coordinations between them. The length and the thickness of the ball are not independent dimensions. There is a necessary statistical connection between them—as one changes, so does the other; that is, they vary together, or co-vary. In other words, they are coordinated. Moreover, this coordination is itself embedded within the coordinations of the action: between the changing visual appearance of the ball and the sense-receptors in skin, muscles, and joints. This is what happens in all our everyday thoughts and actions like digging, riding bikes, and lifting objects.

The essential idea is that only by internalizing coordinations—not elements—can an individual conceive the invariant structures in objects and events against constant variations in appearance. And only with such conceptions can cognitive systems operate in the real world. Being neither prestructured adaptations nor simple copies of experience, in the form of simple associations, Piaget stressed that these mental structures develop as coordinations are assimilated.

For example, the primitive perceptual and conceptual routines available at birth suffer constant disturbances. So they must constantly readapt (Piaget used terms like "compensation" or "re-equilibration"). In the process, Piaget said, these coordinations become ever deeper and more inclusive of the dynamics of the world, slowly developing into the "logico-mathematical" structures of adult cognition.

Piaget struggled to describe the nature of logico-mathematical structures in formal terms, but never satisfactorily.[9] Some further grounding of these concepts in the real informational structures of experience is obviously needed, and I return to that problem below. Obviously, there is affinity here with the covariation parameters and structural grammars I have described as the basis of intelligent functions at other levels, from molecular networks to brain networks.

■ ■ ■

These are the three sets of principles to be found underlying models of cognition. However, it is worth mentioning in passing that these ideas have not formed in psychologists' heads as if in a vacuum. Instead they have been aspirated out of the experience of those psychologists in their own social and political circumstances. Psychologists, after all, are recruited and paid for by special institutions—employment, education, law, mental health—dealing with the needs and problems arising from the maintenance of a particular social order.

So they have been associated with different ideological and political frameworks. From the time of ancient Greece and Plato's *Republic*, nativism has been associated with authoritarian regimes with preformed knowledge structures and social ranks. Associationism has been associated with seemingly benign periods emphasizing learning and the training of minds, albeit through shallow, utilitarian, principles. Constructivism has tended to reflect deeper principles of renewal and revolution.

COGNITION TODAY

The field of cognition has produced a wonderful array of findings. But they have been mostly confined to highly specific examples of cognition related to highly specific models or theories. There has been no agreed-on perspective except in very general terms. This needs to be remembered in discussions about measuring cognitive abilities and grading people by them.

To illustrate the problem, consider cognitive theories of thinking. It is probably agreed that thinking is really the core of cognition. IQ test constructors and education theorists claim to be in close touch with it. As with nearly all cognitive theory, though, we are not told exactly what thinking is. We are given lists of problems that thinking seems to solve, but not how it does it. For example, in their leading textbook, *Cognitive Psychology*, Michael Eysenck and Martin Keane tell us that thinking is a set of functions, including problem solving, decision making, judgment, reasoning, and so on. A vast variety of flowchart models of these processes has arisen from an impressive array of ingenious studies on animals and humans. Some of these distinguish between different kinds of thinking, such as declarative (used in speaking) and procedural (used in doing things) knowledge. But they are described by their ends, not their means, which tend to be represented only by boxes and arrows.

The most popular recent notion—reflecting the dominant computational model—is that thinking consists of operations on internal cognitive symbols, or representations, using some kind of rules. The difficulty lies in specifying the nature of those representations, how they are set up and manipulated, and by what kind of rules. The tendency has been to break inputs down into ever-smaller if-then associations. On encountering a problem, these associations are thought to be configured into appropriate sets and sequences by the rules built into the system or learned in some way. According to theorists like Steven Pinker, any problem can be broken down into a series of steps like those forming the basis of a machine "that thinks," and so explain the vast repertoire of human thought and action. So-called artificial intelligence grew out of the computational assumption that machines can be made to think like humans in the same way.

The view has been increasingly criticized for the implausibility of many of those assumptions. Real inputs to real cognitive systems do not come in the form of discrete elements, as already mentioned. They are continuous and coordinated, and usually novel, fuzzy and imperfect. Moreover, although internal representations are usually described by analogy with tangible images, like pictures, maps, three-dimensional models, or even linguistic propositions, they cannot literally be like that in the soft tissue of the brain. And it is doubtful whether thinking really consists of sequences of simple if-then rules. That assumes a repetitiveness of experience that is rare in the real world. So the rules of real cognition must be of a different form—ones that regularly need to be updated.

This may explain why the implementations of computational models in artificial intelligence have produced machines that do simple, routine things (like select and count items according to fixed criteria) very well. But they cannot deal with fuzzy, dynamic inputs like moving images or human speech, except by breaking down such inputs into simpler steps. Discrete characters and steps suit a computer program, but this is not what real experience is like. So those programs are unlikely to yield truly realistic models of thinking.

Similarly, for the purposes of cognitive research, very simple, well-defined, stationary (nondynamic) tasks devoid of context have been resorted to. But the strategy has probably been self-defeating, because most everyday problems are dynamical, ill defined, and intimately related to context. Consequently, we seemed to have learned a lot about the difficulties humans have with unreal cognitive tasks in an unreal world, but not so much about real thinking in the real world.

The strategy has also encouraged a remarkable underestimation of human thinking/reasoning powers in the real world. The pessimism about the cognitive ability of both adults and children in cognitive testing (as in IQ testing) was mentioned earlier. In contrast, studies of problem solving in real-life contexts reveal complexities of thinking that greatly exceed those manifested in such tests (see chapter 3).

We have probably also underestimated the cognitive abilities of nonhuman animals.[10] Research over the past few years has revealed previously unsuspected complexities of cognitive processing in invertebrates such as bees and flies, as well as vertebrates, such as fish and jackdaws.[11] It seems clear that current models of thinking are failing to capture the true qualities of cognitive operations in complex, dynamic situations, from the foraging of the honeybee to humans carrying out cooperative, constructive acts over long periods.

In sum, a vast variety of work is going on in a wide variety of special cognitive topics. But the attainment of an agreed-on model of what cognition is and how it works remains very much a work in progress. Instead of characterization, nearly all textbooks of cognitive science start with a list of approaches and then present (more-or-less confused) students with a theoretical salad, as if from a grocery store. The resulting frustration explains why many psychologists are turning to neuroscience for the insights that cognitive science has failed to deliver.

So it is perhaps hardly surprising that psychologists who claim to measure cognitive potential have only a very rudimentary cognitive theory, and are not sure what they are measuring. Although some specialized tests may be useful for identifying categorical cognitive disorders (which will manifest in many other ways) there is no firm basis in cognitive theory. This has serious implications for the notion of cognitive ability, the nature of individual differences in it, and for expensive searches for the genes and brains that "determine" those differences.

In my view, the difficulties describing cognitive functions arise because the nature of experience is misunderstood. As mentioned several times already, cognitive psychologists, imbued with mechanically and ideologically derived metaphors, tend not to have appreciated the complexity of the structure of natural experience. So they fail to connect cognitive functions in explicit terms with evolution, including what the brain and cognition are really for.

A DYNAMICAL SYSTEM

In my view, these problems can be ameliorated by realizing that what has evolved is a system of dynamical, nonlinear processes that deals in patterns or structures, not specific contents (e.g., symbols, features, codes, or images). The patterns/structures are condensed as the statistical parameters (or grammars) assimilated from experience. They reflect the deeper correlations among interacting variables, not simple associations. Because experience consists of ever-changing environments these are rules that can be constantly updated. This is life's universal grammar, through which predictability can be tracked in the midst of change.

EVOLUTION OF COGNITION

Cognition-like interactions are apparent in metabolic and developmental systems, as mentioned earlier. But brain-based cognitive functions probably emerged in evolution as soon as brains evolved. They did so as a means for dealing with an enormous leap in environmental change. And that arose as soon as animals started to move around more, in a world full of other organisms and objects.

To the individual organism, those "others" are experienced as moving, transforming, fleeting, and usually incomplete, impressions. Animals need the ability to integrate these unreliable impressions into some more definite internal representations of that turbulent world outside. They needed to obtain predictability of the unfolding behaviors of other objects, both animate and inanimate; and also to fashion fruitful responses to them, even while the scene is changing and being changed by the individual's own activity.

The unique function of cognitive systems, then, is to go beyond the information given: to disambiguate confusing, incomplete, and rapidly changing data while generating equally unique, yet adaptive, responses.

Such functions arose very early in the evolution of nervous systems. Even the soil-dwelling nematode worm *Caenorhabditis elegans* has cognitive abilities. Although barely one millimeter long, and with a brain of only 302 neurons, it can learn the smells, tastes, temperatures, and oxygen levels that predict aversive chemicals or the presence or absence of food.[12]

Cognition in flies and bees is quite sophisticated by comparison. The honeybee has a brain of less than a million neurons. But it exhibits complex concept formations that go beyond what Martin Giurfa calls "elemental associative learning."[13] For example, bees learn categorizations based on generalized properties such as vertical and horizontal, a basic cognitive function in advanced systems. Moreover, after foraging in their usual zig-zag fashion, bees do not find their way home by simply retracing their outward meanderings: they memorize and integrate the directions and distances experienced to construct the most direct route. In doing so, the bees internalize the higher-order relations of the landscape topography. In other words, they have assimilated the statistical structural parameters.

In vertebrates, from reptiles to fish to apes, the same story of previously unrecognized cognitive abilities has been unfolding. And, again, it is being realized that these are abilities based on deeper structures than simple cue-response associations. The standard computational model cannot explain how we deal with real, dynamic environments. So let us see what such structures are and how they are connected with the brain. Being more clear about this is crucial to an understanding of what varies in cognitive systems, even at a rudimentary level.

FROM BRAIN TO COGNITION

The distinctive demand on cognitive systems is that they first need to be very good at assimilating those structures. This is not only because environmental structures are deep in the statistical sense, but also because they change frequently—more quickly than can be dealt with by epigenetic, developmental, or physiological processing. Cognitive functions must be able to use such structure to predict immediate and distant futures from inevitably partial current information.

As seen in chapter 6, the brain is superbly evolved for supporting just such functions. Incoming stimuli are highly variable and noisy, and so are the responses of single neurons to them. Yet cognitive experience of the environment is much more stable and consistent. How is this achieved?

We have already seen that the brain is a dynamical system, quite distinct from the linear processor in the computational model of cognition. It assimilates the structure of experience at various levels, but not in the form of direct sensory records, such as snapshots or movies. What must be internalized are the parameters describing the statistical relationships in experience. These comprise the grammars of experience. This is what the vast numbers of connections in extensive neural networks are for.

By virtue of such grammars, activity in the network becomes both constrained and generative. As with the whirlpool in the river (or Bénard cells, or a million other natural structures), they are attractors or basins of attraction. Inputs consistent with the cognitive grammar—like the order of words in language grammars, or characteristic movements of an object—will tend to be pulled into corresponding attractors and become more predictable. The attractors take on myriad forms in the brain networks to form a constantly developing attractor landscape. It may be better, in fact, to liken them to a seascape, because the medium in which they exist is more turbulent than a landscape and because they are being constantly updated with experience in a domain.

By the nature of experience, then, attractors in brain networks are being constantly perturbed by streams of inputs. Sometimes these inputs will create temporary distortion, causing activity (patterns of nerve impulses) to cycle around a central tendency. These are limit cycle attractors, as with a swinging pendulum bumped off its regular swing into a distorted loop. In contrast, activity often can be stretched to a critical state by inputs with highly unusual values. The activity in the attractor can flip out of its regular constraints to a more searching state, in which new structures and responses can emerge.

This has been called a "chaotic" state. Generally speaking, it has been found that attractors in the brain are maintained on the edge of such criticality by constant perturbations from inside and outside the brain.[14] Sudden switching to chaotic states helps find optimal resolutions to current inputs very rapidly.

Just how such brain states create cognitive states is illustrated in Walter Freeman's studies. He has studied electrical activity in the brain (electroencephalograms) during olfactory experiences in mice. These suggest that previous experiences with smells will have led to the emergence of attractors, as just mentioned—one for each kind of smell. And these will reside in the first subcortical level in the brain, the olfactory bulb.

Novel experiences perturb a corresponding attractor, as just mentioned, but not with a fixed response. Each new smell, Freeman reports, produces "a global wave of activity" in those attractors. But failure to reconcile the current with past experience creates chaotic activity in the attractor landscape. This chaotic activity quickly resolves into an updated attractor by adjusting parameters, as recorded in cell connections. The new resolution—the newly defined smell—is quickly passed on to the cortex for further response.

As a consequence of such dynamics, Freeman and colleagues suggest that it is not external smells per se that animals respond to, at least directly. Instead they respond to internal activity patterns created by the dynamics in the olfactory bulb. The nonlinear, chaotic dynamics of the system seem to be an essential part of the process. This is because "neither linear operations nor point and limit cycle attractors can create novel patterns . . . the perceptual message that is sent on into the forebrain is a construction, not the residue of a filter or a computational algorithm."[15]

Freeman cites evidence that similar kinds of nonlinear dynamics hold for visual, auditory, and somatosensory inputs. But the key point is that such a construction is now a cognitive agent rather than a mere neural one. Just as the combinations of atoms form a molecule with properties not envisaged in the components and free to interact with other such molecules, so this cognitive agent enters into a new emergent level of regulations with other such agents, creating new properties of life in the process. It can do this because it is not a singular entity—a token, or symbol—but a unique spatiotemporal output of an attractor basin and its structural grammar.

As with the acoustic forms in speech, such forms can now interact with other such constructs to form new levels of information with emergent properties. That is how cognition emerges in the brain: to become something different from the brain, while remaining attached to it. Let us now consider how that new level works in broader perspective, again using visual experience as an example. This perspective will be crucial for understanding individual differences in cognition.

THE GRAMMAR OF COGNITION

Visual experience starts as objects traverse the visual field and/or as we move around them, handle them, and so on. In effect, this experience consists of a rapidly changing mosaic of light energy on the retina: a cloud of moving points exciting areas of photoreceptors in the retina in seemingly disorderly succession (see chapter 6).

But the cloud is not really disorderly. By the nature of the structure of the objects it emanates from, this changing two-dimensional mosaic of light on the retina is not a random pattern. Statistical analyses of natural stimuli, mentioned in chapter 6, have revealed numerous correlations among points at various levels or depths.

Just imagine any three points on a line—say, the edge of a book—as you move around the object. The apparent movement of any pair of points in space and time will be correlated (they vary together, or co-vary). But these correlated movements will co-vary with any third point, and so on over innumerable points, across increasing depths of co-variation. The precise values of that set of parameters will be characteristic for any line. Curved lines, corners, edges, and myriad other features will have different, but equally characteristic, sets of parameters.

As described in chapter 6, it is such parameters, not isolated stimuli or features, that the brain is interested in. They are assimilated into networks by modifying the network connections. Through them, the network can compute predictability from the onslaught of spatiotemporal change in sensory inputs. This is the informational grammar through which endless novel forms make sense.

And that is what the concept of a line really is: an internalized set of statistical parameters abstracted across innumerable such experiences. You see three light spots moving toward you; the pattern enters your visual networks and gets pulled into the attractor comprised of those stored parameters. After a flurry of further activity—as described above for odors—the whole pattern gets filled in as a line and passed on for further processing. Indeed, when viewed on a computer screen, such a pattern of only three light spots quickly becomes reported by viewers as a line.

The important point is that the line is now a cognitive entity, over and beyond a neural one. This is because it will merge with myriad other sets of parameters it co-varies with, at increasingly higher levels or depths. For example, the co-variation patterns that separately define sides and edges of the book you were looking at will co-vary together at a deeper level. As you move around it, they move together and exhibit a structure peculiar to that object or class of objects.

Likewise, as we move around a chair, the line cluster that is picked up from the front will co-vary in space and time with the line cluster emerging from the edge or back of the seat. The shared, overlapping, sets of parameters form an attractor at a higher level. This, in my view, is how object concepts are formed as the basis of knowledge. They furnish tremendous predictability when dealing with the fuzzy nature of experience in changeable environments.

Viewing concepts and knowledge in this way, rather than as stored, ready-made images, suggests that what we perceive can be different from what we sense. We use the stored grammars to go beyond the information given. This is confirmed in the fascinating research on visual illusions.

We have experienced that strange perception of wheels appearing to rotate backward when the vehicle they carry is clearly going forward. In the well-known Kanizsa triangle, we add lines where the stored parameters in our networks predict them to be. (It has also been shown, incidentally, that neurons in the visual cortex fire as if the lines really were present). The fluidity of the dynamics may also explain the possibility of two or more rival attractors competing for the sensory input. For example, the Necker cube illusion appears as an image that flip-flops between two or more equally suitable possibilities, first one face at the front, then the other (figure 7.2).

Richard Gregory argued for many years that visual illusions arise from the restructuring of current experience from stored knowledge. He showed, for example, that patients recovering from lifetime blindness are not susceptible to typical visual illusions.[16] Presumably, this is

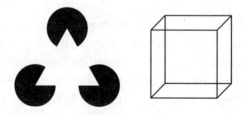

FIGURE 7.2

Visual illusions: the Kanizsa triangle and the Necker cube.

because they have not acquired the deep structural parameters that, in the rest of us, tend to generate the illusory outputs.

COGNITIVE INTELLIGENCE

Like singers moving together in harmonies, this merging of structures continues so long as there is mutual co-variation to pick up from them. Thus, attractors in different sense modes will share a number of their parameters or correlations. For example, the movements of a singer's lips will tend to be in sync with the sounds. So they will tend to become integrated into more inclusive, multimodal attractors, furnishing further powers of predictability.

In this way, information through one sense mode could readily disambiguate uncertain information through another, just as lip-readers do. A sound heard by predator or prey can create a visual expectation of what is in the bushes. Tactile impressions help us visualize the texture of a familiar surface. Smells convey information about possible tastes to follow.

The attractors can sometimes fool us, as when the patter of the ventriloquist appears to come from the dummy's mouth. Even young infants learning a language will be perturbed by any sudden asynchrony in sound and lip movements (cunningly contrived by an experimenter on a computer screen). All this is well attested in the multisensory cells and regions of the cerebral cortex. As Marc Ernst has said, the "strength of coupling" of signals from different modalities "seems to depend on the natural statistical co-occurrence between signals," and "perceptions are tuned to the statistical regularities of an ever-changing environment."[17]

Those points seem more obvious when we remember the dynamic aspects of experience. Objects are never experienced in perfectly static poses but as events in spatial transformation over time. Whereas any particular view may be unfamiliar, it usually becomes recognizable through the grammar in the networks.

In addition, of course, most objects are usually experienced as events with other objects—birds with nests, bats with balls, people with tables and chairs, and so on—in sequences of familiar spatiotemporal association.

Cognitive psychologists have used terms like "action schema" or "event structure" to reflect such general patterns with varying specific contents. Such general patterns permit further predictability in changing environments.

Most importantly, event concepts will be integrated with patterns of action or response. For example, the moving visual sensations of cycling will be integrated with familiar patterns of information from somatosensory receptors in muscles and joints. Approaching a hill will duly predict the strain about to follow in those impulses. The cumulation of an individual's specific experiences shapes the neural networks through which actions are expressed. That is, cognition shapes brain networks rather than vice versa. That gives rise to enormous individual differences, as well as other interesting properties.

EMERGENT FUNCTIONS

So far we have been talking about the merging and integration of network attractors. It has long been known, however, that ensembles of such attractors do not merely cooperate in an additive fashion, like cogs in a machine. They mutually reorganize in the process to exhibit emergent properties. Such properties would not be expected from the behaviors of the separate attractors. This should not be too surprising, as it happens in metabolic networks of single cells, in development, and in physiology, as mentioned earlier. At the level of cognition, however, emergent properties become even more important.

It is important to note that these complex traffic flows are carrying cognitive constructions over and above the neural activity. They are the outputs of attractor landscapes, but as cognitive interpretations, not passive patterns of neural activity. The nervous system receives two-dimensional light patterns. It is the cognitive system that converts these into meaningful three-dimensional objects and other constructs useful for making predictions. These properties are emergent and cannot be reduced to the properties in the separate networks.

This becomes clear from a number of observations. Take, for example, the effects of a novel stimulus on the amygdala, part of the brain involved in interrelating cognition and feeling. It has been shown that its responses are more closely related to the outputs of an attractor—what a stimulus means, or predicts, to the individual—than to the physical characteristics of the stimulus itself.

Likewise, in numerous areas of brain, neuronal activity has been found to vary according to the size of the reward being expected from an object or event rather than its physical form. Activity level is also related to the current degree of motivation. Reciprocally, anticipation of a highly valued reward is reflected in measures of arousal, attention, and intensity of motor output. As Luiz Pessoa put it, "Complex cognitive–emotional behaviors have their basis in dynamic coalitions of networks of brain areas, none of which should be conceptualized as specifically affective or cognitive."[18]

Some further evidence has come from experiments with artificial neural networks. Self-supervised networks were mentioned earlier. Even quite simple versions can form attractors from statistical associations and interaction parameters in the inputs. Of particular interest are experiments in which two or more artificial neural networks are coupled (i.e., have reciprocal connections), much as different regional networks are in real brains. It has been shown that as the numbers of units in such networks increases, clustered and hierarchically organized activity emerges that was not present in, and could not have been expected from, the separate networks individually.

Such coalition-based emergent properties are reminiscent of Jean Piaget's theory of cognitive development. Although intended as a model of human cognition (of which more in chapter 9), it probably also applies to less-evolved animals, at least up to a certain level. Piaget described how, in human infants, the coordinations in primitive sensory-motor activity become integrated. The predictabilities so furnished then became expressed in behavior: for example, in predicting the future location of a moving object from its current trajectory, as in catching a ball.

These primary coordinations then become nested into more inclusive coalitions (coordinations of coordinations) as experience and development proceeds. From such coalitions, all the major abilities of conservation, number, classification, seriation, logic, and so on, emerge. Whereas the infant's first schemas are simply compressed copies of reality, the more developed schemas allow representations beyond immediate appearance and thus anticipation of novel possibilities, as in math and theories of science.

This is Piaget's important concept of reflective abstraction. It reflects his findings that knowledge can emerge from interaction among the schemas to obtain a more abstract level, detached from immediate experience. One example is the formation of higher inclusion classes (e.g., fruit) from simple concepts, such as apples, oranges, strawberries. The latter are empirical abstractions, in that they form from what is directly experienced. But the inclusive class demands a higher self-organization, over and above what is experienced.

Another example is that of transitive inference. The conception that object A, say, is bigger than object B, may be gained from direct empirical experience. Likewise with the experience of B being bigger than object C. But appreciating that object A is *necessarily* bigger than object C requires reflective abstraction and the emergence of a new schema that provides an important system of logical thought.

Both examples demand abstraction of the more implicit structure, beyond what is available in direct experience or appearance (or in low-level attractors representing experience of them). They yield knowledge about the world deeper than that in immediate experience and must have been a tremendous boost to the evolution of predictability, and adaptability of behavior, in all animals. Both examples have in fact been demonstrated in nonhuman animals.

Finally, the assimilation of external structure into network coalitions also brings the outside world into the brain and cognition in ways not usually envisaged. The system allows us to sense far more of the world than is in current sense data, just as, in the visual system, we "see" far more than any current image. Mark Johnson has called this "embodied understanding."[19] It "is not merely a conceptual/propositional activity of thought, but rather constitutes our most basic way of being in, and engaging with, our surroundings in a deep visceral manner."

Of course it also takes cognition out into the world by providing much more scope for adaptable action. The basic message is that cognitive systems provide a deeper synergy between the inside and the outside than is usually imagined.

In humans, cognition evolved in further spectacular ways, as we shall see in chapter 9. In the meantime, we should not forget that all this emergent activity influences physiology, cell metabolism, and gene transcription. Together these activities make up nested systems that furnish adaptability by transcending direct experience. They further indicate the dangers of collapsing such multi-level systems into the rules governing only one of them, as in attempts to reduce cognitive functions, and individual differences in them, to neural activity per se.

PROPERLY DESCRIBING COGNITION

Real cognitive systems, then, consist of coalitions of attractors, from sensory reception to higher cognition. What are usually identified as key components of cognition are all emergent properties of such coalitions at various levels—perception, conceptual categorization, learning, knowledge, memory, thinking, and so on. There has been a tendency to describe these as independent systems with genetically determined processing. But they are aspects of self-organizing systems based on shared co-variations, the structures of experience. Let us take a quick look at some of them.

Learning

I described developmental plasticity in chapters 5 and 6. It evolved to provide the adaptability in changeable environments not possible from gene-determined functions. Learning is lifelong developmental plasticity (in the brain) for dealing with lifelong environmental change. In dynamical terms, we say that learning is the continuous updating of attractors in extensive attractor landscapes.

This updating happens even in single cells and in development and physiology, as we have seen. In cognitive systems, as Walter Freeman explains, "learning establishes attractor landscapes in the sensory cortices, with a basin of attraction for each class of stimuli that the animals have learned to identify. The basins of attraction are continually re-shaped by experience, and each attractor is accessed by the arrival of a stimulus of its learned class."[20] The most basic attractors are those assimilating the structure of sensory inputs. But these integrate as coalitions to yield higher conceptual learning and the basis of reflective abstraction.

Across the cognitive system, the basins of attraction are continually reshaped. This is possible because the modifiable connections in neural networks can assimilate structural relations. But there are also regions of brain in which the intrinsic wiring is biased for handling some informational structures better than others. Such biases will be ones shaped by canalized development, as described in chapter 5. For example, in the mammalian brain, the initial configurations of the hippocampus seem to be specially adapted to integrate spatial information and to assist memory (because memory will be in the form of spatiotemporal parameters, not sequences of "stills," as often assumed). Areas in the infero-temporal cortex seem to have evolved architectures that support the abstraction of finely timed movement and acoustic information (and, later, human speech).

In contrast, most regions of the cortex—those most recently evolved—are remarkably plastic. For example, research has shown that blind people tend to be more sensitive to differences in auditory pitch and touch than people who are sighted. And individuals who are born deaf may be better at detecting motion and "seeing" in their periphery than individuals with normal hearing.[21]

Such findings raise questions about the phenomenology of experience and consciousness, such as what makes one set of stimuli visual and another set auditory. But the important point for the moment is that what is learned—the updated network configuration—is much more than a mirror reflection of the outside world. It captures the nonobvious depth and structure behind it that furnishes predictability. It includes action and affective components, and, of course, emergent aspects that have not been directly experienced.

Note, also, that learning can be shallow or deep, depending on the depth of statistical structure assimilated from experience. It will be as well to remember this when we turn to learning and the school curriculum in chapter 11.

Memory

Memory is a consequence of learning. However, nearly all cognitive models of memory treat information storage and retrieval as of fixed-point attractors (discrete files, sequences, or traces). Again this is a resort to inappropriate metaphors. It should be clear by now that neither the cognitive system nor the brain is fashioned to handle information in that form. Rather they evolved to handle abstract statistical structures, the grammars, from which most likely specific instances can be generated from constantly novel inputs. Memories, as dynamic entities, reflecting dynamic experience, are enshrined in such grammars (or attractors).

This explains several prominent findings in memory research. Semantically related memories will tend to have related parameter structures. Accordingly, a person being asked to recall items in a category like "animals" will tend to do so in groups of related species: all the birds, first, say; then all the fish; and so on, rather than by recency of experience. It also explains why memories become distorted over time, but in particular ways. For example, Frederick Bartlett in the 1930s described studies in which participants were asked to reproduce stories and pictures experienced earlier. The participants typically elaborated or distorted the originals in ways that imposed their personal and social experiences over the original story line (another example, in fact, of the workings of ideology).

This makes sense in a dynamical perspective, because events tend to become assimilated into more general parameter structures— attractors—with experience. Over time, these also come to include personal and cultural ideals, as in the developing child. In other words, with continuing learning in a domain, memory attractors become revised and reorganized in relation to one another. This is a well-established finding in research and is reflected in the nature of knowledge. (In fact, in recent years, most cognitive models have been concerned with working memory, which I scrutinize below).

Knowledge

In the cognitive sense, knowledge is obviously the product of learning (indeed, the word "cognition" stems from the ancient Greek for "to know"). For the past few decades, at least, it has been assumed to reside in patterns of neural connections. As David Elman and colleagues put it, "Knowledge ultimately refers to a specific pattern of synaptic connections in the brain."[22]

In the standard approaches, however, it has been a confusing area of research and theory. In his book, *How the Mind Works*, Steven Pinker mentions how psychologists feel perplexed and baffled about the nature of knowledge. In a review of the subject, Emmanuel Pothos says that "overall, there has not been a single dominant proposal for understanding general knowledge."[23] In the popular textbooks of Michael Eysenck, Mark Keane, and others, knowledge appears to be treated as an ingredient in a vast diversity of cognitive processes but not as a subject in its own right.

In some theories, some or most of an individual's (animal or human) knowledge will be deemed to be innate. To most this usually means genetically determined, or coded in genes, and therefore in the form of if-then (cue-response) rules. The same idea is applied to higher concepts of cognition, like our concept of space and number, and also in the form of constraints on learning that limit what can be learned. Just how codes in linear strings of DNA can specify such complex structures as object concepts is not described. Other theories have viewed knowledge as catalogues of simple associations (as in the concept of dog mentioned above); still others are based on the rather vague notion of constructed schemas.

I discussed problems with these views earlier. For example, it is impossible to find a convincing model of what form such associations and concepts actually take in the mind/brain. Symbols, icons, pictures or other representations are mentioned, but these are just tentative labels. Accordingly, it is difficult to describe individual differences in knowledge.

In a dynamical perspective, each domain of knowledge is an emergent attractor consisting of relational parameters. These are the experienced co-variations between variables—the structural parameters—interdependent at many different levels. They furnish predictability in changing conditions.

General knowledge consists of attractor landscapes—coalitions of the more specific attractors for each domain of experience. The attractor landscapes consist of relational parameters abstracted from each domain of experience and condensed into neural network configurations. Any individual is knowledgeable in a domain when it is possible to utilize those configurations to generate predictions about current or future situations.

However, as described earlier, local attractors tend to form coalitions, giving rise to emergent knowledge not present in the original attractors. These include hierarchies of superordinate concepts, multimodal concepts, novel abstractions, and logical structures. Their formation reflects the abilities that Jean Piaget called "reflective abstraction." These powers stem from the nature of a system that is supersensitive to structural co-variations.

Unfortunately, these structures are sadly overlooked in assessments of learning ability through knowledge tests. These tests overlook, for example, how most knowledge is much deeper than psychologists seem to suspect. It is not captured in tests and exams that seek mere regurgitation of shallow—factual—knowledge. Moreover, in humans, the deeper layers can vary drastically across cultures, even when superficially similar (see chapters 10 and 11).

Finally, because of reflective abstraction, knowledge seems to snowball once established in a domain. That, too, reflects the value of cross-domain fertilizations, as in learning from analogies and the use of metaphor. And it sometimes emerges in the form of sudden new insights. The way that knowledge is bound up with feelings and motivations, forming beliefs, is also often overlooked, as in much education. As we shall see in chapter 9, these emergent aspects of knowledge became much expanded in humans as the basis for cultural learning.

Thinking and Problem Solving

Thinking is considered to be the core of cognition. Above all else, when psychologists and behavioral geneticists refer to individual differences in cognitive ability, they are implying the ability to think competently or effectively (and that such variation will have something to do with variation in genes and brains). However, as we have seen, thinking has not been well understood in the traditional frameworks.

In a dynamical systems perspective, thinking is the spread of neuronal activations through the attractor landscape. The spatiotemporal pattern of that activity determines its compatibility with an existing basin of attraction, into which it may then be drawn. Some further output may then ensue: another form of activation with various possible consequences that depend on circumstances.

On the one hand is aimless "mind wandering" through the attractor landscape, perhaps prompted by random stimulus inputs. (Dreaming is probably just such random activations through the attractor landscape during sleep—or even daydreaming). On the other hand is motivated searching that recruits attention and turns thinking into problem solving.

For example, hunger may initiate a planned search for food. Dissonant or contradictory current attractors may need to be resolved (as in making choices). Or a means of escape from danger rapidly devised. These all demand thinking. Responses may then emerge as reinstatements of activity patterns from memory, novel patterns of motor action, some retuning of networks, and gain in knowledge, or some other resolution of the perturbation. Most likely, each immediate response guides activity to the next attractor basin in a series of activations that lead to a final resolution.

Such processing is achieved with amazing speed and productivity. Unlike the relative slowness and irreversibility of genetic selection or even developmental plasticity and physiology, dynamic cognitive systems can respond quickly and adaptively to deal with rapid environmental change. These are adaptations (or "species") arising on scales of milliseconds rather than centuries or epochs.

In contrast, some decisions can be deliberately delayed by interactions among centers. This allows responses to emerge over longer time periods, as in stalking and hunting in animals, or most important decisions in humans. Thinking activity may involve gradual retunings, as when we slowly change our minds; or sudden resolutions, as when we gain insight into a situation, understand the meaning of a novel signal from context, or classify an object or event. Activity can also oscillate between rival states in situations of uncertainty, distraction, or weak motivation.

A crucial aspect of thinking with dynamic systems is its creativity. By definition, changeable environments demand creativity of responses from living systems. But flexibility based only on innate knowledge or processing, or shallow associations, would inevitably be rather limited. As mentioned above, constantly changing environments demand constantly novel responses. That challenge can only be met by a special kind of chaotic dynamics, so that thinking is almost invariably a kind of constructive imagination.

Creativity is also enriched by attractor coalitions that involve affective states, or feelings (see chapter 6). Thinking, as Jerome Bruner once said, involves a search through feelings as much as through independent cognitions. This meshing of cognitive activity in feelings—and, through that, their grounding in the outside world—is what distinguishes animals from robots. It resolves as the deeper phenomenon of *belief*, which might be defined as knowledge with feeling. We can talk about robots learning, memorizing, making decisions, having knowledge, and so on. But we do not talk about robots having beliefs. Without beliefs, they remain much less creative than the humans who invent them.

This attractor landscape combining cognition and feeling is what forms individual identity—in humans, emotional intelligence, personality, and the concept of self that is known by the self and others: our minds.

EVOLVING COGNITION: A SUMMARY

Darwin understood that natural selection, as the primary architect of evolution, would only work if there were some system of variation production in living things. He did not understand how the variation was produced. But he recognized its importance. What he also did not understand was that survival in changing environments required different systems of variation production. The standard neo-Darwinian view is that all or nearly all variation arises from random gene mutations. This is what the behavioral geneticists of IQ take to be the basis of the heritability of IQ and the grounds for gene hunting.

Complex changeable environments, however, require adaptation on a different time scale from Darwinian natural selection. That is, it requires *intelligent* (adaptable) variation rather than random variation. Variation arising from intelligent systems far outstrips that possible from random gene mutations. Yet the latter is still claimed by behavioral geneticists as the main source of differences even in human intelligence.

The origins of life probably entailed random variations among components of molecular ensembles. By the time of yeasts and bacteria, however, living organisms had evolved into self-organized intelligent systems involving a variety of variation-producing processes (usually summarized by the term "epigenetics"). They survive today, in changing environments, through sensitivity to the available statistical structure. Moreover, the variation so produced can help steer future evolution in more adaptable, intelligent, directions. That is, intelligent systems are active creators of variation; variation is not the passive outcome of genetic and environmental forces as in the standard behavioral genetic equations.

That early intelligence became incorporated into multicellular organisms and became extended and amplified in developmental systems. It is exhibited in the vast variety of cell types and functions generated from a single precursor. Intelligent development is also exhibited in the emergence of body form, in two ways. First, different members of the same species develop highly uniform basic characteristics even when they have different genes. This is canalization, as described in chapter 5. Second, very different body forms can develop even in a group of animals with identical genes, or between parents and offspring, in response to environmental change. This is developmental plasticity. Again, we get adaptive creation of variation, predicting and anticipating change to a far greater degree than is possible by genetic variation.

The communication systems coordinating cells subsequently evolved as physiological systems. They are highly sensitive to changes both inside and outside the multicellular body, creating further variation. In turn, physiology became embedded in nervous systems with a further leap in scope for individual differences, now in behavior. The further evolution of brains produced the emergent properties of cognition, as described above.

I hope to have shown you the tremendous further scope for variation in cognitive systems, as in learning, memory, knowledge, thinking, and so on. In turn, what has evolved is an efflorescence of adaptive variation that is woefully underestimated by behavioral geneticists of potential.

Let us now look at the implications of all this for individual differences.

WHAT VARIES IN COGNITIVE ABILITY?

The various models of cognitive functions permit individual differences to appear, theoretically, in a variety of ways. Under strict nativist assumptions, there will be few individual differences: natural selection of its critical survival qualities will have ensured common genes and common functioning. In contrast, a less-strict nativist would argue that differences in cognitive hardware will result from random gene mutations. And these will put individuals on a genetic ladder of ability. This is the model of the behavioral geneticist who adopts metaphors like speed and power to describe the hardware differences.

However, nearly everyone agrees that the gene-directed variation in that model will be attenuated by the environment and experience. Usually, some level of associationism is incorporated in the form of learning ability. So we get a mixture of sources of variation, as in the contemporary nature-nurture debate.

Under constructivism, individual differences emerge from experiences in development in a way that transcends simple associations and built-in nativist functions. This does not preclude the possibility of individual differences forming from genetic and environmental differences. It is just that their respective effects will be more concealed in the workings of the system.

In this chapter, I have put forth another point of view: that of active organisms incorporating environmental structures and changing their own structures and functions to accommodate them. This view reflects the extensive evidence that systems have evolved with many buffering mechanisms and ways of varying in adaptive, creative ways. Nearly all individuals will have good enough systems for assimilating structures of experience and actively generating individual differences accordingly. After that, to quote Eric Turkheimer again, "behaviour emerges out of a hyper-complex developmental network into which individual genes and individual environmental events are inputs. The systematic causal effects of any of those inputs are lost in the developmental complexity of the network."[24]

Of course, rare deleterious gene mutations—the absence of a crucial ingredient—might result in disorders at any of these levels. Absence of specific environmental components, or toxic insults, may do likewise (see chapter 10). Such mutations and insults result in distinct categorical variants, not extremes of a normal distribution, or a ladder of ability.

In general, however, causal explanations of individual differences in complex traits are not going to be found in individual genes or environments. Attempting to do so results from misunderstanding those systems and how and why they have evolved. This is why the literature that portrays individual differences merely as a resolution of genetic and environmental forces contains little on cognitive theory as such and relies heavily on mechanical metaphors, like speed, capacity, power, or efficiency.

Similar arguments apply to attempts to find causes of individual differences in brain differences. Since the brain and cognitive systems evolved to abstract external structures and assimilate them in neural networks, associations are more likely to come from the outside in rather than the other way around. This is a point that researchers are inclined to overlook.

COGNITIVE MODELS AS UNCONSCIOUS IDEOLOGY

As I have mentioned, vague concepts in psychology soon become vehicles of ideology. To further illustrate the problem in the science of cognition, let us consider working memory. It is said to (probably) exist in other animals as well as humans.[25] As mentioned in chapter 3, aspects of working memory have been proposed recently as the basis of individual differences in human intelligence. Moreover, investigators consider these differences to be rooted in genetic differences, explaining the alleged heritability of intelligence.

The bases of such claims are correlations between working-memory capacity scores and IQ scores. Patricia Carpenter and colleagues (in a 1990 paper) proposed that the ability to maintain a large set of possible goals in working memory accounts for individual differences in IQ. Peter Carruthers says that working memory abilities "account for most (if not all) of the variance in fluid general intelligence, or g."[26]

The notion of working memory was first proposed by Alan Baddeley and Graham Hitch in 1974, but has been much updated since. It is generally conceived as a process for holding representations "in mind," either input from sensory receptors or retrieved from other "stores." The central function appears to be control of attention. As Peter Carruthers explains, "It is by targeting attention at representations in sensory areas that the latter gain entry into WM [working memory]." There they are "held in an active state for as long as attention is directed at them," that is, for other cognitive processes to operate on them.[27]

As just mentioned, individual differences in working memory—and intelligence—are said to stem from differences in its capacity. Attempts to measure that capacity duly consist of tasks that require holding something "in mind" while attending to a concurrent task. Especially popular are so-called n-back tests. Participants are presented with a string of items (e.g., letters or digits) on a computer screen. They have to decide whether each stimulus matches the one that appeared n items ago (where n may be, e.g., 2, or 3, or 4, as dictated by the investigator).

How is this model said to work? Well, through our old friend the "executive system." Baddeley himself used the metaphor of a company boss to describe it.[28] The company boss decides which matters deserve attention and which can be ignored. The boss then gathers information together from other departments, to integrate and clarify the problems, and then selects strategies for dealing with them. These departments recruit two "slave systems," a verbal store and a "visuo-spatial scratchpad" that deliver information in the required format.

Huge amounts of research and mountains of texts have been done and written on working memory, especially as a source of cognitive potential. And the models are at least logical interpretations of the data gathered. But they bear little resemblance to how a dynamic cognitive system operates. As we have seen, there is no evidence of an executive system in the brain. Instead, cognitive functions are emergent properties of a self-organizing attractor landscape. Likewise, no evidence exists for a coherent "store," as a discrete place varying in capacity or some other easily quantified dimension. And, there are no representations in sensory areas in the form of coherent symbols or tokens: certainly none that gain entry into working memory, in the sense of parcels in a letter box.

Investigators often acknowledge, indeed, the problems of interpretation in the area. As Susanne Jaeggi and colleagues say about the n-back test, "Little knowledge is available about the cognitive processes that mediate performance in this task and consequentially, about the processes underlying n-back training."[29]

In other words, it seems that individual differences in cognitive ability are being ascribed to variables that do not actually exist in reality. It is perhaps hardly surprising that the group set up by the American Psychological Association to investigate intelligence concluded that the definition of working memory itself remains vague and relations with IQ or *g* are still uncertain.

It may seem churlish to object to attempts to describe extremely complex functions by breaking them down into smaller steps or chunks. But such objections are warranted when strong and influential statements about the genetics of individual differences are based on such models and metaphors.

That the latter have their roots in social ideology is clearly demonstrated in the one-sided, causal, interpretation of correlations. For example, little consideration is given to the fact that correlations between test performances can arise from joint association with emotional factors, such as self-confidence and motivation in testing situations. Indeed, Carruthers remarks in an aside that "attention is quite sensitive to interference, so sustaining a representation in WM [working memory] for an extended period is by no means easy."[30]

Interpreting score differences as definitely one thing and not the other can be dangerous in such circumstances. But much the same can be said for differences on all the other so-called tests of intelligence or potential—especially when they emerge in social contexts: the subject of the next chapter and beyond.

8

POTENTIAL BETWEEN BRAINS

Social Intelligence

OF INDEPENDENCE AND INTEGRATION (AGAIN)

For some reason, the typical mindset of psychologists in the Western world cannot deal very easily with integrations of living components. We have seen this problem in the nature-nurture debate with the notion of additive genes and environments, and with individual differences as the sums of independent effects. Yet, from the origins of life itself, so much of what has evolved has depended on the interaction and integration of components. The evolution of intelligent systems is the most conspicuous result of such integration and interaction. Those systems became even more complex with the amalgamation of individual organisms into social groups.

The advent of social aggregation, in fact, became a game changer for living things and for sources of individual differences. Encounters with other "objects" that are themselves animate, mobile, and unpredictable are already rather challenging. But when relations among them became collaborative, still further demands were made on intelligent systems. Yet social cooperation emerged quite early in the evolution of species. It first appeared as occasional behavior patterns for dealing with changing environments. But then, with advantages manifest, it quickly took off; it allowed organisms to evolve more adaptable life styles and to survive in less hospitable circumstances. As a consequence, for at least two billion years, most of life has not existed as individual units but as social conglomerates.

That has enormous implications for understanding intelligent systems and the nature of individual differences. In this chapter, I examine those implications. I show how different social forms have evolved, with different implications for intelligent—including cognitive—systems. This examination provides some of the conceptual grounds for better appreciating human cognitive systems in subsequent chapters.

FROM SINGLE CELLS TO SOCIAL BEINGS

The benefits of sociality are often described quite simply as twofold. First, it imparts a wider effective awareness of things going on in the world (two pairs of eyes are better than one, etc.). Second, it creates a wider range of possible responses (two pairs of hands, etc.). Both are particularly valuable in changeable environments. Such are the conditions that first brought many bacteria and slime moulds into social groups. The theoretical implications for intelligent systems explain why such groups are studied so intensively.

One subject of study has been the soil organism Myxobacteria. If nutrients become depleted, individuals start to secrete a mutually attractive substance. Small molecules called auto-inducers are synthesized in individual cells in response to the stress warnings. These are then released into the local medium and bind to receptors on individuals in the neighborhood. When the concentration of inducers reaches a certain threshold, they trigger internal signaling and movements that bring the cells together.

In response to this quorum sensing, approximately one hundred thousand cells aggregate to form a single fruiting body. Continuing waves of intercellular signaling orchestrate recruitment of numerous genes to produce new cell components. Extensive morphological and biochemical changes follow in the cells, and thick-walled spherical spores are formed on, and shed from, the fruiting body. These are sessile and resistant cells that can withstand starvation until nutrients reappear.

This is, of course, analogous to the use of morphogens in multicellular organisms for coordinating differentiation and development. As with development, the functioning of those signals depends on the statistical structure—their distribution in space and time—not just concentration. Indeed, up to twenty chemically distinct signals are used in the microorganism's response to stress. And these signals must be integrated to create the divisions of labor and their coordination. As Karen Visick and Clay Fuqua put it, "The regulatory cascades involved in the response can be staggeringly complex."[1]

In other words, the intelligence that emerges among cells, coaxing differentiation and cooperation, is as dependent on environmental structure as any that evolved within cells. There is no distinct executive or supervisory agent. Individual cells do not bear a genetic code for the form of a fruiting body or for spore production. The process is self-organizing, the effect emerging from interactions among the individual cells, themselves responding to a changing environment.

Such occasional sociality, of course, evolved into the permanent cooperation of multicellularity. That entailed wider differentiation of cells together with increased complexity of cell signaling and physiology. The upward spiral in changeable environments eventually turned these multicellular conglomerations into individuals that behave as one. And with that came nervous systems and brains and rudimentary cognitive systems.

Evolution thus laid the ground for another form of cooperation, at a higher organizational level, to face even more challenging conditions. This is the cooperation among behaving individuals, now through brains and cognitive systems rather than quorum sensing and physiology alone. Obviously, that turn of the spiral provided even greater scope for individual variation in behavior, but only if those individuals could be coordinated through an even more complex form of regulation. Let us call that "epicognitive" regulation, meaning at a level above or beyond the cognition of individuals. It is important to ask: What is the nature of that regulation? And what are its implications for understanding individual differences?

SWARM INTELLIGENCE

Answers to those questions have been explored in social insects through models and theories of what is now called "swarm intelligence." Wikipedia provides an interesting definition: "[Swarm intelligence] is the collective behavior of decentralized, self-organized systems . . . interacting locally with one another and with their environment. . . . The agents follow very

simple rules, and although there is no centralized control structure dic-
tating how individual agents should behave . . . interactions between
such agents lead to the emergence of 'intelligent' global behavior, unknown
to the individual agents. Examples in natural systems of [swarm intelli-
gence] include ant colonies, bird flocking, animal herding, bacterial
growth, fish schooling and microbial intelligence."

The absence of a centralized control structure does not mean that there
is no regulation at all. Local perturbations are perceived by individuals,
and they respond to them. But they do so in a way conditioned by the
responses of other individuals, which the first individuals, in turn, also
perceive, and so on. So feedback and feedforward cycles of stimulus and
response are set up, and a harmonious pattern of intelligent behavior
somehow emerges.

One of the objectives of this book has been to explain how such emergent patterns have been important throughout the course of evolution.
How intelligence emerges among cooperating individuals is another turn
of the evolutionary spiral. The details have been revealed mostly in the
study of social insects, such as ants, termites, and bees. Although hugely
successful as life forms—in terms of biomass, they probably outweigh
all others on the planet—they live in challenging conditions. As Duncan Jackson and Francis Ratnieks point out about food supply alone, "Social insect colonies live in a dynamic, competitive environment in which
food sources of variable quality are constantly changing in location.
Most ant species are dependent upon ephemeral food finds."[2] Other daily
hazards include intrusions from predators and clumsy passersby, weather
fluctuations, water incursions, and earth movements.

In responding to these conditions, remarkable degrees of cooperation
are exhibited. The building of intricate nests; defense against predators;
food foraging and transportation; route marking; laying, clustering, and
sorting of eggs; brood care; dealing with sudden obstacles; urgent nest
repairs and maintenance; and so on, are all tightly regulated. This is
achieved most conspicuously through development of individuals into
anatomically, physiologically, and behaviorally specialized roles. Reproduction (egg laying) is confined to a single queen. Defense is by soldiers
and other specialized individuals. Foraging and food collection is done

by workers. Brood care, nest building, and cleaning are all done by other specialized forms.

It might be tempting to think that these different castes develop through different individuals having different genes. But there is no such correlation: the extreme phenotypic specialization arises through developmental plasticity that is itself under the regulation of the emergent behaviors of the group as a whole. Fertile queens develop from larvae that have been fed a sugar-rich substance called "royal jelly." The jelly is produced by workers and contains the hormones that turn larvae into fertile queens rather than sterile workers. One or more of these bloated queens subsequently produces all the eggs cared for by workers. During this stage, the level of care and nourishment the larvae receive determines their eventual adult form: workers, soldiers, nurses, winged nymphs that become reproductive adults, and so on. These are sometimes called "polyphenisms" and are excellent examples of developmental epigenetics.

Bees, ants, and termites are famous for their striking organization and highly efficient responses to everyday challenges. Much of that is achieved through chemical messages called "pheromones." Depending on species, there are about ten to twenty different pheromones, and all individuals are sensitive to them. Some pheromones can quickly summon up thousands of ants to a food discovery. Others can signal attacks on prey, the defense of the colony, or even the relocation of the colony. Still others help ants distinguish among different family members, nest mates, and strangers. The queen also has special pheromones that influence workers' behaviors.

Pheromone following has been best described in ants' trail laying to food. When a forager finds a food source, it heads back to the nest, leaving a trail of pheromone that other ants can follow. After feeding, in turn, each recruit reinforces the pheromone trail. The result is a nonlinear increase of ants around the food source and a more efficient utilization of it. Such self-organized responses are even more impressive in activities like nest building, with a complex topology of underground galleries housing treelike networks of tunnels and chambers.

So how does the result become greater than the sum of its parts? After all, the complexity of nest form is not within the cognitive potential of individual ants. As Mehdi Mossaid and colleagues put it, "The contrast between the limited information owned by single individuals and the 'global knowledge' that would be required to coordinate the group's activity is often remarkable."[3]

No one is yet very sure. Individuals may appear to be responding on the basis of simple if-then rules. Information picked up by individuals is responded to in stereotyped ways that then become information for others to respond to. It all suggests little in the way of cognitive resources. Computer modeling—so-called individual-based modeling—has suggested that much can be achieved from individuals using simple response rules reciprocally. Indeed, the models have been copied in artificial intelligence programs for commercial delivery systems, telephone companies, and even air traffic control systems.

A closer examination suggests that the cognitive resources of ants may be a little more complex than that. Even the following of a pheromone trail requires integration of positive and negative feedback involving a number of variables. These variables include the freshness and density of the volatile pheromones, the quality of the food source, crowding at feeding sites, direct bodily contact with other nest members, and changes in the environment itself (including the original perception and then the depletion of the food source).

Ants also communicate in several other ways: vibrations through antennae and legs, food or liquid exchange (trophallaxis), mandibular contact, and direct visual contact. Research has also revealed how the different volatility of different pheromones, together with these other forms of communication, and used in different combinations (multi-modal communication), achieves remarkable sophistication of communication in all key tasks.

Such a multiplicity of variables suggests dynamical principles rather than cue-response (if-then) rules. In the group, the flow of communication results in streams of positive and negative feedback loops with nonlinear relations. These self-organize dynamically, as attractor basins, reflecting the statistical structure of those interactions (the relational parameters being the "rules"). It is the outputs of these attractors that are then refracted through individuals. As Claire Detrain and Jean-Louis Deneubourg, put it, "Fluctuations, colony size and environmental parameters act upon the dynamics of feed-back loops between interacting nestmates and hence shape the collective response of the whole ant society."[4]

Such emergent systems of positive and negative feedback loops has been studied in a number of colony behaviors, including nest building, vacating, and cleaning, as well as food foraging. As described by Detrain and Deneubourg, "While excavating, the insect adds pheromone to the cavity walls and the laid pheromone, in turn, stimulates other nestmates to dig at that site. As the nest volume increases, the density of insects and so the frequency of their visits to the digging sites decrease what ultimately leads to a self-regulation of the excavated nest volume."[5] Similar self-organized dynamics have been described when dealing with dead bodies, forming nest clusters, and defending against predators.

Individual ants are not entirely if-then machines; they can exhibit some complexity of cognition and of individual learning, even with a brain of only about a quarter of a million neurons. For example, at least some ants do a "learning walk" to integrate positional coordinates of the nest site. When foraging and exploring, they also integrate a variety of cues, such as sun position, polarized light patterns, visual patterns, odor gradients, wind direction, and step-counting, to navigate and compute a "bee-line" home. By tuning their responses to the emergent patterns of interactions in the group, they are doing more than following built-in rules. And they are obtaining greater adaptability than they could achieve alone.

Thus the cognitive systems, even of ant brains, are already coalitions of attractors (i.e., attractor landscapes), as explained in chapter 7. In ant colonies the individual ants are now forming coalitions at a higher level. As well as the coalitions of attractors in brains, there are "coalitions of coalitions" across brains. Higher-level attractors emerge among brains, just as they do among individual neurons within them. These are new statistical abstractions that go beyond the information given. They do not, as it were, hang in cyberspace. They are dispersed among individual brains—through learning—to regulate individual behaviors.

The point is that this new level of attractors, based on deeper statistical patterns in the here and now, provides far greater adaptability than built-in rules or individual learning, possibly could. And that applies to the group as a whole, as well as to its individual members.

Having said that, we should not lose sight of the fact that individual ants are developmentally, cognitively, and behaviorally limited. Considerable adaptability of the whole is achieved from limited adaptability of individual members. They interact on the basis of a limited range of individual behavior patterns. Stephen Guerin and Daniel Kunkle refer to such systems as "thin" agents, with limited depth of cognition. They contrast with "fat" agents capable of internal reasoning in more complex social systems, to which we will turn below.

Accordingly, my task is to get you to imagine such coalitions where the "units" of cooperation have, for other evolutionary reasons, already developed advanced cognitive systems: that is, they have much greater individual adaptability. It will be like comparing an orchestra of three-hole hornpipes (each with limited repertoire) with one of eight-hole hornpipes. Far more interesting things can happen.

Before that, though, it is instructive to look at some other forms of social living in species more evolved than social insects.

SHOALS, FLOCKS, AND HERDS

The Cambrian explosion of about 550 million years ago produced an enormous variety of species. That, in turn, produced more intense predator-prey interactions and thus improved hunting and defensive systems. Rapidly moving among objects, some of which are themselves moving, creates high levels of unpredictability: sensory inputs are largely fleeting, fragmentary, and form novel combinations. Among these changing environments, the bony fishes emerged with great maneuverability and speed, new sensory and motor systems, and bigger brains and more advanced cognitive systems to regulate them. Some of them also evolved to form social groups, or shoals and schools.

About a quarter of fish species form shoals. The advantages seem fairly obvious: enhanced foraging success, better chances of finding a mate, and even some hydrodynamic efficiency. But the overwhelming advantage is defense against predators. This happens through better predator detection and by diluting the chance of individual capture.

Fish social groups do not have the extensive and close-knit organization of colonies of social insects. This is probably because their need is mostly limited to the benefits of foraging and predator avoidance. Nevertheless, the dynamics of their coordination are fascinating and have been widely researched. Again, there has been much debate about how it actually happens. As Iain Couzin put it, "Decision-making by individuals within such aggregates is so synchronized and intimately coordinated that it has previously been considered to require telepathic communication among group members or the synchronized response to commands given, somehow, by a leader."[6]

Some shoals are quite loosely aggregated. Others are more tightly organized, all moving at the same speed and in the same direction. Then the fish are said to be schooling. Fish schools move with the individual members precisely spaced from one another and undertake complicated maneuvers, as though under top-down regulation.

As with the social insects, though, there are no executive or supervisory agents at work. All the group choreography is achieved by mechanisms in individuals interacting with the structural dynamics emerging in the group as a whole. The results include patterned behaviors for foraging and more complex ones for avoiding and confusing predators. In the process, they manage to avoid collisions between individuals, collisions with obstacles, or breakdown in the coherence of the overall patterns. The speed of reaction to predator attack suggests rapid transmission of information from one group member to another, initiating changes of direction. They then turn in concert, resulting in escape waves, sometimes fanning out or using other tactics to confuse the predator.

As with the ant research, rather simple if-then rules of individual behavior have been proposed to account for the composite integrity and success of shoals or schools. For example: keep an eye on your immediate neighbors, move in the same direction as your neighbors, and remain close to them (but not too close). These have formed the bases of "individual-based modeling." In truth, though, the actual behavioral rules followed by individuals in fish shoals are still poorly understood. Simple rules, like those just mentioned, have only been partially successful in accounting for school behavior. So some researchers have proposed that the "telepathy" emerges from self-organizing dynamics, as in ant colonies. As Naomi Leonard argues, "The dynamics of collective animal behavior are typically nonlinear due to nonlinearities in individual dynamics, nonlinearities in interaction dynamics, nonlinear coupling between the individual dynamics and the interaction dynamics, and nonlinearities in the configuration space."[7]

DOMINANCE HIERARCHIES

Of course, forming shoals or schools is not the only form of social behavior in fish. Even in loose aggregates, individuals learn from the successes and failures of others in such activities as food foraging, nest building, and route finding (this is called "observational learning"). Perhaps the most obvious social formation in at least some species of fish (as well as other animals) is the dominance hierarchy. In such groups individuals are clearly differentiated in a way that regulates priority of access to food, mates, or other resources. Studies of such hierarchies have been revealing.

Deterministic models have generally assumed that rank status is due to some individual attribute, such as size, aggressiveness, superior genes, physiology, or other indication of "biological fitness." Experiments by Ivan Chase and colleagues, however, question that idea. They brought groups of fish together in a tank to form a hierarchy. The fish were then separated for a time before being brought back together again. In those reunions of exactly the same fish quite different hierarchies tended to form. Although individual attributes did seem to have a minor role, social dynamics seemed to be much more important.[8] This result is confirmed by other studies in a wide range of species.

Such findings suggest that social hierarchies are self-organizing, self-structuring, dynamical phenomena. Their function may be that of reducing uncertainty in social access to resources, where "queuing" is more mutually beneficial than a scuffle. As Chase and colleagues glean from further research, behavior is highly context dependent and is conditioned even by numbers in the group. Individual attributes have little or no ability "to predict the outcomes of dominance encounters for animals in groups as small as three or four individuals."

Instead, they go on, "We review the evidence for an alternative approach suggesting that dominance hierarchies are self-structuring. That is, we suggest that linear forms of organization in hierarchies emerge from several kinds of behavioral processes, or sequences of interaction, that are common across many different species of animals from ants to chickens and fish and even some primates."[9]

In other words, individual variation and group variation are functions of the dynamics of the system. This has implications for attempts to delineate the origins of individual or group differences. The research effort, Chase suggests, should be to concentrate on rules governing the dynamics of interaction rather than to focus on individual-level explanations.

Even in fish, then, where the functions of social grouping are narrower than in the social insects, much of the intelligent system of the individual is a refraction of the dynamics of the group. There may be individual "constitutional" variations, but almost all members will be good enough for full group participation and its benefits.

BIRD FLOCKS

Much has been learned about the dynamics of collective intelligence from bird flocks. The coherent flight of bird flocks is one of nature's most impressive aerial displays. These are really spectacular, and video footage is readily available in the social media (e.g., see www.youtube.com/watch?v=8vhE8ScWe7w). They show rapid changes of speed and direction involving thousands of birds in a complex aerial choreography, all without a conductor and without collisions.

The dominant function of flocking again seems to be predator avoidance. During a predatory attack, waves of agitation are observed originating from the point of attack and propagating away from it. The wave velocities are often faster than the velocity of the birds themselves and are highly confusing to predators. Studies have shown that the higher the velocity of those waves is, the lower the predator success will be.

As with studies of ant colonies and fish schools, attempts have been made to explain the dynamics of bird flocks in terms of a few individual-based rules. One popular theory has suggested that flock behavior is successful if individual birds only interact with a few neighbors (the so-called influential neighbors). As with fish schools, the interactions follow three simple rules: get close to neighbors; align speed and direction; avoid collision. Any response in one of the birds then quickly ripples through the flock.

However, empirical data show that the total response is faster than such ripples: changes of speed among different birds are correlated beyond the influential neighbors. The changes seem to be cast near instantaneously through the flock as a whole. This obviously enhances a predator escape maneuver, but its mechanism is not clear.

Irene Giardina and colleagues studied flocks (or murmurations) of common starlings. Using multiple video cameras, they tracked and analyzed flock movements from second to second. They showed that changes in the velocity of any one bird affected the velocity of all other birds in a flock, regardless of the distance between them. That sort of relationship is known as a scale-free correlation, and is seen in systems poised at the edge of criticality. Such dynamics are typical of attractor landscapes, like the metabolic interactions in cells and processing in nerve networks, as described in previous chapters.[10]

As with players on a soccer team, properly aware of one another, "the change in the behavioral state of one animal affects and is affected by that of all other animals in the group, no matter how large the group is. Scale-free correlations provide each animal with an effective perception range much larger than the direct inter-individual interaction range, thus enhancing global response to perturbations. Our results suggest that flocks behave as critical systems, poised to respond maximally to environmental perturbations."[11]

Remember from chapter 4, and the Bénard cells, that a critical point is where distant elements of a system become correlated with one another, far beyond the range of local interactions among the individual elements. As William Bialek and colleagues say with regard to bird flocks, "The critical point is the place where social forces overwhelm individual preferences;" it is where "individuals achieve maximal coherence with their neighbours while still keeping some control over their speeds. . . . Away from criticality, a signal visible to one bird on the border of the flock can influence just a handful of near neighbors; at criticality, the same signal can spread to influence the behavior of the entire flock."[12] In other words, the dynamics of the flock is a far better way to make predictions from fleeting signals than the use of built-in rules or personal learning.

The study of flocks is thus instructive for understanding group dynamics. Birds are more adaptable than ants as individuals. But they bring only a small repertoire of behaviors to the group, sufficient for realizing the narrower range of benefits. The group dynamics are consequently less complex even than those of the ants. Nevertheless, to participate at all, individuals must exceed a rather demanding threshold of cognitive ability. It seems likely that strong natural selection would ensure that. Through a combination of canalization and developmental plasticities, the vast majority of individuals must be "good enough" to meet such demands and ensure that coherence of social behavior among hundreds or thousands of birds is maintained.

SOCIAL COGNITION

However, flocks are not the only kind of social life experienced by birds. As with fishes, many flocks are only temporary and many bird species alternate between solitary and social lives (even if that consists only of mating and rearing offspring).

A widely discussed hypothesis in recent years has been that an animal interacting with others can create a cognitive model of another's cognitions and so be able to predict the behavioral intentions of others. This has been called the "theory of mind" theory. It is controversial, because it has been thought by many theorists to exist only in humans—indeed to require the "advanced" brains of humans.

The suggestion, and evidence to support it in birds, comes from an increasing number of studies. For example, Milind Watve showed how bee-eaters (birds) would not return to their nests if they saw a potential predator in the vicinity. They would, however, be less cautious if the predator could not actually see the nest. In contrast, they were very cautious if the predator had, in fact, already seen the nest before, even if it could not see it at that moment. Watve and colleagues argue that to behave in this way, the birds must have formed an impression of what the predator knows about the nest—that is, have some knowledge of the other's knowledge.[13]

In another study, Nathan Emery and Nicola Clayton noticed that jays that had buried food for later retrieval (as they habitually do) when other jays were around tend to return later, alone, and hide the food in a different place. They were even more likely to do so if they had experience themselves of stealing other jays' food (as they also habitually do). These observations have been confirmed in replication studies. As before, the jays re-cached a much higher proportion of the peanuts if another bird could see them.[14]

Of course, these abilities have less to do with social cooperation than with general (nonsocial) interactions with other animals. But they may be indicative of ways in which social interactions could encourage the evolution of even more complex cognition. For example, it has been claimed that birds that have experienced social hierarchies are more capable of transitive inference of the kind:

A is higher than B;
B is higher than C;
therefore, A is higher than C.

However, there have been doubts, mainly to do with the interpretation of observations and lack of experimental control. I will not go into these details here. However, these studies may indicate that birds have cognitive intelligence more advanced than previously thought.

SOCIAL COOPERATION IN MAMMALS

So far we have discussed the nature of cognition in social groups and inquired how individual differences might be related to them, at least in ants, fish, and birds. Variation seems to depend on the actual degree of cohesion of the group. And the cohesion varies with the range of advantages gained over the kinds of environmental changes to be contended with.

The social insects, facing unpredictability of many kinds, exist entirely in highly cohesive groups and cannot exist outside them. Individual differences are achieved through developmental plasticities, affecting anatomical and physiological, as well as cognitive and behavioral, differences. That is, the dynamical rules emerging in the group as a whole are refracted through the already-developed biases in individuals to set up structured feedforward and feedback loops.

The ancestors of species that form fish shoals, bird flocks, and mammal herds had already evolved improved cognitive systems (and bigger brains) for reasons other than social existence. The dynamics of these social groups are less encompassing than in the social insects. They organize a relatively restricted range of individual behaviors into a narrower range of group behaviors, based mainly around passive food foraging and prey avoidance.

Important exceptions to such limited, rather passive, patterns of social behavior are those where individuals fully cooperate in hunting. It has been suggested that even some fish engage in cooperative hunting. In a review, Redouan Bshary and colleagues say that "co-operative hunting in the sense that several predators hunt the same prey simultaneously is widespread in fish, especially in mackerels, which have been described [as] herding their prey." It is observed that "individuals play different roles during such hunts (splitting the school of prey, herding the prey) and refrain from single hunting attempts until the prey is in a favorable position."[15] Similar behavior has been reported in other fish species.

Such reports are startling, because the cognitive demands of true cooperation are very great. Acting jointly in a dynamic activity like hunting requires a rapidly updating shared conceptualization of a fleeting target. That alone places high demands on attention, both to the target and to other pack members. In addition, individuals need to form momentary sub-conceptions corresponding with their complementary place in the global perception—a kind of dual cognition. Finally, the coordination between global and individual cognitions must give rise to individual motor responses that are also globally coordinated. In that coordination, such objects as bodies of water, trees, rocks, and ground undulations need to be negotiated, all on timescales of seconds or even milliseconds.

It suggests highly refined regulations emerging among the individual cognitions of participants. Again, let us call that behavior epicognitive by analogy with the epigenetic regulations among cells in development (and likewise based on emerging attractor landscapes). Although probably rudimentary in fishes, such regulations appear more strikingly in certain mammals.

In mammals, however, there is a big difference. Mammals generally evolved from reptiles as terrestrial beings. To survive on land, they had already evolved highly complex cognitive systems for dealing with non-social environments. Social groupings of cognitive systems, already so advanced, generated possibilities for social cognition far greater than in the ants or birds. The results depend on how much of that adaptability is actually brought into group dynamics.

The answer seems to be quite a lot in some species, not so much in others. Social cooperative hunting as a major survival strategy is prominent in a number of species of canids (wild dogs) and felids (wild cats). The main advantage is better success at catching bigger prey, which would be far beyond the capabilities of individuals. Among the social canids, cape (African) hunting dogs have been well studied for their habit of running down prey through divisions of labor, surrounding the prey, strategic delays, and other tactics of organization. Similar patterns are observed in wolves and lions.[16]

Above all, social cooperative hunting in mammals brings individually adaptable cognitive systems into joint action. That means far wider possibilities in the dynamics of the group. For example, the dynamics can incorporate deeper environmental structure, because the individual brains can cope with it. That implies epicognitive regulations far more complex than in ants, fish, or birds.

Whether this entails increments of cognitive ability over and above that in nonsocial species is, however, uncertain. One approach has been to examine whether there has been some reflection in brain size and structure. Even in wolves and lions, however, teasing out the effects on brain size of social factors from those of nonsocial factors is not straightforward.

For example, across species generally, brain size increases with body size and increased innervation. It is also related to other aspects of habit, such as diet, and also behavior. However, mammal species with larger brains tend to survive better when introduced to novel environments.[17] Carnivores generally have increased size in the frontal cortex compared with other mammals. This is thought to reflect the extra demands on cognition of even solitary hunting. Socially cooperative carnivores have been reported as having even bigger brains, but not vastly different from corresponding nonsocial species. We can presume that such an increment, if it exists, would have involved changes to affective or affective-cognitive regulatory networks as well as cognitive networks alone. It is also well known that domestication in any species drastically reduces brain size over generations, indicating effects of ecology and lifestyle.

However, an important point is that there may have been an evolutionary limit to further brain expansion in quadrupeds, because a horizontally extended neck would have been unable to bear the greater weight of a bigger brain in a bigger skull. Of course, many quadrupeds manage to carry large horns or other protuberances on their skulls. But they are not the species engaged in social hunting. It may be that further evolution awaited the emergence of a bipedal (upright) gait, as described in chapter 9.

PRIMATES AND "FREE MARKET ECONOMICS"

The primates (monkeys and apes) are usually considered to be the most evolved species of nonhuman mammals, and also the most human in terms of anatomy and aspects of behavior. Similarities to humans include occasional bipedalism, manual dexterity of hands, and varieties of gestural and vocal communication. Probably for that reason, there has been a tendency to assume that they are more "intelligent" than other mammals, with richer cognitive systems and bigger brains. So they are presented as a model of human ancestry.

However, studies of primate cognition and its origins also tend to have been infused with anthropomorphism and the liberal use of metaphors: in some cases drawn from the neoliberal economics of Western societies. In this section, however, we must first ask: To what extent do primates have bigger brains? And to what extent are primates more "intelligent" than other mammals?

Answers to the brain-size question, as evidence of cognitive ability, are not straightforward. Primates generally have relatively bigger brains than other mammals because of their finer visual discrimination and digital dexterity. Those that forage for fruit have bigger brains than those that eat leaves, perhaps because they need more detailed cognitive maps for more scattered locations. And those that eat insects have still bigger brains, because they need even finer sensory discrimination and motor skills.

Bigger brains or not, a popular idea is that primates have enhanced cognitive abilities because of their enriched social lives. Robin Dunbar says that "primates have unusually large brains for body size compared to all other vertebrates: Primates evolved large brains to manage their unusually complex social systems."[18] Living in primate groups certainly involves intense interactions among group members, requiring a wide range of gestural and vocal communications. The downside, it is said, is that they have to compete with one another for food and other resources, often forming temporary partnerships to do so. Observers frequently report enhanced ability to predict others' intentions, constructing feints and deceptions, cheating, and so on.

Somewhat paradoxically, then, it is such within-group competition that (it is argued) demands sophisticated social cognition in primate groups. This argument was put forth by Nicholas Humphrey in the 1970s.[19] He claimed that the need for individuals to negotiate tricky rivalries and rank orders has demanded a new kind of smartness. A "Machiavellian hypothesis" became widely accepted, based on observations of deception and cheating among members of a few primate groups. By such cognitively clever means, individuals are said to reap the benefits of social life, while minimizing the personal costs.

In what has become the dogma of a paradigm, this kind of cooperation is defined purely in terms of individual gain, dictated by personal perception and cognition. Thus Bshary and colleagues have attempted to model primate behavior on a market philosophy and game theory with fixed rules.[20] A more recent article in *Nature* (May 26, 2015) cites Ronald Noë, a primate behavioral ecologist, who has "come up with a biological market-based theory of cooperation. It proposes that animals cooperate to trade a specific commodity—such as food—for a service that would promote their survival, such as protection from a predator." Again, cognitive abilities are reflective of the need to "deceive" and "cheat."[21]

So has this evolutionary "market economy" produced more complex cognitive skills situated in bigger brains? Let us address the brain question first. In fact, the evidence is indirect and thin. It has mainly been couched in terms of whether those living in more complex groups have bigger brains. However, complexity has not been analyzed except in terms of group size. It has simply been assumed that bigger groups will demand more complex cognition for managing more relationships. Such sociality, even if accurately described, is a far cry from the kind of complexity required in, say, group hunting among wolves.

Anyway, correlations have been reported between size of cortex and group size across different primate species. The trouble is that many other attributes correlate with group size, such as body size and general (nonsocial) learning ability. Christine Charvet and Barbara Finlay suggest that, when account is taken of these, the correlation is unlikely to be significant.[22] We still cannot be sure whether group living in primates is such as to require greater cognitive abilities than everyday, "ecological," problem solving.

So what about cognitive ability in primate groups? This, too, has been measured indirectly, in terms of the incidence of identifiable practical skills. Simon Reader and colleagues assessed the degree of social learning, tool use, extractive foraging, and tactical deception in sixty-two primate species as reported in published papers. The statistically converted data suggested low to moderate correlations with brain volume.[23]

Interpretation is not straightforward, however. First, compiling data from published sources requires copious adjustments and many assumptions. More importantly, the actual definition of these behavioral flexibilities as indices of cognitive ability is by no means clear. Social learning is simply defined as learning through observation of, or interaction with, others. Such learning has been reported in fish and birds, as mentioned above. And as already implied, there is a big difference between observation and interaction, depending on how the latter is defined. Unfortunately, in this area little micro-analysis has been done of cause and effect in the way that the microbiologist has revealed the intelligence of the cell, or the embryologist the intelligence of development.

Finally, considerable variation exists in social cohesion across different primate groups. The chimpanzee is considered the closest to humans in terms of cognitive ability and brain size. But as Craig Stanford put it, "Chimpanzee society is called fission-fusion, to indicate that there is little cohesive group structure apart from mothers and their infants; instead, temporary subgroupings called parties come together and separate throughout the day."[24]

Some of the problems may be those of studying small samples of a species over limited time periods. There may be failure to recognize unexpected degrees of cognitive flexibility even in seemingly fixed hierarchies. For example, baboon troops exhibit strong dominance hierarchies. Yet observations in the field also reveal highly collective decision making at times, in which high-ranking males are sensitive to the preferences of others: "These results are consistent with models of collective motion, suggesting that democratic collective action emerging from simple rules is widespread, even in complex, socially stratified societies."[25]

Such wider observations suggests that many, if not all, of the conclusions of the "selfish individualist" paradigm might be explained by self-organized group dynamics rather than individual constitutional differences. As Claire Detrain puts it, regarding studies of macaque monkeys, "Self-organization . . . even alters common beliefs about the origin of highly complex structures such as hierarchy in primate societies which ranges from 'despotic' to 'egalitarian.' . . . It is suggested that the numerous behavioral differences between egalitarian and despotic macaques automatically emerge within the society and can be traced back by simple differences in group-members interactions."[26]

THE PATHWAY TO HUMANS?

Much of the research on primates has been motivated by the belief that it will tell us more about human culture and human nature. But some myths and distractions have been created in the process. For example, it seems to be assumed that many primates live social lives a lot like those of humans, that they have bigger brains because of such social lives (and that explains why humans have big brains), and they exhibit human "culture."

However, if there are increments in brain size attributable to social life at this (prehuman) level, they are not huge (compared with increments due to demands of the physical world, including nonsocial interactions with other animals). In any case, as Harry Jerison has said, much social complexity can be achieved without much increase in brain size. More important than sheer size and (assumed) processing complexity may have been modulation of emotional and motivational expression by cortical centers.

An alternative suggestion is that the social lives of primates are not nearly as complex as those of humans. Correspondingly, the cognitive demands of the social lives of primates may be a little more complex than the demands of life in general, but not greatly so.

A closer look at the culture of primates, for example, suggests that it has little correspondence with that of humans. In the primate literature, "culture" tends to be defined as merely the product of individual observational learning. For example, chimpanzees will start to use a stick to dip for termites or monkeys might start to wash potatoes on the seashore, after observing similar behaviors in peers. The claim is made that this amounts to cultural transmission of knowledge and cognitive abilities across generations, similar to that in humans.

As should become clear in chapter 9, this is quite different from culture in humans, which is the product of intense social dynamics with emergent properties. The study of primates is an intrinsically fascinating and important field. But those who hope to find in it a better understanding of humans may be disappointed. Conspicuously, another sort of social life is largely absent from primates: one that really is more demanding cognitively. This is true cooperation—including cooperative hunting—as described above.

ABOVE "MARKET COMPETITION"

Indeed, complex cooperative activity, as in hunting, foraging, defense, and so on, is rare in pre-human primates. Although living rich social lives in other respects, monkeys and apes rarely help group members other than close family, and joint action and teaching are also rare. There is little evidence, even among chimpanzees, of agreement to share or of reciprocation.

For example, chimpanzees do not take advantage of opportunities to deliver food to other members of their group. In experiments reported in 2013, Anke Bullinger and associates found that bonobos, chimps, and marmosets all prefer to feed alone.[27] Observers have reported similarities in the form and function of many gestures produced by the chimpanzee, bonobo, and human child. But within a few months, human infants are developing turn-taking patterns and coordinated vocal interactions.[28] Moreover, multimodal expressions of communicative intent (e.g., vocalization plus persistence of eye contact) that are normative even for human children are uncommon in apes.

As a result, there are also differences even in gestural communication. "Unlike the gestures of human children, the majority of ape gestures are dyadic, intended to draw another's attention to oneself, rather than triadic, intended to draw another's attention to an external entity. . . . Also unlike most humans, ape gestures are frequently imperative (requests) and less frequently declarative (attempts to share experience with another)."[29]

In fact, the primate mode of existence—their ecological niche—has probably never demanded more than a restricted form of cooperation. As with fish and birds, the primary advantage of social life in apes is protection against predators. Otherwise, as Michael Tomasello has pointed out, "the vast majority of nonhuman primate cooperation is in the context of intragroup competition." As he goes on, such a lifestyle works "against the evolution of cooperation in these species, as individuals who are given favorable treatment by conspecifics are those who are best at competition and dominance." Because of that, nonhuman primates lack "shared intentionality." As Tomasello puts it in his book, *A Natural History of Human Thinking*, they do not self-monitor and evaluate their own thinking with respect to the normative perspectives and standards of others or the group.[30]

These views of Tomasello are an echo of the other side of Darwin, as emphasized by Peter Kropotkin in his book, *Mutual Aid: A Factor in Evolution* (1902). Kropotkin reminded us that Darwin did not define the fittest as necessarily the strongest, or individually most clever; instead the fittest could be those who cooperated with one another. In many animal societies, Darwin noted that struggle is replaced by cooperation, amidst warnings, in his *Origin of Species*, about the dangers of interpreting his theory too narrowly.

Later, in the *Descent of Man* (1871), Darwin pointed out how, in many animal societies, the struggle between separate individuals for the means of existence disappears and is replaced by cooperation. That creates the conditions for the evolution of greater "intellectual faculties," he suggested, in turn securing survival. He emphasized again that the fittest are not those with the greatest physical strength or most cunning, but those who learn to mutually support one another for the welfare of the group as a whole.

Kropotkin complained that, "Unhappily, these remarks, which might have become the basis of fruitful researches, were overshadowed by the masses of facts gathered for the purpose of illustrating the consequences of a real competition for life."[31] This seems to have been forgotten in today's haste to reduce humans to the Machiavellian individualist in a free market economy.

As it is, among nonhuman primates, chimps alone seem to exhibit occasional cooperative hunting. Reported examples include raids on neighboring groups for killing some of the males and capturing some of the females; or opportunistically predating on other primates, small pigs, and so on. It is difficult to be sure from sporadic observations what the intentions, cognitions, and motivations behind such rare behaviors really are, but they may well indicate rudimentary social cooperation. As such it suggests a kind of flicker of true cooperation at this level of evolution. As we have seen, though, that behavior occurs in other social mammals to a much greater extent.

Since Darwin, the intellectual gap between humans and other primates has remained the most challenging problem for a theory of evolution. Perhaps we need to ask: Is there some other factor? Perhaps the answer has already been hinted at in this chapter—and even by Darwin himself. Indeed, the argument to follow is that the evolutionary pathway to humans took another course, based on the cognitive demands of true cooperation and other aspects of social organization—that is, the further evolution of intelligent systems rather than fixed adaptations. That is the subject of the next chapter.

9

HUMAN INTELLIGENCE

INDIVIDUALISM

Since Darwin, the intellectual gap between humans and other primates has remained the most challenging problem for a theory of evolution—and, we might say, a theory of intelligence. It has been described as the human paradox: how can humans be so biologically continuous with other species yet be so different, especially in their potential? The paradox haunts evolutionary psychology and sociobiology and other evolutionary accounts of human potential. And it leaves the field of human intelligence in a state of increasing confusion.

I argue that the question is a misleading one. It arises from attempting to impose a narrow "biological" model on human potential, without due regard of why or how the human species evolved. So the field is pervaded with tensions, doubts, and not a few myths, of which the nature-nurture debate is only the most prominent. And it has thwarted a deeper understanding of human intelligence.

One of the tensions is about the often-felt need to head off allusions to human uniqueness. The priority has been how not to threaten Darwin's famous claim, in the *Descent of Man*, that the difference in mind between humans and higher animals is one of degree and not of kind. So care is taken to ensure that no Rubicon has been crossed. We must keep our Darwinian integrity; humans are products of the natural selection of genes; and so on. I have already suggested how the apparent evolutionary gap might be filled: not by subverting biology but by revising and enriching it.

A major obstacle to such a revision is understanding how culture distinguishes us from apes (assumed to be the nearest biological model) and thus what kind of system we really have as humans. The Rubicon hangup has put many in a state of denial on that subject. Yet many bridges have been crossed in the course of evolution, with major leaps in variation-creation and adaptability. The crossing from unicellular to multicellular forms was one. The evolution of brains and cognitive systems was another. And humans are starkly different from apes in many other ways.

These include huge brains, bipedalism, manual dexterity, tool making, and more complex cognitive systems, for a start. Humans also have complex technologies, cooperative systems of production, and a system of communication radically different from any other in the animal world. Most striking is that all other species—including apes—are specialized for a particular environmental niche. Not so humans. We have no particular niche, colonizing every corner of the world: we adapt the world to ourselves rather than vice versa. All this has created confusion about how evolution—and what kind of conditions—could have produced such an animal.

Most investigators acknowledge that our capacity for culture is the major factor underlying human accomplishments. But then to keep it tethered to a Darwinian mast, a special definition is imposed on it. Culture is reduced to "social learning," an ability to acquire knowledge and skills from other individuals. So culture "can then be modeled as a Darwinian process comprising the selective retention of favourable culturally transmitted variants."[1]

In that definition, culture comprises discrete packets of information, like genes, that can be naturally selected. In *The Selfish Gene*, Richard Dawkins called these packets "memes" (from the Greek for what can be imitated). Although originally referring to things like tools, clothing, and skills, the idea has been generalized to include ideas, beliefs, and so on, as well as specific behaviors. So—with some relief, it seems—culture can be brought into the neo-Darwinian framework after all.

In this picture of culture, of course, human cooperation also has to be specially defined. As with the Machiavellian hypothesis mentioned in chapter 8, cooperation is defined as a cognitively clever means by which individuals can reap the benefits of social life while minimizing the personal costs. So cooperation, too, falls into the selfish gene scheme of a gene's way of making copies of itself. That, in turn, means a particular definition of intelligence: being smart in that selfish sense, as a result of a lucky deal of genes in the poker game of life.

And yet, in such schemes, confusion remains. Most behavioral geneticists, evolutionary psychologists, and sociobiologists probably share the exasperation of Mark Flinn and his colleagues. They agree that humans have evolved intelligence that is extraordinary: that it is "a spectacular evolutionary anomaly" and "the evolutionary system behind it as anomalous as well." In keeping with standard Darwinian logic, they look for a solution in "competition among conspecifics" in some specific environment. Unfortunately, they go on, "it has proven difficult to identify a set of selective pressures that would have been sufficiently unique to the hominin lineage."[2] Similarly, Frederic Menger senses the puzzle that "Modern humans are smarter than what is demanded by our evolutionary experience as hunter-gatherers."[3] We have already noted Michael Tomasello's observation that natural selection for competitiveness hardly explains the evolution of cooperation.

Otherwise theorists opt for the sweeping, despairing tone of Edward O. Wilson who, in his book, *The Meaning of Human Existence*, describes humans as an "ensemble of hereditary regularities in mental development that bias cultural evolution in one direction as opposed to others and thus connect genes to culture in the brain of every person." We are all "saints and sinners," he says, "because of the way our species originated across millions of years of biological evolution."[4]

I suggest that all this stems from inadequate microscale analysis of the way our species did originate at the cognitive level. The obvious presupposition just mentioned is that humans evolved and exist—and their cognitive apparatus is made—for a state of competitive individualism. That assumption, of course, serves a class structure in a free market economy (the winners are the winners because they are the "best," most intelligent, and so on). But it also handicaps attempts to understand human brain and cognition, distorting even the methods of inquiry.

On this point, I can do no better than quote Uri Hasson and colleagues: "Although the scope of cognitive neuroscience research is vast and rich, the experimental paradigms used are primarily concerned with studying the neural mechanisms of one individual's behavioral processes. Typical experiments isolate humans or animals from their natural environments by placing them in a sealed room where interactions occur solely with a computerized program. This egocentric framework is reminiscent of the Ptolemaic geocentric frame of reference for the solar system. . . . Despite the central role of other individuals in shaping one's mind, most cognitive studies focus on processes that occur within a single individual. . . . We call for a shift from a single-brain to a multi-brain frame of reference."[5]

Of course, the same thing may be said about intelligence testing and the research and models surrounding it. In this chapter, I show for human potential what Raymond Tallis said about consciousness in his book, *Aping Mankind*: It "cannot be found solely in the stand-alone brain; or even just in brain; or even just in a brain in a body; or even in a brain interacting with other brains in bodies. It participates in, and is part of, a community of minds built up by conscious human beings over hundreds of thousand years."[6]

In other words, the splendid variation of humans only arises—people only individuate—in the context of a culture and its history. Through that framework I explain that there is no "evolutionary anomaly," no ghostly "sinners" in the machine, no fanciful bell-shaped curves in our genes or our cognitive abilities. There is only the continuity of a largely overlooked evolutionary logic that has produced spectacular changes in the complexities of living things.

I explain how the intense selection for a cooperative lifestyle helped overcome complex changeable environments by shaping a unique kind of intelligent system—continuous with but overshadowing all others. I discuss how this has evolved as a quite different kind of relation between individual cognition and culture than that portrayed in other primates. Finally, I discuss its implications for the creation of individual differences in intelligence today.

BECOMING HUMAN IN EVOLUTION

The human evolutionary story goes back some seven or eight million years, according to fossil findings in Africa. The story remains rather hazy, but another ratchet upward of environmental complexity and uncertainty was certainly involved. The older picture (now becoming more complicated) was quite neat. The climate dried, forests thinned, and former forest dwellers were forced onto the forest margins and open savannah. Without natural defensive equipment and deprived of traditional food resources, those species became extremely vulnerable. They had to face unreliable food supplies, which resulted in wide ranges and new diets, exposure to large predators, and so on.

They responded by cooperating with one another to a degree far outstripping that of any of their primate ancestors. But doing so presented new demands on a cognitive system. The micro-environment of true cooperation is vastly more intricate, and more demanding on cognitive systems, than the macro-environment of climate, geophysical changes, and the social groupings of nonhuman primates.

This was the context for the appearance of hominids (The word "hominid" usually refers to members of the family of humans, *Hominidae*, which consists of all species on our side of the last common ancestor of humans and living apes). Some new adaptations were physical, such as the adoption of bipedalism, present in our earliest ancestors of 7.5 million years ago. But the cognitive demands are much more important. As mentioned in chapter 8, they are present to some degree in most primates. Only in this new branch, however, did they really make a difference. Defense, hunting, and foraging became vastly more effective as an organized group than as a mere collection of individuals. Also, reproductive relations, child rearing, divisions of labor, sharing of products, and so on, became less fraught when cooperatively ordered.

As also mentioned previously, some of that lifestyle is foreshadowed in cooperative hunters like wolves and African hunting dogs. It was also foreshadowed by a suite of pre-adaptations already evolved in some form in the apes. Bipedalism occurs sporadically in some monkeys and apes to provide height to see over tall grasses and shrubs and to wade in rivers. It was further selected and improved in the hominid ancestors. Freed hands permitted rudimentary tool use, which, together with the longer visual perspective of open spaces, would have improved visuo-motor abilities. These factors may well have added the first increment in brain size. That, too, would have been anatomically fostered by the new bipedalism and the ability to balance a heavier skull on top of the now-vertical spine.

However, it seems to have been the benefits of true social cooperation that we need to keep coming back to. More refined cognition, communication, and epicognition impelled more rapid brain expansion. That started a virtuous evolutionary spiral in cognitive systems. Early hominds were gradually able to release themselves from the confines of a single habitat (to which even our nearest cousins are still tied). With deeper abstractions of the dynamics of the world, they were better able to anticipate and change their environments. By thus adapting the world to themselves, rather than vice versa, they were able to expand into many less-hospitable environments.

The old story, though—or, at least its chronology—has been complicated by more recent fossil finds. As with any incursion of species into new habitats, it turns out that early hominid evolution was not so much a ladder as like a bush. It consisted of diverse "experiments" with different permutations of features, such as bipedalism, diet, manual dexterity, and bigger brains.[7] Up to about four or five million years ago, they include varieties with a braincase similar in volume to today's chimpanzees (340–360 cubic centimeters). Other features, such as hands and arms designed for climbing, are indicative of life in thin forests.

From about 4.5 to 2 million years ago, hominids are represented by a diverse group of undoubtedly bipedal, socially living creatures with slightly bigger brains (about 450–500 cubic centimeters). They had humanlike teeth and hands. These have been called *Australopithecines* (the most famous being *A. afarensis*, which existed between 3.9 and 3.0 million years ago). Fossil deposits suggest that they lived in small groups of twenty to thirty individuals; lake margins or river floodplains may have been their habitat, with the availability of trees for escape from predators, especially at night. But every now and again another report appears in *Science* or *Nature* of more fossil discoveries suggesting further revisions to the story.

The newer finds do not negate the general picture of evolution of cooperative lifestyle described above; it is only more staggered in time. From about 2.5 to 1.6 million years ago, remains of a variety of other species appear in the fossil record in East Africa, including one with more human features, originally called *Homo habilis*. (The picture is complicated by the report in 2015 of the discovery of another new species; named

H. naledi, it exhibits humanlike skeletal details but has a small brain, although as yet it is undated). These are the earliest known members of the genus *Homo*; that is, the first human species. They lived in a period of further forest thinning when traditional forest fruits must have started to become scarce. *Habilis* sites are associated with a particular tool industry characterized by crude stone flakes, rounded hammer stones, and possible stone weapons. They are regularly found near dismembered animal remains.

A review in *Science* in 2014 reported that "over the past decade, new fossil discoveries and . . . new environmental data sets suggest that *Homo* evolved against a background of long periods of habitat unpredictability that were superimposed on the underlying aridity trend." It goes on, "From ~2.5 to 1.5 million years ago, three lineages of early *Homo* evolved in a context of habitat instability and fragmentation. . . . These contexts gave a selective advantage to traits, such as dietary flexibility and larger body size, that facilitated survival in shifting environments."[8] That picture is largely confirmed in more recent reports. However, they still tend to stress macro-environmental factors (dietary flexibility, unpredictable environments, range expansion, etc.) rather than cognitive demands of cooperation per se.

The scatter of animal remains around *Homo* sites suggests living in large groups with cooperative hunting, or possibly scavenging, as a habit. But the most important feature is an enlarged brain, in a cranium of about 650 cubic centimeters. This is some 30 percent bigger than the *Australopithecine* brains and is much bigger than that of chimpanzees (when allowance is made for body size). In addition, they had lighter skeletons and skulls, with dentitions much reduced in size, suggesting less reliance on tough vegetable matter in the diet. All this suggests some parallel between cooperative hunting and living in larger social groups, on the one hand, with brain size and cognitive complexity, on the other.

A new species, *Homo erectus*, first appeared about 1.8 million years ago. It is associated not only with another large increment in brain size but also with continuing brain expansion across the period of its existence: from 850 cubic centimeters at 1.5 million years ago to 1,200 cubic centimeters at 200,000 years ago (modern *Homo sapiens* has an average brain size of 1,400 cubic centimeters). *Homo erectus* probably originated in

Africa but soon migrated across wide areas of Asia and Europe. They encountered some of the harshest climatic conditions on the way. This is what probably, in part, demanded an even closer cooperative lifestyle (reflected in the brain expansions just mentioned) and associated cognitive and practical abilities.

Liane Gabora and Anne Russon have reviewed some of the evidence as follows: "*Homo erectus* exhibited many indications of enhanced ability to adapt to the environment to meet the demands of survival, including sophisticated, task-specific stone hand axes, complex stable seasonal home bases, and long-distance hunting strategies involving large game."[9] By 1.4 million years ago, the species was also producing stone tools requiring considerable cognitive imagination and dexterity of hand and eye.

The uniformity of shape evident in this tool manufacture also probably reflects a group design, and thus a truly intercognitive culture of shared imagination and action. Sites occupied by *Homo erectus* also indicate large social groups, and their hand axes and flints have been found around the remains of very large prey (elephant, bison, and so on), suggesting advanced cooperative hunting skills. In addition, abundant evidence at campsites indicates the use of fire and perhaps cooking.

It seems safe to assume that such organization would have required other cultural innovations such as rules for group behavior and a system of communication more sophisticated than the thirty or so fixed signals of the chimpanzee.

MODERN HUMANS

Anatomically modern *Homo sapiens* probably originated in Africa around one hundred and fifty thousand years ago. By seventy thousand years ago they had spread across Europe and Asia, replacing existing Neanderthals that had already been there for hundreds of thousands of years. Fossils indicate sophisticated, adaptable, and variable stone cultures.

Home sites suggest large social bands, with alliances among them, and extremely successful social hunting and home-making skills. Fossil skulls from that period also indicate a huge increment of brain expansion to the modern human average of 1,400 cubic centimeters. This is about three times bigger than the average for chimpanzees, even when body size is taken into account.

As explained earlier in chapter 6, simple volume obscures a vast increase in folding of the cortex to increase functional tissue volume. So correlation between volume and function is not always clear. It seems reasonable to suggest, though, that most of this enlargement is to support the more complex cooperative lifestyle. Cooperative hunting on that scale would require advanced cognitive skills at an individual level but also ability to share epicognitive regulations at the group level. Cooperative hunting would also demand intimate knowledge of the prey and their typical behavior patterns, local and wider ecology, terrain, predator risks, weather patterns, and so on.

The new cognitive flexibility is emphatically indicated in the surviving material culture of stone tools. The stone culture of *Homo erectus* and the Neanderthals, over a period of a million years, tended to conform to a stereotypical pattern or style. The period from about seventy thousand years ago, however, is marked by a spectacular creativity and variability of styles. The culture includes complex bone technology; multiple-component missile heads; perforated seashell ornaments; and complex, abstract, and artistic designs. The distributions of this culture also suggest trading between groups in which technological exchange accelerated cultural development: the more contact there was between groups, the more rapidly technology developed.[10] This is an important additional layer in a hierarchical dynamical system that would engender further creativity.

EVOLVED FOR COOPERATION

This, then, was the context for joint evolution of superb cognitive and epicognitive systems, including the genetic resources to support it and the brain to house it. Human social cooperation was not a voluntary process, in which a collection of "brainy" individuals chose to work together. Instead, it emerged as a solution to stark environmental dynamics. In other words, social cooperation was the *context*, not the result, of rapid cognitive evolution. This has important implications for understanding human brains and cognition.

After years of making comparisons, Michael Tomasello says that humans are evolved for collaboration in a way that the apes are not. That is, richer connections between cortical and affective centers in brains have fostered deeper awareness of others through relations among brains. Not only can humans think what others are thinking, but they can also feel what others feel.

Tomasello's studies show that, although chimps may collaborate sporadically, human children collaborate obligatorily. For example, in a joint activity producing rewards, either chimp is likely to purloin all the rewards, whereas children share equally. Children prefer to work jointly rather than individually, whereas chimps rarely show intention to collaborate. Children reject unfair rewards in joint activities and will not tolerate free riders.[11]

True cooperation permitted habitation of increasingly marginal territories, and co-evolutionary effects made the biology of humans unique in other ways. Dangerous living, widespread migration, and constant testing of new habitats meant that our ancestors passed through some periods of intense biological selection. Such periods, which would have been ones of rapid elimination of the less fit—in this case fit to participate cooperatively—are referred to as ecological bottlenecks. This would have further ensured the good-enough fitness of survivors.

As a result, genes have followed culture—selected to support socially devised innovations—not vice versa. Our human ancestors migrated from the African savannahs, across the Alps into Northern Europe, and northeastward into Asia between 150,000 and 60,000 years ago. In doing so, they would have encountered a colder climate, different food sources, and new predators. Genetic selection supported some specific adaptations, such as skin lightening, digestion of novel food sources, and immunity to novel pathogens. But *adaptability*, as in food procurement, the invention of warm clothing, use of fire, robust dwellings, and so on, took place at an entirely different level—a cultural one.

As a consequence, humans are remarkably alike genetically. But by being individually good enough to have brains to participate at that cultural level, there is also a new effervescence of variation. Humans have only a quarter of the genetic variation of apes, our nearest biological relatives.[12] In spite of that, human behavioral and cognitive diversity far outstrips that of any other species. Let us now see how that complexity with diversity may have come about and is expressed in cognitive abilities.

THE MUSIC OF (HUMAN) LIFE

Survival in unpredictable environments is only possible through systems of pattern abstraction. Adaptation by genetic selection is too slow to keep track. Simple cues are unreliable. So living things are intelligent systems based on pattern abstraction (in this book, I have given it various other names as well). In a sense, pattern abstraction is intelligence, and intelligence is pattern abstraction. Life itself originated with it. But as organisms inhabited more complex environments, their powers of pattern abstraction had to increase.

Cooperating with other individuals in survival and other tasks presents the most complex of all environments. Having evolved to deal with it explains why humans are now compulsive pattern abstractors. We not only revel in finding it in nature, we also celebrate the ability by creating such patterns. That is what music, dance, art, and beauty are all about. We also contrive pattern abstraction, and what to do with it, in a vast variety of games and sports. Encountering patterns in any way makes us feel better, because it makes the world feel more predictable, more secure. Discovering and creating patterns makes a fine form of therapy.

Humans, in other words, are pattern junkies. But other lessons can be learned from that. Why humans so exuberantly engage in activities like art, dance, and music has been a major puzzle in the evolutionary psychology/cultural biology tradition. In *The Descent of Man* (1871), Darwin wrote that musical ability must be among the most mysterious with which humans are endowed. In 2008, the journal *Nature* published a series of essays on it. The authors of these essays agreed that "none . . . has yet been able to answer the fundamental question: why does music have such power over us?"[13]

In a narrow Darwinian tradition, in which only rigid adaptations make sense, what kind of adaptations can these be? It is possible that they are happy side effects of abilities that evolved for other purposes, as proposed by Steven Pinker. However, there may be a better explanation.

These forms of structure may be reflections of our more general capacity for structure abstraction. In his book *Musicophilia*, the late Oliver Sacks says, "In all societies, a primary function of music is collective and communal, to bring and bind people together. People sing together and dance together in every culture, and one can imagine them having done so around the first fires, a hundred thousand years ago." Sacks adds, "In such a situation, there seems to be an actual binding of nervous systems accomplished by rhythm."[14]

"BINDING" OF NERVOUS SYSTEMS

Such "binding" of nervous systems is exactly what happens in the cognitive dynamics of the social animals (from ants to wolves). Those self-organizing dynamics have amazingly creative results for groups and individuals. But it depends on underlying pattern abstraction. Indeed, the capture of structure in experience, from the molecular ensembles of the cell to advanced cognitive systems, is what makes a system intelligent. It leads from a state of uncertainty to one of predictability. When humans find it or discover it—in nature, pictures or human faces, for example—they describe it as beautiful and harmonious and they feel better about it. Humans deliberately contrive such structured experience in simplified forms in music (as well as art, dance, etc.).

It is hardly surprising, therefore, that Oliver Sacks's account should lead Holger Hennig to the question: Is there any underlying and quantifiable structure to the subjective experience of "musical binding"? From what has been said about the nature of cognition, we would surely expect so. Accordingly, Hennig examined its statistical nature by getting pairs of humans to play rhythms in synchrony. The results provide further clues to the true nature of human culture and human intelligence.

It has been known for a long time that anyone playing an instrument gets the timing of each beat very slightly wrong—a few milliseconds before or after the metronome. It turns out, though, that in any individual sequence, the deviations are not random. They correlate with each other over long intervals in the sequence. That is, they reveal an underlying statistical dependency, or structure (also known as fractal structure).

This is what Hennig found in each of his volunteer musicians, tapping out their beats on keyboards. But, after a few minutes, he also observed something else: that their deviations from the metronome, though initially different from one another, soon started to correlate *with each other*. They not only failed to follow the metronome rhythm, they also started to follow each other's deviations from it.

There seems to be only one explanation for such convergence of patterns. All along, each player must be tracking the other's history of rhythmic deviation and then assimilates it so as to finally predict each next stroke. Even between only two collaborating individuals, the abstracted structure seems to achieve "an actual binding of nervous systems," Hennig suggests.

Remarkably, listeners significantly prefer music exhibiting such structured deviations over independent fluctuations: perhaps because such music has a richer, deeper, structure. Now we know why: structure provides predictability and evokes the good feelings just mentioned. Such structure is, of course, the kind already discovered among components in cell metabolism, physiology, and brain functions; and among individuals in ant colonies, fish schools, and starling flocks. Forms of "interpersonal brain-to-brain coupling" are also being demonstrated through new techniques of brain scanning.

Such binding of nervous systems also suggests that human culture consists of much more than what the individual can take from social cooperation as individuals. The reality is of far more complex, epicognitive, dynamics: something happens that is more than passive social learning. Whether in, say, cooperative hunting of prey or making music, each member of a group is registering, and adapting to, the structure in the behavior of one or more others: each is complementing, augmenting, and compensating for the behaviors of others. There is a new emergent layer of regulation.

This seems to be the case with humans from a very early age. For example, infants between seven months and one year of age acquire turn-taking behavior when vocalizing with others; that is the cognitive binding that characterizes interactions between them and caregivers at that stage of life. And to begin to learn language properly, infants need to detect and assimilate the statistical structures that segment words in the highly complex acoustic streams of speakers. As we know, they do this with great facility so long as a speaker is playing the game: vocalizing with clarity and emphasis, anticipating difficulties, and "scaffolding" the experience for the learner.

The immediate result in all cases is a compound spatiotemporal structure, full of nonlinear feedforward and feedback loops among, not merely within, brains. As with the dynamics of the ants, then, culture is a self-organizing structure. The great advantage is that such a dynamical culture is far more representative of complex environments than any one brain alone could achieve and is far more adaptable to environmental and social change. By the same token, the dynamics can generate well-targeted, harmonious, responses far more quickly and creatively and can refract them through individual participants.

Joanna Sänger and colleagues refer to these shared cultural structures as "hyperbrain networks." They arise among and then modulate the "within brain networks" of individuals.[15] Others have referred to a "super-brain," the "invisible social brain," or a "distributed" social cognition. Uri Hasson and colleagues put it as follows: "Cognition materializes in an interpersonal space. The emergence of complex behaviors requires the coordination of actions among individuals according to a shared set of rules. . . . Brain-to-brain coupling constrains and shapes the actions of each individual in a social network, leading to complex joint behaviors that could not have emerged in isolation."[16]

This is what human culture is and where culture resides. It is not the pale extension of ape mentality and social life but a whole new intelligent system. As Gabora and Russon explain so well, "One does not accumulate elements of culture transmitted from others like items on a grocery list but hones them into a unique tapestry of understanding, a worldview, which . . . emerges through interactions among the parts."[17]

COOPERATION, DEMOCRACY, AND HISTORY

Here we may reflect on these points a little by comparing and contrasting human social intelligence with the social intelligence of ants, and fish schools, and bird flocks. The ant brain exhibits relatively little plasticity.

Individual inputs to the group are limited in range, although the social dynamics add much to adaptability in challenging circumstances. Likewise with fish schools and bird flocks.

In contrast, the social mammals, like all mammals, inhabit more demanding environments. Their brains have already evolved considerable plasticity independently of becoming "social." When they did, their contribution became so much greater to the micro-demands of social hunting and other social structure.

The early hominids seemed to have experienced even greater unpredictability of environment, which demanded greater cooperation. In other words, in human evolution, the ability to participate in collective intelligence itself became a precondition for survival and natural selection. The biological response was vastly bigger brains with far more extensive neural networks for social participation. In consequence, vastly greater contributions could be made by the individual brain to the collective intelligence. That furnished the ability to abstract the statistical structure of cooperation to a far greater depth.

So humans cooperate in a different sense than do ants, fishes, and apes, establishing a far more creative and adaptable intelligent system. In the case of ants, as described in chapter 8, interactions among them lead to the emergence of intelligent global behavior; but the consequences of individual contributions are *unknown to the individuals*. The opposite is the case in humans. Participants in human joint behavior need to be mutually aware of the global objectives and consequences of individual contributions—if, that is, individuals and group are to maximize both the benefits of cooperation and individual development.

But these are the biological, cognitive, and social foundations of democracy, defined as "a system in which all the people of a polity are involved in making decisions about its affairs" (from Wikipedia). Therefore, far from democracy being an unnatural state, as some seem to suggest (see chapter 1), human cognitive functions seem to demand it. Social and individual abnormalities emerge when, for various reasons, individuals are excluded from such awareness and such democracy (discussed in chapter 10).

However, there is another big distinction with all previous forms of social cooperation. We know that, from very early in human evolution, humans went further than the isolated group in their collaboration. Hunting and gathering bands collaborated in larger coalitions, sometimes sporadically, and at other times more permanently. Such arrangements are observed among hunter-gatherer tribes today (and in their successors in modern societies). These additional coalitions present still further levels for interaction—reciprocal feedforward and feedback loops, group dynamics, and behavioral innovation. In other words, they provide the scope, through a further level of reflective abstraction, for even deeper intelligence about the changing world and how to manage it.

Again, these dynamics would not be possible without the co-evolution of interdependencies across levels: between social, cognitive, and affective interactions on the one hand and physiological and epigenetic processes on the other. As already mentioned, the burgeoning research areas of social neuroscience and social epigenetics are revealing ways in which social/cultural experiences ripple through, and recruit, those processes.

For example, different cognitive states can have different physiological, epigenetic, and immune-system consequences, depending on social context. Importantly, a distinction has been made between a eudaimonic sense of well-being, based on social meaning and involvement, and hedonic well-being, based on individual pleasure or pain. These different states are associated with different epigenetic processes, as seen in the recruitment of different transcription factors (and therefore genes) and even immune system responses.[18] All this is part of the human intelligence system.

In that way human evolution became human history. Collaboration among brains and the emergent social cognition provided the conceptual breakout from individual limits. It resulted in the rapid progress seen in human history from original hunter-gatherers to the modern, global, technologiocal society—all on the basis of the same biological system with the same genes.

Let us now look at some of the psychological implications, including those for human potential and individual differences. I hope to show that the "cost" (if we can call it that) of such a dazzling system is that it is quite impossible to separate the potential (or lack of it) of any individual from that of the culture and the circumstances in which it operates. Human intelligence only individuates through these dynamical cultural processes.

HUMANS ONLY INDIVIDUATE
THROUGH CULTURE

Since ancient Greece, philosopher-psychologists have known that we can only become recognizable humans through social interaction. That also seems to have been recognized in tribal societies. In an interview in *New Scientist* in April 2006, Desmond Tutu told us that "in the African Bantu language, the word *ubuntu* means that a person becomes a person only through other people."[19] That recognition has become lost in the methodological individualism of the Western world. But real understanding of cognitive potential and cognitive differences becomes difficult without it. So let us remind ourselves of that.

True cooperation is seen when even as few as two individuals act jointly on a task. This is already a challenge for perception and cognition. Each partner needs to form momentary perceptions corresponding with their complementary place in the global perception while holding on to a shared conception of the final goal. Then the coordination between global and individual perceptions must give rise to individual motor responses that are also globally coordinated. And all this is rapidly changing in seconds or even milliseconds.

In the process, the partners jointly reveal a depth of forces in nature, and in each other, that would not be experienced by either one operating alone. We have already seen what happens even with two musicians coordinating key-tapping for a few minutes. With more realistic tasks like moving an object, such coordination is already much more demanding. Each has now to take account of that new complexity of forces—of mass, gravity, shape, and friction—for the joint action to become coordinated. In jointly lifting an object, say, the natural relations between forces of each individual and the actions of one participant all become conditioned by the actions of the other. In addition, of course, all these forces have to be geared to a *shared* conception of joint purpose.

In hunting and defensive actions against predators, the task is even more complex, because the object is itself active, and reactive, against the joint action. The dynamics of action and reaction are not even remotely experienced by noncooperating animals. They cannot be regulated through a narrow range of stereotyped chemical, gestural, or other signals or the networks of lesser brains.

As Patricia Zukow-Goldring noted, ordinary experience involves a multitude of signals from a multitude of variables, through various senses, constituting what might be a multitude of objects (animate or inanimate) undergoing rapid spatiotemporal transformations. But in social cooperation, these inputs need to be coordinated between individuals so that "perceiving and acting of one continuously and reflexively affects the perceiving and acting of the other."[20]

CULTURAL TOOLS BECOME COGNITIVE TOOLS

What such demands imply for human intelligence has been much neglected in psychological theories. Most of the evidence for it has come from the two giants of developmental psychology Jean Piaget and Lev Vygotsky. Their interpretations were different in some ways but also complementary.

Piaget viewed social engagement as a potent source of perturbation or disequlilibration in the developing mind. For example, by telling stories about toddler's misdemeanors and then asking who was "naughtiest," he learned how children's thinking changed over time. He agreed that coordination of viewpoint in the social world was more demanding than any problem in the physical environment. In that way, he said, cognition unavoidably develops from interrelations with others. As he put it, "The individual can achieve his inventions and intellectual constructions only to the extent that he is the set of collective interactions," and "there are neither individuals as such nor society as such. There are just interindividual relations."[21]

Piaget concentrated on what changes emerge in the individual cognitive system during development. In contrast, Vygotsky focused on what goes on in the dialectics between collective and individuals, and the intertwining of changes in both. These seem to follow the dynamical models described above.

Thus, joint perceptions and joint actions create rapidly changing feedforward and feedback loops that emerge as agreed-on regulations—or attractor states. Examples include conventions like turn taking, queuing, agreed-on signals and gestures for demanding joint attention, pointing (to focus joint attention), means for indicating the goal of an action and routes to it (strategies and "maps"), agreements about divisions of labor, signals for monitoring and modulating progress, and the use of a unique language. Just think of helping someone lift a long table through several doorways, or a wardrobe downstairs: a cascade of cognitions through a mutual attractor landscape ensues—rules and conventions we tend to take for granted.

Other regulations emerge in the fashioning of shared mechanical devices and artifacts. These range from simple stone tools and shelters to emergent technologies, all literally designed with others in mind. The critical design of a wrench is not merely matched to a hand. It derives from an abstract statistical structure of its use—that is, an attractor basin—shared in the mind of the designer.

The self-organization of larger coalitions of humans has also emerged in the form of highly structured institutions. They issue overtly stated, or legalized, obligations and conventions, as in marriage rules, kinship identities, asset ownership conventions, and so on. These define and legitimize more explicitly what is expected of individuals, often in stated laws, shaping cognitions in the process.

Human language, with its complex structure, unique productivity, and speed of transmission, evolved specifically for coordinating social cooperation. And, as mentioned above, other sets of rules (contrived attractor basins, as in a square dance) are devised for simply celebrating and enjoying expressions of synthetic structure. All these forms of regulation shape and cement cooperative social relationships, most of which can vary radically from group to group or place to place. But their detailed cognitive structure, even as used by young children, is almost incredibly complex.

Vygotsky referred to these rules and devices (or, as we would now say, shared attractors) as "cultural tools." They all reflect a depth and detail of informational structure not experienced in, or demanded of, previous species. But they also become internalized as psychological tools. It is through them that individuals think and act. Accordingly, the cognition of individuals becomes fashioned by its sociocultural tools, just as the activities of single neurons are fashioned by the global activity of their networks or as the behavior of ants is fashioned—albeit in a much more restricted sense—by the emergent dynamics of the collective.

As Vygotsky put it, "By being included in the process of behavior, the psychological tool alters the entire flow and structure of mental functions. It does this by determining the structure of a new instrumental act just as a technical tool alters the process of natural adaptation by determining the form of labor operation."[22]

Vygotsky concluded that human thought and activity is embedded in social life from the moment of birth. He argued that the entire course of a child's psychological development from infancy is achieved through social means, through the assimilation of cultural tools. The highly complex forms of human cognition, he argued, can only be understood in the social and historical forms of human existence. From toilet training to becoming an airline pilot, a scientist, or a bus driver, socially devised tools become individual cognitive tools. This is the basis of his "law of cultural development": "Any function in the child's cultural development appears on stage twice, on two planes. First it appears on the social plane, then on the psychological, first among people as an inter-psychical category and then within the child as an intra-psychical category."[23]

It now seems certain that it is precisely for the assimilation of such tools that the extreme plasticity of the human brain evolved: a brain for the internalization of socially devised apps rather than one reliant on built-in routines. The big difference is that, unlike the apps we download for phones and computers, the brain constantly refashions them and feeds them back into the collective. Nevertheless, we now know from brain imaging and other studies how background experience with specific cultural tools or procedures results in corresponding changes in brain networks.[24]

For example, such experience often reflects occupational specialization. London taxi drivers, who are required to develop a detailed memory of street layouts, show enlargement of that part of the brain involved in spatial memory (the posterior hippocampus). Likewise, the part of the cortex involved in the sensorimotor aspects of finger coordination in violin players is expanded on the corresponding side of the brain, but not on the other side. Another study showed changes in cortical thickness and hippocampal volume following intensive foreign language learning in army trainees. In an article titled "Culture Wires the Brain," Denise C. Park and Chih-Mao Huang review studies suggesting that even subtle differences in wider culture, such as tending to be individualistic or collectivist in outlook, are reflected in brain networks.[25] Of course, these differences will emphatically not be ones of learning ability as such. Yet that is how they will appear in the single cultural yardstick of an IQ test.

One of the key products of such interactions among brains, and between individuals and culture, is the creativity it has fostered, from stone tool making to modern technology and artistic culture. This is the manifestation, on an interindividual level, of the coalition of attractors and reflective abstraction described in chapter 8. A simple example—but one far beyond the capability of other species—is the creative imagination required to make even the simplest stone tools like those in widespread use in early human evolution. An individual could pick up a rough stone as an implement. But it takes the conceptions of a group, presumably with much intercommunication, to imagine a purpose built hand-axe to be used by all. And it takes a dynamical relationship to translate that purpose into hand-eye coordination, visual imagination, in the brains of individuals. These collective dynamics put tool making way beyond nest building by weaver birds, dam building by beavers, and the use of straws for termite dipping by chimpanzees.

None of this should imply that individual minds are passively programmed by social forces. The dynamics of social life are constantly refracted through individuals with their own creative cognitive systems. For example, development in expertise (at all ages) lies in the "collision" (as Vygotsky put it) between mature forms in the cultural network with the less-developed forms in the novice. The need to resolve incongruities often results in the individual creating novel conceptions that can feed forward to broader cultural change. Far from culture being merely transmitted or copied, as Tim Ingold says, "Knowledge undergoes continual regeneration. . . . It is impossible to separate transmission from knowledge generation."[26]

The makers of stone tools were the world's first product designers. But we all follow them. Everyone is inventive on a daily basis. Whether it is suddenly imagining a new recipe, a sneaky way through city streets, or clothing design, we pass such ideas on to others, who refine them and pass them back to the collective. And so on, so long as individuals have the freedoms and avenues for doing so.

Another major feature of these sociocognitive dynamics is that, as for hunting in wild dogs or wolves, the switch of cooperative pattern can be almost instantaneous, even in quite novel situations. This was demonstrated rather dramatically when a bus ran over a cyclist in a London street in June 2015. Within seconds, dozens of former bystanders got together and coordinated their actions. They actually lifted the heavy bus, in concerted effort, at least enough for the victim to be pulled away.

"With so many strangers suddenly and rapidly trying to work together, it's difficult to discern how much coordination there was. But at the very least, some kind of collective understanding of what they were trying to achieve began to develop. . . . [It was] almost like working towards a mutual goal but unknowingly, without actually communicating it across."[27] An understanding of the dynamics of social cognition helps explain that collective understanding.

The power of those dynamics also explains why human culture is not merely an efflorescence of human cognition but is its very medium and constitution. As anthropologist Clifford Geertz explained, culture does not merely "supplement, develop and extend organically based capacities," as the notion of observational learning might suggest. Instead it is an ingredient of those capacities. A cultureless human being, Geertz suggested, "would probably turn out to be not an intrinsically talented though unfulfilled ape, but a wholly mindless and consequently unworkable monstrosity."[28]

In a culture of individual competitiveness, of course, such ideas may seem odd. As Daniel Siegel pointed out in his book, *The Developing Mind*, the idea that we do not "own" our minds, independently of others, tends to make people feel uncomfortable. It particularly contrasts with the idea that intelligence, or cognitive functioning, in humans reflects some basic nonsocial functioning that varies among individuals like personal strength or power. Such is the power of ideology.

We can, perhaps, now see why attempting to summarize people's abilities with a simple yardstick, like an IQ test, rudely cuts across what it means to be human. To illustrate this point, I next discuss some of the "cognitive abilities" claimed to be measured in IQ tests. I discussed some of these in chapter 7 as products of dynamical systems generally. Here I put them into human cultural context.

IQ IN CULTURAL CONTEXT

Thinking or Reasoning

Thinking, reasoning, and problem solving are considered to be at the heart of intelligence. Above all, IQ tests are considered to be measures of "good thinking." But, as mentioned earlier, there remains considerable confusion about what thinking is. Mainstream investigation has spun off into numerous specialized topics. Research has mostly consisted of presenting generations of participants (usually long-suffering students) with banks of esoteric puzzles mostly detached from real-life problems. In these contexts, thinking is sometimes described as judgment or decision making. What has been repeatedly reported is that humans are not good thinkers, and frequently make shocking errors! A number of best sellers and thousands of scholarly papers have entertained us with the numerous ways in which human thinking is "irrational."

As a consequence, psychologists have spent much time considering why most—otherwise seemingly "intelligent"—people make errors with simple logical problems devised and presented in the laboratory. For example, participants have been presented with problems like: A bat and ball cost a dollar and ten cents. The bat costs a dollar more than the ball. How much does the ball cost?

Most people confidently blurt out that the ball costs ten cents, which is the wrong answer. A little reflection should tell you that the ball costs five cents. Studies presenting participants with large numbers of such problems show that average scores are unrelated (or even inversely related) to IQ test performance. The findings have led a number of psychologists to suggest we have deficiencies in our basic cognitive "machinery," or the way the brain is "wired." Our rationality is said to be "bounded." Daniel Kahneman—author of one of the bestsellers—describes laboratory errors of thinking as deficiencies in the basic "architecture" of the cognitive system.[29]

This again, it seems to me, is a problem of engaging human thinking out of context. The reason that some people fail such problems is exactly the same as the reason some people fail IQ test items like the Raven Matrices tests, described in chapter 3. It is simply not the way that the human cognitive system is used to being engaged.

As described in chapter 6, cognitive systems evolved to deal with deeper structures in experience. Learning is much more than a mirror reflection of the outside world. It captures the subtle depths and structures behind it that furnish predictability. Knowledge also includes emergent aspects that have not been directly experienced. These are the cognitive structures that make familiar problems predictable and solvable. The Raven tests are widely accepted as tests of "pure reasoning" (i.e., of individual reasoning strength or power, detached from experience). But, as explained in chapter 3, they simply measure the extent to which cultural tools of text management and the like have been assimilated by some individuals more than others. It is not that others cannot deal with that kind of trick problem; they are just not used to such problems.

It is a relatively simple matter to illustrate how human thinking arises in a cultural context. Consider the case of multiplication in what is, to most of us, simple arithmetic. It emerged as a cultural tool thousands of years ago; we all now use it to some degree in our thinking; and its development in children is well described in Piaget's studies. Children generally develop the concept of addition from direct experience, for example:

$$3 + 3 + 3.$$

That concept is one of "empirical abstraction," as Piaget called it. We can presume that it was present in early humans. But the reflective abstraction generated from communication with others in a group turned it into the concept of multiplication:

$$3 + 3 + 3 = 3 \times 3.$$

So, as Piaget insisted, the entire history of the development of mathematics from antiquity to the present day may be considered as an example of the process of reflective abstraction. But that was only possible through the mind sharing of cultural dynamics. He showed how the cognitive demands result in new syntheses vastly extending individual powers of calculation and predictability.

A long tradition of research, especially in child development, has revealed an almost universal dependence of reasoning on background knowledge. Everyday reasoning is based on culturally specific knowledge. For example, the logical abilities of individuals have been questioned when they have difficulty with laboratory problems like:

> All As are Bs.
> Are all Bs As?

Yet the difficulties evaporate when essentially the same problems are presented in socially relevant contexts, such as:

> All humans are mammals.
> Are all mammals humans?

In other words, the idea that individual differences in reasoning are ones of personal cognitive power is too narrow and mechanistic. Lambos Malafouris is only partly correct in saying that "the tool shapes the mind."[30] In truth it is collective cognition, with the help of individual minds, that shapes the tool that reciprocally shapes those minds.

Knowledge

It is usually assumed that intelligent systems must operate on the basis of knowledge, as opposed to fixed cues or signals (which is what distinguishes intelligence from instincts). The nature of knowledge, therefore, is supposed to be at the roots of cognitive theory. Many items in standardized IQ tests are based on simple knowledge questions. Yet in his book, *How the Mind Works*, Steven Pinker mentions how psychologists feel perplexed about the nature of knowledge, describing it as one of the problems that continue to baffle the modern mind. Like other psychologists, Pinker acknowledges that knowledge is involved in nearly all cognitive processes, but he fails to be explicit about what it is.

Again this seems to be a problem that stems from disregarding the way that the human intelligence system is part and parcel of a self-organizing culture. It misses the point that culture extends and amplifies human knowledge in a way quite different from the database of a computational machine.

Take basic knowledge of objects, for example. In prehuman cognitive systems, co-variations picked up from objects—how aspects of them change together over space and time—self-organize into object concepts. These concepts are co-variation clusters that share features and thus predictabilities. (I have also referred to them as attractors and grammars). If a rabbit spots a bushy tail and a red coat in the bushes, it "knows" what can be predicted and acts accordingly. That is the rabbit's knowledge of foxes.

However, human use of objects virtually never consists of isolated cognitions around isolated entities. We know objects—their properties, uses, and so on—through the patterns of social relationships of which they are an integral part. Social action with objects reveals far more than purely personal encounters possibly could, going much deeper than surface appearances and superficial physical properties. This is especially the case for artifacts that are produced precisely for the use of others. The infant's developing knowledge of spoons and forks extends far beyond their immediate utility and includes patterns of social relations surrounding their use.

Such socially embedded knowledge of objects transforms and extends individual cognitions. For example, social cooperation requires us to be jointly explicit about our knowledge, as in sharing it, teaching it, working together with it, and communicating about it. But to share with others our experience of an object and what it predicts, it must be ascribed to a category, defined by a word (e.g., "fox," "table," "furniture," "artifact"). Through the discourse tools of speech and writing, these more explicit knowledge structures can then be publicly reworked and extended into more complex forms. They can be further interrelated, used as metaphors for exploring new domains, and so on.

The same applies to knowledge of all experience. In fostering reflective abstraction, social engagement radically alters personal cognition, a process that creates enormous cognitive potential in people. As concepts become extended in vast knowledge networks, predictabilities about the world become much deeper, and far more potent, than any found in non-human animals. Asking children, as in typical IQ tests, about the capital of France, or the boiling point of water, completely misses this quality of knowledge. Suggesting that such questions indicate a child's true learning potential or knowledgeability is grossly misleading.

Memory

Memory figures prominently in IQ tests as the content of many simple items, such as digit span or short-term recall. In the form of working memory they are thought to reflect the more general cognitive, and even neural, "strength" of individuals. I criticized that view of working memory in chapter 7. But theories of memory, in general, assume that it resides primarily in the individual brain.

In humans, though, memory also exists in epicognitive forms with the aid of cultural tools. Even in preliterate societies, memory is, or has been, expressed in shared verbal and graphic forms (drawings), and in song, dance, story, and legend. At some time in human history memories started to be shared as written symbols, such as marks on sticks and later, pictograms. Even the rudimentary forms of these new tools fostered a crucial medium of social cooperation, as well as extending individual memory capacity.

As Vygotsky and Luria pointed out, the written forms of communication vastly expand the natural memory function, transforming it into a new medium of cognitive organization and planning. Such auxiliary memory tools became vastly augmented by printing, libraries, calculators, computers, and the internet. Internalized, they all form the cognitive tools of individuals in a position to assimilate them. There will be (and are) individual differences in such memory, but again, apart from demonstrable pathology, it is impossible to separate the individual from the social sources.

Science

This interdependence of individual and culture has reached its most sophisticated form in the cultural tool we call "science." The history of science and of scientific thinking is that of a shared toolkit for confirming knowledge. The implicit knowledge found in all other species is, of necessity, made explicit in humans when we need to cooperate to acquire knowledge and, therefore, communicate about it. As with refractive abstraction, found in all cognitive systems, deeper layers of structure are exposed that defy surface appearances.

Since the time of ancient Greece, it has been the task of scientific research to systematize that confirmatory process through scholarly analysis and discourse. Even prior to the more recent scientific revolution, this process radically altered some of our interpretations of everyday observations. For example, it revolutionized perceptions of a flat earth, or the geocentric structure of the universe. In this way, predictabilities of future states from current conditions become ever more precise, permitting more intelligent interventions.

It is now well documented how the great flowering of modern science from the seventeenth century on occurred through the devising of new knowledge-sharing procedures. Observations became more systematic, more experimental, and better recorded for sharing and attempted replication. Observed co-variations in natural phenomena could be shared by inviting others to check findings. Theoretical models could be jointly constructed such as to be more explicit about constituents and how they operate together. Predictions (hypotheses) could then be made and subsequently tested in controlled experiments.

The need for all steps to be made explicit, so that they can be shared and replicated by others, determined the logical structure of modern empirical science. Indeed, it has been argued that the upsurge in the social processes of conferencing, collaboration, and publication among scientists in the seventeenth and eighteenth centuries was the single most important step in the historical development of the scientific method.

Thus the current methods of science grew naturally out of epicognitive dynamics, and they perfectly illustrate the fruits of a cultural tool. But the rational social procedures are also internalized by individuals as a personal psychological tool. Doing so permits us to think scientifically as individuals. This is another example of the transformation and extension of otherwise limited cognitive abilities by cultural tools.

One obvious implication is that school tests that ask individuals to memorize and reproduce details of such methods miss the point. They are merely describing individuals' "schooling" and not successful participation in real science. As a consequence, as mentioned in chapter 3, there is probably little association between such test scores and future success as scientists.

Scientific knowledge and reasoning have, of course, been much assisted by other cultural tools. In particular, mathematics was developed precisely as a language for expressing the deeper structures and patterns in phenomena—patterns that cannot be expressed very easily in ordinary language. It is through such means that scientists reveal structures of nature that remain hidden to other species and to prescientific humans. It is the revelation of hidden structure, in all their harmonious composition, that led Albert Einstein to declare that "there is nothing more beautiful than a good theory." However, as mentioned in chapter 1, it is also part of the culture of science to constantly identify and reassess its presuppositions, especially their roots in social ideology.

INDIVIDUAL DIFFERENCES

Of course, science is itself a lesson about the potentials of humans. Who would have thought that the beings of a hundred thousand years ago, hunting with crude weapons; butchering with stone tools; and gathering roots, plants and fruits by hand could transform themselves into exactly the same species with the same genes we see today? What Vygostsky was pointing out in his law of cultural development was that what a child can do now, alone—often seen as an index of the child's potential—can be transformed by the cultural tools provided by parents, teachers, or more capable peers.

In other words, potential is created by dynamical systems, not merely expressed by individuals. Current status is not necessarily a good predictor of future status—unless of course individuals are artificially constrained to the same circumstances (and, under the power of ideology, they usually are). If, as behavioral geneticists claim, the human intelligence system can be modeled as a simple quantitative trait, like height or weight, then humans would still be confined to narrow niches, like the apes.

GIFTED, OR WHAT?

It is, of course, wonderful how the dynamical attractors of a culture are assimilated into individual minds and then turned around as something fresh and original. It is remarkable to see the speed and extent to which individual humans can develop within a specific domain when committed to it. Noteworthy is the remarkable speed of learning in all human children as they acquire the highly complex human language at a very early age.

However, what many psychologists and others get most excited about is the unusual extent of development of particular children in particular domains. Such individuals thereby become referred to as "gifted." Again the subject becomes shrouded in a mystique with attributions of (entirely unproven) "genetic" superiority. How inert chemical strings of DNA can come to enhance development in, say, mathematics or music is never explained.

Nevertheless, for parents who generally want such distinction for their offspring, this can be an emotive issue. That makes them more gullible to "messages" from scientists and their implications. So a letter in *New Scientist* looked forward to the day when "a screening test for sperm might one day be used to screen for genetic markers, thereby at least increasing our chances of producing brilliant offspring." And, in a more popular piece for parents, Linda Gottfriedson says, "Gifted children unmask the fiction that we are all born equally intelligent. . . . Mother Nature is no egalitarian; she grants gifts and talents in differing amount."[31]

Some studies over the past century have tried to prove this general idea by following up groups of gifted children in adulthood. Indeed, the IQ test designer Lewis Terman did that with a large sample of high-IQ individuals. Of course, the studies are fatally designed, because the children mostly continue to enjoy the favorable social circumstances they started with.

Not surprisingly, then, most did quite well in later life. But few achieved national recognition in their field of choice. In a review in 2006, Joan Freeman points out how Lewis Terman's famous "geniuses" enjoyed well above the population norms of childhood privilege and circumstance. In their seventies and eighties, however, they were no more successful in adulthood than if they had been randomly selected from children of the same socioeconomic backgrounds—regardless of their IQ scores. Similar results have emerged from more recent studies.[32]

Governments are also snared by the mystique, attributing national economic success at least partly to such individuals. That is why they fund the search for genes for "giftedness." I mentioned in chapter 1 how groups in Britain and China are scanning DNA from blood samples from the one in a thousand individuals with the highest IQ scores. They hope to find the magic genes prominent in such individuals that can tell us how they became so gifted. Others have suggested that this could one day help parents select embryos with genetic predispositions for high intelligence. This seems to me to be highly superstitious.

More interesting than such (almost certainly futile) searches are the conditions in which "giftedness" develops. If there is one certain conclusion coming out of research into gifted people it is the way they have been deeply immersed in a very specific aspect of culture, either by their own motivation or that of parents or other patrons. One thing I mentioned earlier about reflective abstraction is that knowledge easily snowballs once triggered in a domain. That is what seems to be confirmed in the research.

In his book, *Outliers: The Story of Success*, Malcolm Gladwell says that few stars of any field succeed without putting in at least ten thousand hours of practice. That includes prodigies, such as Mozart. Also, he says, few achieve great success without being handed privileged experiences early in life. He mentions specific characters in that category, such as Bill Joy, founder of Sun Microsystems, and Bill Gates, creator of Microsoft. Similarly, in his study of high-ability children, Michael Howe reported how "high achievements always depend upon diligent efforts . . . demanding thousands of hours committed to training and practice."[33] He also noted how chances are increased by early engagement in the domain.

Interestingly, one of the conclusions of Joan Freeman's longitudinal study (which did include control groups) is that "in terms of conventional success in life such as high examination marks, rising up the corporate ladder or making money, the primary building blocks were always keenness and hard work, allied with sufficient ability, formal educational opportunity and an emotionally supportive home . . . these factors are found over and over again."[34] I would, of course, interpret "sufficient ability" as itself a product of development in a culture, and not a mysterious constitutional entity.

The same applies to the "heights" of knowledge, as in scientific discovery. Effort is crucial, as is social context and the cultural tools provided by others. We justly praise the special efforts and achievements of individuals, but most will readily acknowledge the role of that very context: ideas do not arise de novo in individual minds. As Einstein himself was always ready to point out, the previous theories of Maxwell and Lorentz "led inevitably to the theory of relativity." He insisted that the work of the individual is so bound up with that of scientific contemporaries that it appears almost as an impersonal product of the generation.[35]

Walter Isaacson points out in his book, *The Innovators*, that collaboration underlies the creative process that has produced virtually all scientific revolutions, especially the current digital one. "Only in storybooks," he says, "do inventions come like a thunderbolt, or a light bulb popping out of the head of a lone individual in a basement or garret or garage." Rather, ideas seem to ferment in the cultural-cognition dynamics before appearing as a new synthesis, as mentioned above. It is such considerations that lead Guy Claxton and Sara Meadows to argue "that both the research base and practical and moral considerations should lead us to exclude ideas of innate and unchangeable degrees of 'giftedness' from our educational practice as incorrect, inhuman, and counter-productive."[36]

In sum, what is described as specific giftedness is really demonstrating something more general: how fast and how far development can proceed in a particular domain through the kind of cognitive-cultural interactions described above. I suggest that most children could become gifted in that sense. But not by the Jesuit boast, "Give me the child for his first seven years, and I'll give you the man." What seems most important is being in a position to access, assimilate, influence and feed off the global strength of a thriving culture. It is obstacles to that kind of process that are, broadly, the subject of the next chapter.

10

PROMOTING POTENTIAL

MODELS OF CAUSES AND EFFECTS

We all operate with mental models of external reality. Usually they are informal and implicit, developed in our own minds from physical and cultural experience: models of people, traffic, language, disease, child development, and myriad others. In virtually every domain of our lives, such models help us predict events and the effects of our actions.

Scientists try to make their models more explicit through systematic observations, constructing cause-and-effect theories, testing hypotheses, criticizing, coming up with revisions or alternatives, and so on. Every step is conducted and reported in ways that permit replication by others. We can be more sure of our models when we all agree that we are "seeing" the same thing.

However distant in time the objective may be, this scientific modeling is nearly always done with the goal of intervention in mind. Funding bodies and governments provide the resources precisely in the hope that the work of scientists will help to devise interventions for medical, social, and other problems. As we have seen, governments have been particularly interested in human potential for both practical and ideological reasons, and such ideology creeps into the science. This has been the case particularly with intelligence and with school attainment.

A good model is crucial for any intervention. It will provide a process account of how a system works; for example, how intelligence develops in childhood and what causes variation in it. So a good model will help us describe the best circumstances for promoting human development, remove obstacles to it, and guide intervention, as necessary. Of course, our models of human development are still unclear and contentious. I have explained this as partly due to vagueness about what develops and how and also due to the influence of ideological preconceptions, especially as they pervade the nature-nurture debate.

In this book, I have been contrasting more recent dynamical models with traditional mechanical (input → output) models. The first thing I do in this chapter is illustrate how the mechanistic input → output model has dominated thinking about both "genetic" and "environmental" causes of individual differences and has shaped interventions accordingly. I then criticize these and discuss alternative, dynamical, perspectives. In chapter 11, I apply the analysis to the institution specially set up to promote the development of potential, namely, education.

GENETIC INTERVENTIONS

Until recently, nearly all conceived genetic intervention in human development has been medical, dealing with single-gene conditions or disorders. These are associated with genetic mutations that either arise in the life of an individual or can be inherited from parents to offspring. There are thousands of known single-gene (or monogenic) disorders, occurring in about one in every hundred births, and many of them have distinctive effects on cognitive functioning.

Research into possible treatments of such disorders has been conducted almost since genes were discovered, in some cases with impressive results. Because of their basis in single genes, and categorical consequences, intervention has been relatively straightforward and uncontroversial, once cause and effect is understood. The standard example is that of phenylketonuria, an enzyme deficiency that, untreated, can lead to intellectual disability, seizures, and other medical problems. An understanding of its genetic basis—or, more accurately, the role of the gene product in metabolism—led to the dietary intervention that prevents the development of the disease.

Such environmental interventions with these inborn errors of metabolism are uncontroversial. Some excitement has been created by the possibility of treating psychological conditions in analogous ways, that is, by assuming that different environments have differing effects on individuals with different genes. This has been called "differential susceptibility" and has been speculated to apply to various conditions, such as alcoholism, smoking addiction, and a range of childhood behavioral problems.

However, the theoretical naiveté of ignoring the many pathway interactions (as I described in chapters 4 and 5) has been pointed out.[1] Failure to account for these interactions probably explains why results have been inconsistent. As Irene Pappa and colleagues point out in their review, pinning discrete functions, or functional variation, on different alleles has been impossible. Studies have given rise to "an overall lack of consistent findings" with "no certainty that these [allelic] differences result in biological, functional effects."[2]

Further difficulties have been encountered when the disease conditions reflect not one, but the combined consequences of multiples of deviant genes. However, it is such a polygeneic model of different sums of more or less "good" genes that has been applied to the normal ranges of complex cognitive traits like human intelligence. The hope has been to find analogous environmental treatments for helping those with poor gene combinations.

This is what seems to be proposed by, for example, Kathryn Asbury and Robert Plomin in their book, *G Is for Genes*. The description gets rather vague, but it seems that each child, after DNA sequencing, will go to school with a learning chip as "genetic predictor" of learning "strengths and weaknesses." Teachers will then devise learning programs to suit, thus ensuring that each pupil gets the best possible treatment. Indeed, in an interview in the *Guardian* newspaper, Plomin says, "It's wholly accepted that preventative medicine is the way to go. . . . Why not preventative education?"[3]

Other inspiration has come from the possibilities of direct gene therapy or genetic engineering, also arising from DNA sequencing. This involves structurally "correcting" a gene, by altering a DNA sequence, to prevent or treat a disease. Several possible techniques are currently being perfected for achieving this. For example, it might be possible to knock out a mutated gene; or it could be replaced with a healthy copy. A new technique (CRISPR, short for "clustered regularly interspaced short palindromic repeats;" it is *Science* magazine's "Breakthrough of the Year" for 2015), allows for DNA to be cut at desired points and pasted with replacement sequences, using selected enzymes, or "molecular scissors." These techniques are currently under study in clinical trials in humans, but only with genetic diseases that have, as yet, no other cure.[4]

However, these developments have led to some rather imaginative construal in psychology. For example, Plomin's colleague Stephen Hsu has an article in the online magazine *Nautilus* with the rather long title: "Super-Intelligent Humans Are Coming: Genetic Engineering Will One Day Create the Smartest Humans Who Have Ever Lived." In the article Hsu says, "If a human being could be engineered to have the positive version of each causal variant, they might exhibit cognitive ability which is roughly 100 standard deviations above average. This corresponds to more than 1,000 IQ points." Suggesting that thousands of genes may be involved, he explains that this "would require direct editing of the human genome, ensuring the favorable genetic variant at each of 10,000 loci. Optimistically, this might someday be possible with gene editing technologies similar to the recently discovered CRISPR/Cas system that has led to a revolution in genetic engineering in just the past year or two."[5]

As mentioned in chapter 1, the Imperial College team is also hopeful that genes have been discovered that could be manipulated to boost intelligence. It is such ideas and models that excite the media and lead some parents to wonder about designer babies. They are classic input → output mechanistic models, and they are hugely naive. Apart from the obvious practical problems (e.g., a teacher being confronted with myriad unique learning chips—how they are to know what will work?), the ideas contain huge genetic fallacies.

Learning, or intelligence, and variations in them, are not due to simple sums of single good or bad genes. They almost certainly involve thousands of the genes, and those that are involved do not "act" as if autonomous agents. Nor are they involved in development and individual differences as if independent of one another, with effects of "good" or "bad" alleles just adding together to form a total gene score. Instead each gene is utilized according to its genetic background—the whole genome—by a dynamical metabolic system, in an ever-changing environment. As described in chapter 4, aside from rare, well-defined disorders, genetic variation is mostly irrelevant. And the output of the genes is highly un-predictable for very good biological reasons.

As also explained in chapter 4, instead of a "dumb" input-output machine, what has evolved, even in single cells, is a highly dynamic intelligent system. Such systems can make do with considerable variation in gene products. Usually, the system can compensate for missing genetic resources by using an alternative product. Or it can find alternative pathways to a desired endpoint, as in canalization (chapter 5). Also, the system can use the same genetic resources to achieve amazing developmental plasticity, even as a lifelong process. Indeed, probably the most interesting aspect of evolution has been the emergence of dynamical systems, at a number of different levels, creating and regulating variation far removed from variation in the genes. And we now know that those intelligent systems can repair or even change the DNA in genes themselves, in "self-therapeutic" ways, as natural genetic engineering.

This is, of course, the problem encountered in genome-wide association studies. The direct associations expected cannot be found, resulting in the so-called missing heritability problem. There are other lessons to be learned there. But it does illustrate the dangers of applying a wholly inappropriate model to highly complex functions and social contexts. In chapter 2, I described the problem as that of applying pretend genes to spherical horse problems.

There have been other suggestions for promoting potential genetically. These are the proposals of the eugenicists, in messages still, alas, not entirely expunged. In imaginings from Galton through a long line of followers, the favorable genes can, as it were, be "gathered together" across generations, and the "bad" genes eliminated. That can be achieved, eugenicists have claimed, by various means: selective breeding, marriage control, sterilization programs, or sequestration (as in asylums). In 1920s America, sterilization was adopted by several states, some of which influenced Nazi thinkers. In Nazi Germany, as well as the horrors of the Holocaust, hundreds of thousands of "mental degenerates" were killed under programs of euthanasia and infanticide because of assumed "bad" genes.

These are stark reminders of how far ideologically driven science can take us. But we also need to be reminded that even the more benign, or therapeutic, plans, can be fanciful—especially when, for the vast majority of individuals, the variation has little, if anything, to do with genetics. Moreover, it is also worth remembering that IQ is not a measure of general intelligence. It is a measure of rather special learning, associated with social class and cultural background. And educational achievement is not a "test" of potential, as I explain in chapter 11.

As explained earlier, though, many of the basic presuppositions underlying the concepts of genes in these proposals are also to be found in conceptions of the environment. These concepts, and alternatives to them, take up most of the rest of this chapter.

THE ENVIRONMENT

Research into environmental effects has been much more abundant than that into genetic effects. This is no doubt because environmental interventions are superficially easier to implement than genetic interventions. Also, in the nature-nurture (part genes, part environments) framework, even the higher heritability estimates for IQ or school achievement leave scope for environmental intervention (however much the heritability concept is misinterpreted).

My first aim here is to illustrate conceptions of the environments thought to cause differences in the development of potential (cognitive abilities and school attainments). I describe the causal model(s) assumed and how they imply the kinds of interventions construed. As we shall see, most studies have been exploratory, with some imprecision about the definition of environments as well as of effects. Some are highly suggestive of real causes and fruitful interventions. Most are revealing about the underlying concepts of the environment (on which I comment below).

Because of the volume and diversity of such studies, this review is far from exhaustive. I merely illustrate studies in a few broad categories according to their scope and specificity: chiefly making a distinction between the simpler material environments and the more complex sociopsychological environments. Cutting across that breakdown is whether the focus is on positive or negative consequences of the environment on development, and whether it affects bodily or mental traits or both. As will be seen, these categories are not entirely mutually exclusive, but my purpose is to illustrate, not exhaustively review.

EFFECTS OF SPECIFIC MATERIAL FACTORS

As with the genes, some research has focused on singular, fairly well-defined, environmental components. Some specific deleterious factors have been shown to affect development. For example, a wide range of toxic agents can perturb cell metabolism with consequences for health and development and, possibly, brain and cognitive functions.[6] These agents include heavy metals (e.g., lead and mercury) in food, water, or industrial wastes and carbon particles and gases from traffic and industrial fumes. Organic pollutants include hormone-disrupting pesticides, paints, and preservatives.

Some incidents arising from side effects of pharmaceutical drugs have been well covered in the media. Smoking during pregnancy has become a big issue since the 1970s, and the abuse or misuse of other substances during pregnancy is suspected to have effects on offspring.

Also in this category are gene mutations that can be incurred through factors like UV radiation or radioactivity (i.e., gene mutations that are themselves environmental in origin). Such cases have arisen from accidents at nuclear power stations. These can have unpredictable, but sometimes highly detrimental effects on wide-ranging aspects of development and subsequent health and vitality. Those effects could be reflected in intelligence test results and school performance.

Conversely, there have been many attempts to identify unitary material agents specifically promoting brain and cognitive development. For example, some studies have suggested that breast-feeding boosts intelligence of offspring in later life, even when other factors are taken into account.[7] Some have reported associations with white matter volume in the brain and IQ (white matter refers to the fatty sheaths around long nerve fibers that act as insulation and speed signal conduction; it contrasts with the gray matter of nerve cell bodies).[8] It has also been suggested that the particular fatty acids in breast milk promote cognitive development, mainly through their effects on neuronal white matter and cell membranes.

However, other dietary factors have been associated with deleterious effects, either as general undernourishment or as malnutrition (absence of specific nutrients). For example, a study of protein-deficient children in India found substantial deficits in cognitive test scores compared with a nondeficient group.[9] Perhaps most famous in this category are studies of the consequences of famines. These include the Dutch Famine of 1944, when a German food blockade affected an estimated 4.5 million people. As many as 20,000 people died. Later the Dutch Famine Birth Cohort Study found that mothers pregnant at the time had children of below average weight. And those children grew up to have children who were likewise affected. Various medical and psychiatric conditions have also been reported in such children. Cognitive effects are, however, more debatable (see below).

Causal pictures of malnutrition are often confused by the timing of effects. In a review in 2001, Janina Galler and Robert Barrett reported that "the brain is vulnerable to the effects of insults during critical periods of brain development from the second trimester of pregnancy until 2 years of age. Malnutrition experienced at these ages will have lifelong consequences that are not reversed by adequate nutrition. Long-term effects of prenatal, postnatal and childhood malnutrition have been reported even after a long period of recovery from the illness itself."[10]

Complementing these findings are studies that have focused on the effects of dietary supplements. Animal and human cognitive studies have suggested that certain micronutrients (metals such as iron and zinc, or vitamins) have specific, critical roles in brain development. Elizabeth Isaacs has suggested that differences in protein and caloric intake can affect the volumes of specific brain regions which (she says) are associated with IQ.[11] Sandra Huffman and colleagues have reviewed research suggesting that omega-3 and omega-6 fatty acids (found in certain fats and oils) play an essential role in the development of the brain and retina: "Intakes in pregnancy and early life affect growth and cognitive performance later in childhood."[12] Since these are highly concentrated in fish oils, the latter have been used in a number of intervention studies, and sold by commercial companies, to "boost" IQ.

It would be rather surprising if nutrition was not related to cognitive functions in some way. There have been some doubts about what mediates these correlations, however. One question is whether cognitive functions are being directly affected or the treatments merely result in changes to general health and vitality. The latter, for example, have significant effects on any demanding task, not just test performances.

And there are some inconsistencies. As regards the Dutch Famine Study, for example, some follow-ups in adulthood have found "little or no suggestion of any impact of exposure to famine on cognitive functioning."[13] In a study reported in 2014, a group of mothers received a protein-energy supplement, in the form of biscuits, from week 20 of gestation to birth. A control group received the same biscuits only for six months, starting from delivery. As Stein says in his review, "In this study, no differences were seen between the groups in several measures of cognitive development measured at the mean age of 19 y[ears]."[14]

These studies suggest that cause and effect are more complicated than expected. Again, it may be that timing is crucial to consequences. For example, the review by Huffman and colleagues suggests there is no evidence for improvements in growth following fatty acid supplementation in children greater than two years of age. Nevertheless, the findings have suggested a fairly simple "growth" model from inputs to outputs and clear implications for intervention. But these concern fairly well-defined, identifiable, material factors.

EFFECTS OF SOCIO-PSYCHOLOGICAL FACTORS

Most research into environmental effects on mental development has prioritized socio-psychological factors, which tend to be less well defined than those just described. Consequently, some studies have been very broad and others more specific. And they involve less well-defined causal models of development and individual differences.

Demonstrating that "global" deprivation in infancy can have long-term effects has been a longstanding research program. One of the best-known contributions has been the English-Romanian Adoption Study. It carried out regular assessments on a group of Romanian orphans who had experienced a profoundly deprived institutional environment. As well as being malnourished, they spent most of the time alone in cribs, lacking physical, social, visual, and auditory stimulation. The study is of those who had been adopted into regular homes in the United Kingdom at various ages.

In fact, most adoptees showed rapid recovery psychologically, although a significant minority, especially of those adopted after six months, continued to experience behavioral problems. The most recent report, into early adulthood, notes "a striking pattern of behavioral impairments, in its core characterized by deficits in social cognition and behavior, as well as quasi-autistic features, often accompanied by cognitive impairment and symptoms of attention-deficit/hyperactivity disorder (ADHD)."[15]

It is known that an adverse fetal environment permanently programs physiology, leading to increased risks of cardiovascular, metabolic, and neuroendocrine disorders in adulthood.[16] High levels of stress during pregnancy have been associated with a number of long-term adverse psychological, as well as physiological, outcomes for both mother and child.

Research has also suggested that stress in the home, experienced during infancy, can result in greater stress reactivity in later life. That, in turn, diminishes confidence and focus in school.[17] As mentioned in chapter 4, stress experienced even before pregnancy can be passed on to the next generation through epigenetic effects, or the way that genes are utilized in an embryo, so that children, and even grandchildren, may suffer.

Understanding such findings is not straightforward, however, as results are often inconsistent. For example, experience of stress and measured circulating stress hormones do not always correlate in the way expected, perhaps due to individual differences in stress management (itself requiring additional explanation).[18] As Kumsta and colleagues report about the Romanian children, heterogeneity in outcome is largely unexplained.

Other studies have investigated effects of "enriched" or "impoverished" environments on brain development and cognitive functions specifically. The environments have usually been conceived in very general terms as ones containing many or fewer stimuli, objects, and opportunities for exercise. For example, studies in the 1960s and 1970s showed that rearing rats in the dark diminishes the nerve cell connections, as well as amino acid and protein production, in visual areas of the brain.

In the 1960s, Mark Rosensweig and colleagues started a long line of research comparing rats reared in normal, sparsely equipped, cages with those reared in cages with toys, ladders, tunnels, running wheels, and so on. They found that enriched early experience improved performance on several tests of learning. Further studies revealed changes in cortical thickness, size, and number of synapses, and extent of dendritic branching.

These findings have been replicated and extended in more recent research. Just a few days of motor skill training, or learning to run a maze, for example, seems to have produced brain structural changes. And it is now known that cognitive stimulation and exercise increases neurogenesis (production of new neurons) in some parts of the brain. The Rosenzweig team concluded that "sufficiently rich experience may be necessary for full growth of species-specific brain characteristics and behavioral potential."[19]

Further studies of institutionalized children have specifically examined outcomes in brain development, with the assumption that it will be reflected in cognitive development. For example, the Bucharest Early Intervention Project found reduced cortical brain activity (as measured by EEG recordings) among institutionalized children compared to never-institutionalized children. However, activity eventually returned to normal in children placed in foster care before the age of two years.

Other studies in this vein have found reduced brain metabolism in parts of the cerebral cortex of institutionalized children and reductions in white matter in various brain regions. There have also been reports of reductions in gray and white matter volumes, and increased amygdala volumes, in previously institutionalized children (the amygdala being part of the limbic system mediating cognitive with affective brain activity, as described in chapter 6).[20]

Studies on humans have been much extended in recent years with the advent of fMRI scans. As mentioned in earlier chapters, some of these have reported experience-dependent changes in brain tissues. Short periods of specific cognitive or skill practice, or general learning or memory training, even in adults, have all been claimed to have structural effects on brain, such as increased regional volumes. Even short periods of aerobic exercise—and of dancing in elderly people!—have been shown to make slight differences.

However, as Martin Lövdén and colleagues point out, many of these studies have serious methodological flaws, and the effects are not usually very big (2–5 percent in most studies). Moreover, changes to cortical volume and thickness may not necessarily reflect learning. The tissue changes may simply reflect the metabolic demands of heightened neural activity.[21]

In an important sense, all assimilation of cultural tools in humans, from early socialization to specific skill learning, is "enrichment training." As mentioned in chapter 6, studies on taxi drivers, violin players, jugglers, and others all found that training was reflected in gray matter changes in cortical areas. It follows that individual differences in brain volumes and associated intelligence might simply reflect differences in access to such cultural tools, not in the potential for learning as such.

These studies all imply direct cause-and-effect relationships arising from environmental deprivation, in a simple model. But they are complicated by other findings. For example, it now seems clear that brain development is specially protected against deleterious effects by physiological buffering. The concept of "brain sparing" is based on evidence from epidemiological studies in humans and experimental studies in animals. Starvation in adults is reflected in loss of body mass and reduction of other organs, whereas the brain and cognitive functions remain relatively untouched. Likewise, studies have revealed that malnourishment or other deprivations during pregnancy yields, as expected, offspring whose physical growth is retarded. Yet the brain growth seems to be much less affected: the ratio of brain weight to that of other organs, and general body weight, increases.

A physiological basis for brain sparing has been well established. It involves neural reflexes that ensure redistribution of blood flow away from other parts of the body to essential organs, including the brain. However, the issue is not free from controversy: some studies have suggested long-term brain and cognitive impairments, at least for severe deprivations in pregnancy. Timing of the deprivation in relation to critical periods of brain development may be crucial.

Finally, this category includes surveys of very general environmental experiences in humans. In these studies, measures of putative environmental factors are taken and then correlated with cognitive and/or educational test scores. The research has ranged from small-scale observational studies to large-scale surveys involving thousands of children. The methods have mostly used parental questionnaires; parent-completed scales; or observational checklists in homes, schools, peer groups, and neighborhoods. A major aim has been to identify, from correlations with test scores, possible targets for future interventions (with the assumption that the correlations might be causal).

The best examples of smaller-scale research are those using the Home Observation for Measurement of the Environment (HOME) scale. As stated on the inventory website, it "is designed to measure the quality and quantity of stimulation and support available to a child in the home environment. The focus is on the child . . . as a recipient of inputs from objects, events, and transactions occurring in connection with the family surroundings."[22] These "inputs"—including factors like parental responsiveness and encouragement; quality of parent-child interactions; and provision of toys, games, and books—are assessed during short visits and recorded on checklists.

Perhaps unsurprisingly, associations between such measures and the children's IQs or school attainments have been reported. Family income, parental education, and neighborhood loom large as associated factors. Accordingly, a big factor is socioeconomic status (SES), usually defined in terms of parents' occupational level or income. It is reported that, among other things, children in lower SES homes have lower relative access to books, games, educational activities, and musical instruments. Their homes are also often reported to be crowded, noisy, disorganized, and unkempt. The children's parents are also less likely to read to them or talk to them, and then within narrow ranges of vocabulary and grammar. So these children tend to experience less teaching of school readiness concepts, such as the alphabet, number concepts, colors, and shapes.

Confirming this general picture are the rather radical environmental changes experienced by adoption in childhood. When this is done from low- to high-SES homes, huge leaps in IQ test scores and school achievement are observed (see chapter 3). With such broad environmental categories, however, it has not generally been easy to establish clear causal pictures: for example, what is it really about SES that retards or promotes test performances and school attainments? I return to this question later.

The other main research strategy consists of large-scale surveys and the national longitudinal studies in several countries. They include, in the

United States, the National Institute of Child Health and Human Development's Study of Early Child Care and Youth Development. In the United Kingdom, the best example is the National Child Development Study. It includes all the individuals—around sixteen thousand—born in one week in 1958 through a number of follow-ups into adulthood.

The aim of these studies is not merely to replicate the already well-known correlations between environmental factors and indices of development, which they do. They are also able, with such large samples, to perform more detailed statistical analyses that might be more indicative of causes, and, therefore, of targets for intervention.

However, that aim has again been mitigated by the need, with such large samples, to use only broad categories of environmental factors (with the exception of some easily identifiable specifics like smoking in pregnancy). For example, in relation to reading and math attainments at eleven years, the National Child Development Study found school performance to be statistically associated with the following variables: SES, degree of parental initiative in contacts with school, housing tenure (owned or rented), geographic region, amenities in the home (e.g., bathroom, toilet), father's education, mother's education, family size, and crowding.

With such a large sample, the effects on attainment of each of these could be statistically "distinguished" from one another. However, although such associations are indicative in some important respects, the true nature of the factors, and how they work, is difficult to establish. For example, one of the biggest effects on gain in reading attainment at age eleven seemed to be having, or not having, sole use of amenities like an indoor toilet. In other words, these factors leave much room for speculation.

In such studies, the implicit input → output model is suggested by the assumption that children in the same home automatically experience the same "environment." Yet children in the same home turn out to be almost as different from one another, cognitively, as those from different homes. In a well-known paper, Robert Plomin and Denise Daniels suggested this was due to children in the same family actually experiencing different environments.[23]

But identifying such "nonshared" environments has been difficult. For example, a meta-analysis of forty-three papers that addressed associations between nonshared experiences and siblings' differential outcomes concluded that "measured non-shared environmental variables do not account for a substantial portion of the non-shared variability."[24] Recent reflections have not changed those overall conclusions.

This evidence suggests problems in the going conception of family environment. It suggests a "missing environment" problem as big as the "missing heritability" problem mentioned earlier: one that might not be amenable to simple correlational analysis. The problem does, however, reinforce Linda C. Mayes and Michael Lewis's acknowledgement, in the 2012 *Cambridge Handbook of Environment in Human Development*, that "indeed, the features of the environment and their various outcomes are poorly understood . . . it is surprising how little systematic work has gone into their study."[25]

I will shortly suggest what this implies for conceptions of environment. But let us first have a quick look at how the simple associations identified so far have been adopted in intervention programs.

APPLICATION OF THESE MODELS

The way in which these factors are conceived is further indicated in attempts to translate the findings into intervention programs. As early as the 1950s, for example, psychologist Alice Heim had reached the conclusion, in *The Appraisal of Intelligence*, that "given better food and more books at home, the poorer children would match, if not outstrip, their more fortunate fellows on the tests."[26]

In the 1960s and 1970s, the model was seized on by state agencies when funding intervention programs. Through it they could demonstrate, they thought, renewed commitment to equality of opportunity—of closing the attainment gaps between social classes and ethnic groups—and making society seem more fair.

The correlational data suggested that the biggest problems arise in the home. So what has been called a "war on parents" was initiated. The interventions have included improving parental encouragement of student effort, improving parent-school relationships, encouraging parental involvement in homework, and improving interest in the school curriculum. And, indeed, at least some of these efforts seemed to result in improved school performance.

However, there has been continual debate about the true nature of these effects, what really mediates them, and how durable they are.[27] For example, it has been suspected that apparent gains simply reflect the sympathetic, but temporary, attention from educational and psychological workers—a different kind of "environment" from that conceived to be operating. So the effects wane when the attention is removed.

The other obvious way of "closing the gap" has resulted in a "war on schools" (meaning teachers and curricula), including the greater demand for preschool experiences. Large and far-reaching compensatory education programs have been created to help disadvantaged children reach their full "potential." In the United States, these include Head Start, the Chicago Child-Parent Center Program, High/Scope, Abecedarian Early Intervention Project, the Milwaukee Project, and the 21st Century Community Learning Center. Parallel efforts have taken place in the United Kingdom and other parts of Europe, and in developing countries.

However, there has been constant debate about their effectiveness. Some evaluations have claimed some success in boosting children's IQs and/or school attainments. Others have claimed that the gains "wash out" over the longer term. Still others report little or no gain at all. For example, an evaluation of Head Start by the U.S. Department of Health and Human Services in 2010 suggested "small" effects. A further evaluation suggested that "on average, Head Start centers perform similarly to non-Head Start centers. Our results suggest that expectations for the Head Start program may be too high."[28]

Some critics warned that the provision of a few hours of home help or supplementary schooling is not getting at the root of the matter. In a review, Lynette Friedrich Cofer describes the efforts as a "powerful example of the unintended consequences of applying a linear, single causal approach to complex social problems." She also warns that "the question of how we conceptualize human development is central to the dilemma."[29]

Finally, an intensive focus on the narrow goal of boosting test scores has become self-defeating, according to increasing numbers of teachers, community workers, and academics. I have much more to say about that in chapter 11.

THE CURRENT CONTEXT

As we moved into the new millennium and beyond, the meritocratic target of equal opportunity and "closing the gap" has become more elusive. On both sides of the Atlantic, social inequality has widened, school performance results have been disappointing, and social mobility has fizzled out. The standard ideology of the child, environment, and individual potential has consequently become more stressed. A report of the American Psychological Association in 2006, specifically on SES, acknowledges reports of the mobility myth and with it the demise of the American dream. Two of the factors often emphasized in these reports are (a) the stagnating or increasing poverty rates, and (b) increasing rates of income inequality.

An Organisation for Economic Co-operation and Development (OECD) report of June 2014 concurs that inequality has increased and social mobility has virtually come to a stop in countries like the United States and the United Kingdom. And in 2015, even the International Monetary Fund says that widening income inequalities is the most defining challenge of our time. Overcoming false barriers to greater equality must be part of the challenge. However, the tendency has not been to look for wider perspectives—to consider that there may be something wrong with the current ideology—but to reinforce the old one in two ways.

One response has been to again start inquiring whether the problems are biological in nature. As described in chapter 1, funds are being pumped into the search for genes for IQ and educational attainment, as if manipulating a few candidate alleles might do the trick, or they may suggest more effective "environmental" interventions. And behavioral geneticists are being called on to give evidence to governments. The subtle difference from historical antecedents is that it is being done under a more benign agenda than in the past: we are told that identifying culprit genes will lead to therapies to relieve their handicapping effects (although it is also claimed that doing so will not change the essential "genetic" inequalities of ability).

The second response has been to reinforce the wars on parents and schools. It is reflected in, for example, programs like No Child Left Behind. Founded in 2002, it was reinforced by Barack Obama in 2012 with plans designed to close achievement gaps, increase equality, improve the quality of instruction, and increase outcomes for all students.[30]

However, new alarms are now sounding around the OECD's new Programme for International Student Assessment (PISA) ratings. The 2013 Education Department report quotes U.S. Secretary of Education, Arne Duncan, as concluding that the performance of American students in the latest PISA evaluation reflected "a picture of educational stagnation." The report goes on: "More resources need to be directed toward disadvantaged students."[31] So it calls for more resources from more taxation to fund public elementary and secondary schools and to improve the quality of teachers available to socially disadvantaged students. It suggests programs for reducing the impact of socioeconomic background on educational outcomes. These are obviously based on manipulating the kinds of environmental factors illustrated above.

Similar measures have been urged in the United Kingdom, when Michael Gove, then education secretary, imposed more testing, inspections, and league tables on schools and more pressure on teachers. In a speech in 2011 he voiced concerns that U.K. education has been "plummeting" down international league tables and spoke of the scale of failure among poor children as "a tragic waste of talent; and an affront to social justice."[32] He suggested that we need nothing short of radical, large-scale reform of intervention and testing, in spite of being heavily criticized by teachers and education specialists.

So the war on parents has intensified, too, particularly in the United Kingdom. Former prime minister Gordon Brown said the problem lies squarely in the backgrounds that (the children) come from. His successor David Cameron has followed suit, claiming the need to address "what is keeping people poor—the family breakdown, the failing schools, the fact that people are stuck on welfare. It's those things that are keeping people trapped in poverty and making them poorer." And Secretary of State for Work and Pensions Iain Duncan-Smith (*Sunday Herald*, April 26, 2009) has more emphatically located our "broken society" in a particular social class, with its problems of family breakdown, debt, drugs, failed education, and so on. He says that if we do not deal with these "causes," things will only get worse. And he puzzles over how the free market principles he has supported produce a "strange phenomenon: growing economies and growing welfare bills."[33]

CONCEPTS OF THE ENVIRONMENT IN THESE STUDIES AND INTERVENTIONS

The above is not intended to be an exhaustive review of a huge, and often highly commendable and well-meaning, literature. My point is to try to identify the more cryptic conceptions of the environment, of the model of human beings and society, and their deeper ideological roots.

I would describe the dominant concept as "elemental": basically, the environment is viewed as a collection of nominal factors, each having some independent size of effect on development. As such, they are conceived as causing individual differences as simple, linear, input → output, relationships (with output indexed by IQ or school performance).

In other words, it is the horticultural view of the child again. In relation to specific aspects of physical growth and specific material factors, the view may have some merit. But, as regards human cognitive potential, I believe it is sadly awry. It invites a "blame the victim" vision of lower class and ethnic minority students laboring under intellectual handicaps because of their genes, parents, family background, language, and culture.

I suggest that there may be a different reason that vision has failed. Our way out of inequality, and "being fair" to individuals and their potential, may not be as easy as a modicum of "horticultural" compensation. A deeper and wider perspective is needed in which individuals are viewed as conscious, reacting systems in a wider dynamic system.

THE REALLY IMPORTANT ENVIRONMENT

In what follows, I suggest that we need a quite different view of the environment and of what actually develops in children. To some extent, it reiterates what has been said in previous chapters about structures rather than elements, so I will be brief.

What have evolved are intelligent, dynamical, self-organizing systems. To properly develop, intelligent systems need many of the ingredients illustrated above, but they need much more than that. They evolved to deal with rapidly changing environments by being sensitive to the statistical patterns or structures in them as the only source of predictability. They are not based on elemental "inputs" involved in assembly-line development for stable functions.

Even at the molecular level in the cell, intelligent systems rely on the abstraction of environmental structure for predictability. In the cell, the abstractions emerge as self-organized structural grammars, also called attractors. They can predict futures from the relations among variables assimilated from past experience.

As you may imagine, in such systems, anything that prevented, or interfered with, the widest possible assimilation of such structure will suppress function. For example, disrupted or blocked interactions in cell signaling are associated with disease, including cancer. Partial, rather than global, receptivity to signals in the developing embryo—perhaps blocked by drugs, toxins, or lack of specific resources—will distort the normal emergence of form and function. Disturbances in the wider environmental—including psychological—context of physiology can lead to disease states (e.g., cardiac arrhythmias and other heart disease).

Evolution in more complex, changeable environments required more powerful abilities for structure abstraction. A most significant leap was the evolution of brains and complex behavior. Through their superconnectivity and supercommunication, brains assimilate environmental structure to extreme depths of abstraction and predictability. We saw in chapter 6 how absence of structure in early sensory experience retards development in respective brain connectivity and function.

Cognitive systems emerge among that structured communication in neural networks. They form powerful attractor coalitions with emergent levels of abstraction (the reflective abstractions described in chapter 7) but are particularly sensitive to structure, or lack of it. Again, incomplete or one-sided engagement results in biased cognitive development, perception, and function.

Investigators find these differences difficult to interpret when viewed in terms of deprivation of elements rather than deprivation of structure. For example, as described earlier, psychologists have been surprised by the way that children in the same family are so different from one another. But group configurations and relationships can be quite different from the various points of view in a family, especially considering birth orders, numbers of siblings, spacing between them, and so on. In dynamical terms, the attractor landscape of the average family can contain many incongruencies (unsynchronized attractors) that refract through different members of the group in different ways.

Siblings share half their variable genes, on average. But as described earlier in chapter 5, even genetically identical mice reared in the same environment develop the full range of individual differences seen in normal mice, with effects reflected in brain networks. Only by attempting to view the "environment" as a collection of disparate elements can we miss such structural effects.

In humans, new levels of coalition emerge through cooperation among brains, forming epicognitive regulations in the form of cultural tools. These are the basis of culture and consist of vastly deeper, more powerful, representations of the world and of action on it (as I mentioned, science is just such a cultural tool). The cognition-culture interactions redefine human potential and intelligence as a consequence.

Culture—including ideas, values, institutions, and ideologies—became the most important environment for human intelligence. But these are not elements to be learned by simple association, as if a shopping list: they are all structural abstractions governed by system dynamics. Failure of access to them, or disengagement from at least the most important of them, can have devastating effects, as I now illustrate.

SOCIAL CLASS STRUCTURES

As with other levels of development, access to system dynamics in societies is crucial for individuals to function fully and to develop individual potential within it. That is, human psychology is only fully realized when fully and equally engaged with the dynamics that govern the whole. Imbalance of access means power for some, subordination for others. This is, of course, just a more sophisticated way of reiterating what has been said before, from Aristotle to John Dewey: namely, that humans are political animals, needing to be socially engaged in a way different from the ant or the sheep.

Early in human evolution, such social engagement was probably the norm. Humans probably developed with a shared conception of the aims and activities of the group as a whole, enjoying its institutions and the shared view of the world. This is the cognitive "binding" described in chapter 9. Fossil evidence from early *Homo* sites suggests living in small bands, perhaps in sporadic coalitions with other groups.

There are a few remnants of such hunter-gatherer groups in fringe areas of Africa and South America and one or two other places. They are considered to be the closest we have to the form of human existence for at least 95 percent of our history. Their psychological and social makeup is very interesting, displaying little rank differentiation, with equitable sharing of resources and little personal property.

A writer in *Science* referred to such societies as "our egalitarian Eden." In his blog on *Psychology Today* (May 16, 2011), Peter Gray says, "Wherever they were found—in Africa, Asia, South America, or elsewhere; in deserts or in jungles—these societies had many characteristics in common. The people lived in small bands, of about 20 to 50 persons (including children) per band, who moved from camp to camp within a relatively circumscribed area to follow the available game and edible vegetation. The people had friends and relatives in neighboring bands and maintained peaceful relationships with neighboring bands. Warfare was unknown to most of these societies, and where it was known it was the result of interactions with warlike groups of people who were not hunter-gatherers."

Of course, we must not get too romantic about information that is skimpy, easily distorted, and far removed from modern circumstances (nor fall into the warm glow of a Rousseau-esque view of the "noble savage"). The observations do suggest, though, the cognitive and social benefits of equal participation in the environmental and social structures.

As always, those original hunter-gatherer structures changed. Populations expanded, and early humans extended their coalition structures ever more widely. Small bands became wider coalitions, and then agrarian "cities," with divisions of labor, then nations and global trading networks. Such cultural inventiveness has been highly beneficial and technologically creative for humanity generally. But it was at the cost of dividing populations into social classes and introducing imbalances of power.

In contrast to the egalitarian hunter-gatherers, organization into so-cial classes introduced inequalities in roles, rewards, and privileges. It also restricted access to cultural tools, including what I have called system dynamics. More simply, institutions enabled the power of a few over the many. Class division also engendered the ideologies required to legitimize such social structure. In that structure, subcultures emerged with different interests and beliefs, and with contrasting conceptions of the world and themselves. Instead of a harmonious coalition, social class systems brought considerable distortion and fragmentation of global cognition, with attendant frictions and conflicts.

Individuals in class societies, that is, do not share awareness and engagement with key social processes equally; the crucial political connectedness is diminished for many. Without that inclusiveness, development and potential will be limited for some.

Putting it like that, I think, better defines the environment and explains the origins of psychological differences more clearly. It is not a question of nature or nurture and their relative contributions to a position on a single ladder, but of place in a culture in an unequal, uneasy, coalition of cultures. Accordingly, the problems of any particular class stem, not from intrinsic properties of the class, but from the dynamics of the class structure as a whole. This deeper perspective is now beginning to be realized quite widely.

For example, it was hinted at by Bernice Lott in an article in the *American Psychologist* in 2012. "In the United States," she said, "one is born into a family that can be identified as working class, middle class, or affluent—divisions that denote status and power, as defined by access to resources." Her article then "explores the relationships between social class membership and a wide array of personal and social daily life experiences." It concludes with "a discussion of classism, which contributes to diminished opportunities for low-income families."[34]

Moreover, the special report by the American Psychological Association on SES (2006) refers to the "network of attitudes, beliefs, behaviors, and institutional practices that maintain and legitimize class-based power differences that privilege middle- and higher-income groups at the expense of the poor and working classes." Or, as Frédérique Autin and Fabrizio Butera put it, "Determinants of inequalities could be better analyzed by considering the way in which the social world is structured and shape people's experiences. . . . Institutions reflect and promote ideas and values (e.g., equal opportunities, meritocracy, etc.), and thereby influence the way people think about themselves, others and society."[35] Most casual observers—and that includes psychologists as well as politicians—seem to be unaware of the deep psychological consequences of a class structure and its durable negative consequences for the development of potential.

A dynamical perspective, then, suggests that the problem of individual differences in intelligence does not lie in the characteristics of a particular class at all. Nor can they be attributed to environments perceived as quasi-horticultural factors. Instead they have their roots in the relations being maintained across classes—and, of course, the ideology that maintains them. It is those social structural relations that constitute the key environments of cognitive development. So let us look at a few examples of those relations and consider their consequences.

WEALTH RELATIONS

"Of course, environments *per se* are not inherited," says Robert Plomin and colleagues in their widely used textbook, *Behavioral Genetics*. Such is the penalty of looking at the environment only in terms of a collection of independent factors "left over" in a statistical model.

Of course, what those authors really mean is inherited "like the genes." But the environment is inherited in many others ways. I mentioned epigenetics in earlier chapters. These are the effects of environmental stress experienced by parents on gene transcription in children. We are as yet uncertain about the extent of such inheritance. But it almost certainly exists and creates individual differences. Yet in the behavioral geneticists' equations, they will be described as genetic inheritance.

Far more conspicuous is the inheritance of wealth. Wealth includes all forms of individually stored up "goods" produced by the society as a whole but unevenly distributed. In the modern world, it includes savings from income, land and property, company shares, cars and boats, art, and so on. It is passed from parents to offspring to create enormous advantages in terms of income stream, power, and privilege, irrespective of actual

merit. Wealth is strongly reflective of historical inequalities and reproduces them from generation to generation, giving "something for nothing" to the beneficiaries. Accordingly, wealth is much more skewed than
income. As mentioned above, the wealth gap between social classes has
been increasing markedly over the past few decades.

The figures are stark enough. A report of the OECD in May 2015 (*In
It Together: Why Less Inequality Benefits All*) says the gap between the
rich and the poor keeps widening. In its thirty-four member states,
the richest 10 percent of the population earns 9.6 times the income of
the poorest 10 percent. As "stored up" income, wealth is then used
(usually by employing others) to make more wealth. So Mark Pearson,
head of the Health Division, told BBC News "It's not just income that
we're seeing being very concentrated—you look at wealth and you find
that the bottom 40% of the population in rich countries have only 3%
of household wealth whereas the top 10% have over half of household
wealth."[36]

The skew is highest in the United States. In government statistics of
2010, Edward N. Wolff and Maury Gittleman found that the wealthiest
1 percent of families had inherited an average of $2.7 million from their
parents.[37] This was 447 times more money than the least wealthy group
of people—those with wealth less than $25,000—had inherited.

Or to put it another way: The richest 20 percent of Americans have
about 85 percent of all the wealth. And the bottom 20 percent have about
0.1 percent.[38] The situation is much the same in Britain. The top 5 percent
control about 45 percent of national wealth. Excluding housing, the figure
rises to nearly 60 percent of wealth in the United Kingdom. A December
2015 report from the Pew Research Center notes that the average income
of the upper tier of society is seven times that of the middle tier. In 1983,
it was merely double.

Obviously, inherited wealth considerably inflates the income streams
of those who benefit, and it creates big individual differences in families
and their children. The benefits are both material and socio-psychological.
Wealthy families have the opportunity to provide every environmental
benefit for their children when they are young. It will be used to create
enriched physical environments from birth and throughout childhood:
good living conditions, stable and predictable circumstances, secure

childcare, cooks and nannies, private healthcare, holidays abroad, second homes, boats, private club membership fees, and so on. All this provides for healthier lifestyles promoting physical growth and cognitive vitality.[39]

Inherited wealth also does much to promote children up institutional ladders to positions of influence and power, irrespective of ability. Paying for attendance at private schools (about 10 percent of school students in the United States and 8 percent in Britain) is a major route to such ends. Such schools, with high expectations, tend to be much more focused on gaining entrance to higher education and occupations. They therefore have more rigorous academic programs geared to coaching for the SAT and other exams.

Perhaps just as important is that private schools ensure networking with people in influential positions. In that way, social class patronage is won, including preferential access to capital and enhanced lobbying power. With that comes a wider perception of society and cognitive engagement with it. In consequence, we get a preponderance of ex-private schoolchildren in powerful positions in all the institutions of developed countries, especially politics.

Perhaps even more importantly, inherited wealth creates powerful psychological effects. Even relatively small amounts, including the expectation of inheritance, can engender a sense of economic security, household stability, and predictability of circumstances. Psychological research in the United Kingdom and the United States has shown how these benefits include increased sense of personal security, self-efficacy beliefs, and self-confidence.[40] These foundations, in turn, promote political interest and participation, future orientation and planning, confident career focus, and secure grounds for risk taking.

Interestingly, in spite of its significance for individual differences, inherited wealth rarely figures as an environmental factor in child development studies or as candidate causes of individual and social class differences (the nearest we get to it is an SES category). I have never heard it suggested that, in the interests of equality of opportunity, inherited wealth should be abolished. Perhaps it confirms that diversions of attention to genes and IQ, the "wars" on parents and schools, and a few compensatory programs have more to do with the ideology of maintaining a class structure than with trying to establish true equality of opportunity. To reinforce this point, let us look at the other side of the coin.

THE PSYCHOLOGY OF DISENGAGEMENT

Cognitive engagement with the system dynamics of a society is crucial for the full development of individuals and the creation of potential. Those on the margins in the social class structure are cognitively disengaged in many ways. And that seriously affects their own potential for learning and that of their children. I considered these matters in the context of IQ test performance in chapter 3. But it is worth taking another view in the present context of causes of individual differences.

The main finding of *The Spirit Level*, by Richard Wilkinson and Kate Pickett (analyses done in a number of countries, including fifty U.S. states) was that the greater the inequality of the society to which a child belonged, the wider the differences in school performance. Why should this be so? Well, it seems that feeling at the bottom of an even steeper social cliff has even bigger psychological consequences. It means more depressing comparisons with those "above," even lower self-esteem, lower sense of security, and a reduced sense of personal control over circumstances.

This view has been supported in other studies. Pamela Smith and colleagues showed how a sense of powerlessness inclines individuals to view themselves as merely "the means for other people's goals."[41] More recently, Candice Odgers has also noted how "among wealthy nations, children in countries with higher levels of income inequality consistently fare worse on multiple indices of health, educational attainment, and well-being. New research also suggests that low-income children may be experiencing worse outcomes, and a form of 'double disadvantage', when they live and attend school alongside more affluent versus similarly positioned peers." Added to little money is what she calls "the role of subjective social status."[42]

Those psychological effects make the material circumstances of lower social classes bite even harder. They are already serious enough. Parents' constant grinding worry about money shortages and future security depletes energy resources for cognition. A report in *Science* (August 30, 2013) notes how poverty-related concerns impair cognitive capacity and other behavior. "Simply put," it says, "being poor taps out one's mental reserves. This could explain data showing that the poor are likelier than others to behave in ways that are harmful to health and impede long-term success—in short, behaviors that can perpetuate a disadvantaged state."[43]

This explains why one of the alleged "weaknesses" of the working class—failure to delay gratification—may, in a world of uncertain futures, amount to quite rational behavior.[44] As Anuj K. Shah and colleagues suggest, "certain behaviors stem simply from having less." They suggest that scarcity changes how people allocate attention: "It leads them to engage more deeply in some problems while neglecting others ... and that can help to explain behaviors such as over-borrowing."[45] That may also explain why average numbers of children in the family decreases with increasing prosperity.

A major issue is that of job security, rarely pressing among the upper class. Among the lower classes, the problem has become worse in recent years. As Antonio Chirumbola and Alessandra Areni point out, "In the USA and Europe, phenomena such as merging, downsizing and reorganization have turned out to be more and more widespread in the last two decades. . . . These transformations have changed the nature of work and caused feelings of uncertainty, stress and anxiety for many workers about the existence and the features of their job."[46] Claims that IQ tests are valid measures of intelligence because they predict job performance have to be judged against such a background, as explained in chapter 3.

Lack of employment security and pressing debt problems, preempt forward planning or thinking beyond current circumstances. They also affect levels of self-confidence, stress, motivation, anxiety, and reduce physical and mental vigor. These all detract from cognitive engagement in wider society and expressions of personal ability. Surveys regularly cite lack of confidence or self-esteem as one of the barriers of escape from low-pay/no-pay situations. For example, the organization Psychologists against Austerity in the United Kingdom identified five severe consequences of lower social class experiences:

- Humiliation and shame
- Fear and distrust
- Instability and insecurity
- Isolation and loneliness
- Being trapped and powerless

Albert Bandura referred to such consequences in terms of "cognitive self-efficacy beliefs." Citing diverse lines of research, he said that they "exert considerable impact on human development and adaptation. . . . Such beliefs influence aspirations and strength of goal commitments, level of motivation and perseverance in the face of difficulties and setbacks, resilience to adversity, quality of analytical thinking, causal attributions for successes and failures, and vulnerability to stress and depression."[47]

However, as more recent studies reported by Willem Frankenhuis and Carolina de Weerth show, children from stressed backgrounds who perform lower on standard tests are not generally impaired in their analytical thinking. Instead they actually exhibit improved detection, learning, and memory of stimuli that are ecologically relevant to them (e.g., dangers) in their particular circumstances, compared with safely nurtured peers.[48]

Perhaps most ironically, lower-class parents, through their own experiences, especially in school, are likely to have reached negative conclusions of their own abilities. This could well have been assisted by media reports of the "genetic" basis of their failures. It is difficult to feel self-confident and aspirational for self or children in a society that has certified you as deficient in brainpower. Children who enter school with such beliefs are less inclined to engage with school learning, become more easily distracted, and even rebel (see chapter 11, especially the work of Carol Dweck).

This helps explain the reduced parental encouragements and reduced inclination to engage their children in activities relevant to future schooling—a regular finding of the surveys mentioned above. Lower social class parents are less likely to have confidence in their ability to help their children succeed in school and have lower academic aspirations for them. Even the fear of being labeled as of inferior potential may impede performance. "Social psychological research suggests that negative stereotypes about women and minorities can create subtle barriers to success through *stereotype threat*."[49] I described in chapter 3 the ways in which these effects impact IQ and educational test performances, resulting in serious misreadings of individuals' cognitive potentials.

Note that these effects do not stem from actual personal potential or quality of childcare. They emerge from perceptions of place in a social order and the extent of control in, and cognitive engagement with, that order.

These are crucial sources of individual differences that are usually ignored. Their effects will not be reduced by simply "lifting" people out of poverty according to some convenient statistical criterion.

Most people acknowledge that the diseased, the hungry, and the homeless are in a very poor position to seize or create opportunities, plan productive activities, be self-reliant, and so on. But few scientists seem to realize that much the same applies to those whose conceptions of their own cognitive ability have been damaged by erroneous assumptions and the criteria of ability that flow from them. These essentially ideological assumptions are serious issues of social justice. But they are also serious impediments to otherwise "good enough" intelligent systems, their development, and their fullest social and democratic participation.

In spite of such conditions, it is worthwhile to point to some other investigations in this area, because they indicate the resilience of humans under duress. In a report in 2010, Paul Piff and colleagues note that lower social class is associated with fewer resources, greater exposure to threat, and a reduced sense of personal control. Accordingly, they suggest, we might expect lower-class individuals to engage in less prosocial behavior and emphasize self-interest over the welfare of others. However, across four studies, they found that lower-class individuals proved to be more generous, charitable, trusting, and helpful, compared with their upper-class counterparts. Further analyses showed that lower class individuals acted in a more prosocial fashion because of a greater commitment to egalitarian values and feelings of compassion.[50]

In this chapter, I have considered some of the possible factors creating individual differences in potential in general, and cognitive intelligence in particular. They already suggest strategies for intervention and the promotion of intelligence better than the piecemeal efforts of the past. However, those forces are played out more strongly at an institutionalized level in education systems around the world. There have been thousands of books and papers on that subject. In the next chapter, I confine myself to a few thoughts on education systems as they relate to the creation of individual differences.

11

THE PROBLEMS OF EDUCATION ARE NOT GENETIC

FUNCTIONS OF EDUCATION

Education systems serve a wide variety of stated functions: equipping children with economically useful mental and physical skills, preparing them for social roles and citizenship, passing on knowledge deemed to be important, transmitting important aspects of the nation's culture and heritage, and so on. But to most people, it is probably the idea of fulfilling children's innate potentials in order to access deserved levels of occupation that matters most.

To most people, too, this seems to be a natural process. Since the institution of compulsory national systems of education in various parts of the world, school achievement has come to be seen by almost everyone as both the development and ultimate test of children's potential. This is why IQ tests are, after all, calibrated against school performance. So to the general public, schooling is perceived as a fair natural selection process: children "prove" their innate potentials by being asked to learn a neutral curriculum in an equal environment. To almost everyone, children are seen to enter the system on an equal basis—and then come out in grades exactly reproducing the class structure of the society from which they entered in the first place.

Teachers, in fact, see this natural selection function as their professional contribution to social and national goals, as well as to realizing children's innate potentials. Indeed, the whole gearing of the school and education system to a hierarchical job market and class-structured society makes it difficult for them to conceive of their role in any other way.

Politicians laud the system as vital to the nation's economy and also welcome its role in the ideology of social inequality, as "an engine of social justice."[1] Children have every chance to prove their inherent potential, it is claimed; if they do not succeed in climbing the social class ladder and obtaining just rewards, it is their own fault.

Research shows quite clearly, though, that schooling has little to do with learning ability, cognitive potential, or intelligence as such. It really consists of a process of psuedo-assessment and attribution of potential based on social class background.

Perhaps Samuel Bowles and Herbert Gintis have put this most clearly. They argue that schools do not supply employers with skills, but with suitably socialized workers. Schools reproduce the values, expectations, and attitudes that prepare people to put up with inequality, accept their lot, and support the system, unequal as it is.[2] This task explains much of what goes on in schools. Behind the nurturant and scholarly façade, that is, something else is going on in education systems—something that en-sures that equal opportunities cannot be equally taken and learning ability for most is underestimated or even suppressed.

Mountains of critiques already exist about this aspect of schooling. David Berliner and Gene Glass, in their *50 Myths . . . in Education*, say that "we keep assessing our students and their teachers with methods best suited for a 19th-century model of education, one based on the simple transmission of knowledge."[3] In this chapter, I simply want to illustrate how typical state education systems, trying to hold together a social class system, fabricate individual differences in potential and intelligence. They do this by demanding processes of pseudo-learning, through what has been called a "hidden curriculum," with outcomes presented as diversity of potential and so reproducing the class structure of society across generations.

SELF-FULFILLING LABELING

Covert selection starts almost immediately. From the findings described in the previous chapter, it is obvious that children do not enter school as equals. Many are already advantaged or handicapped by the social inheri-tance of preconceptions of their own likely abilities, as well as aspects of physical and mental health. And, of course, they exhibit cultural aspects—language, dress, self-presentation, and so on—of the social class to which they already belong.

When they enter school, children quickly become labeled by teachers as "bright" or "dull." These are not technical terms, although they are used regularly by educators, psychologists, and politicians. They are also much used by behavioral geneticists, who argue that they reflect underlying genetic variation.

Pygmalion in the Classroom was a 1968 book (updated in 1995) by Robert Rosenthal and Lenore Jacobson summarizing the effects of teacher expectation on children's classroom performance. They had conducted an experiment in a public elementary school, in which teachers were told that certain children could be expected to be "growth spurters" based on the results of a psychometric test. In fact, the test was nonexistent, and the children were chosen at random. Nevertheless, those labeled children made massive gains in IQ scores over an eight-month period compared with controls. The expectancy effects have been confirmed in much additional research since.

Research in the 1970s and 1980s suggested that such attributions were on the basis of language accent, self-presentation, parental priming, and even facial appearance—that is, corresponding closely with social class background. Guy Claxton and Sara Meadows have summarized more recent research indicating a range of other criteria, including the following:

- Physically alert and energetic
- Strongly oriented to adults and alert to their presence
- Facial expressions
- Sensible responses for the classroom context
- Ability to maintain focus
- Articulateness
- Quick on the uptake
- Ability to sit still and listen to adults
- Greater ease and fluency with peers
- Ability to remember and make links to what has happened
- Proactive and inquisitive
- Greater perceptiveness about sensory details and patterns

From this, they suggest, it is clear that "bright" is a portmanteau word: "Being 'bright' is not a single thing; it is woven together from a number of separable developmental achievements, some social, some perceptual, some cognitive and some linguistic."[4]

It seems likely that, because they have read some genetics—or at least popular and media accounts—teachers will readily assume that those pre-education differences are at least partly innate. So they will behave toward children in different ways according to those preconceptions. Children soon assimilate the subliminal messages to reinforce self-efficacy beliefs already obtained from preschool experiences in family and neighborhood. So the early perceptions become self-fulfilling. But the process continues throughout schooling and higher education.

Carol Dweck has shown in her research that teachers readily and easily transmit the idea to the children themselves that achievement is due to being smart. Accordingly, children who fail tend to think that it is because they lack learning ability. Various studies have demonstrated these subtle effects of rank ordering in schools, even among children previously assessed by other teachers as of equal ability. One of these showed that students allocated to a higher rank have higher perceptions of their intelligence, higher expectations about their future careers, and receive more support from their current teachers. "If two students with the same ability have a different rank in their respective cohort, the higher-ranked student is significantly more likely to finish high school, attend college, and complete a 4-year college degree."[5]

As Jo Boaler similarly observed "Whenever ability grouping happens— whether students are told about the grouping and its implications or not—students' beliefs about their own potential change in response to the groups they are placed into."[6]

The self-fulfilling prophecy hits math-related subjects—and females— particularly. In one study, Donna K. Ginther and Shulamit Kahn concluded that "the core reason that women do not enter math-intensive fields . . . is the stereotypical beliefs of the teachers and parents of younger children that become part of the self-fulfilling belief systems of the children themselves from a very early age."[7]

This appears to be compounded later, in that females' entrance to math-related careers is discouraged by the beliefs of tutors that "raw" innate aptitude is required. Data from the Early Childhood Longitudinal Study in the United States has suggested that underrating girls' mathematics potential accounted for a substantial portion of the development of the mathematics achievement gap between boys and girls who had performed equally in the early grades.[8] Such gender prejudice may have been changing in recent years, but the point is the potency of the attribution processes.

In contrast, banishing "genetic" prejudices has been highly effective. As Boaler also says in his review, when schools abandon grouping by assumed ability and move to mixed or heterogeneous grouping, achievement and participation improves significantly. Likewise, in her book *Mindset* (that quickly became a *New York Times* bestseller), Carol Dweck summarizes her research showing that changing teachers' and children's "mindsets" about ability can dramatically improve motivation and achievement.

In seeming agreement with this view, the U.S. Department of Health and Human Services made changes to the Head Start programs (June 19, 2015). These are designed, they say, "to better support the ability of programs to serve children from diverse economic backgrounds, given research that suggests children's early learning is positively influenced by interactions with diverse peers."

There seems little doubt that these expectancy effects will be due, at least in part, to the culture of fatalism and pessimism created by psychologists of cognitive ability through the instrumentation of IQ. And they will contribute markedly to individual differences in test performances and school attainments. But now let us look at effects of the curriculum; what is being taught, and how; and how it reflects increasing anxiety, not about genuine learning but about maintaining an ideology of children's potential and a bogus equality of opportunity.

AN ALIENATING CURRICULUM

Covert selection in schools also takes place through a special brand of learning in the form of a set curriculum. That basic model of the school curriculum has permeated the developed world. But it is as ill designed for real education as taking fish out of water to teach them to swim, or birds out of the air to teach them to fly. In his *Autobiography* (1876), Charles Darwin notes that mathematics was repugnant to him from being devoid of meaning. School students have much the same problems today with so many areas of the curriculum.

The problem is that the curriculum—what is taught, and the teaching-learning process—is set up as a "test" of children's general learning potential when it is really a test of social background. Researchers have noted this immediately from the way that knowledge is specially prepared, with special features, to be learned in certain ways, for convenience of testing and then largely forgotten. Designed both to be of some obscure future use, and as a way of sorting out children on the basis of their basic learning potential, it boils down to a test, above all else, of their motivation, perseverance, and belief in their own ability.

One clue to this is the way that so much school knowledge is shorn of meaningful reference to the environments, economics, and social structures of children's actual communities, in forms that might help their development as responsible citizens. It is abstracted, refined, and packaged in a way that seems to minimize its immediate social and historical interest. Learning theory in education—how children are thought to learn—is haphazardly laced with a few child-centered concepts from developmental psychologists. But it is utterly dominated, however thinly disguised, by the traditional learning model in which individual minds are required to memorize knowledge "chunks" to be regurgitated in tests and exams.

School learning involves what educational theorist Jerome Bruner called "artificial made-up subjects." These come in specially packaged forms, in carefully controlled time slots, modules, and formats, and in fixed chunks and sequences that can be reproduced in multiple-choice or short-answer tests.

This form of learning is a very unusual, quite remote from knowledge as learned and used in the social and practical world. School knowledge is not "knowledge" as we know it in the everyday sense—as the knowledge we use in work and social intercourse all the time. Nor is it knowledge as we know it in the academic or scholarly sense. As a specially packaged variety, its learning occurs as disjointed fragments and is motivated, on the part of pupils, by long-term status goals, rather than current intrinsic interest.

But it is just right for testing dogged motivation for rote learning and regurgitation, valued not for its intrinsic worth as socially useful learning, but as a means to an end. In that tedious grind, as Berliner and Glass put it, schools come to assume "that learning must be incentivized, and issue rewards and punishments to this end . . . [so that] students are motivated to perform, either by avoiding sanctions or garnering positive reinforcement from their teachers. School then becomes not a place where creativity can flourish, but a place where creativity is extinguished."[9]

In a chapter in the *Encyclopedia of the Sciences of Learning* (2011), J. Scott Armstrong notes, likewise, how "tasks are often of little interest to students, feedback focuses on content (facts) rather than skills, and application is seldom addressed. Motivation is based on extrinsic rewards and punishments in a competitive environment."[10]

Most teachers do their best, against the institutional grain, to instill some social relevance into the process. But "subjects," especially math and science, become notoriously difficult and laborious in the context of the school curriculum. In his best-selling work, *Dumbing Us Down: The Hidden Curriculum of Schooling*, John Gatto shows how even reading and writing are made more difficult than they should be.

More than anything else, the process depends on the preconceptions of learning ability borne by pupils and their parents, and the motivation and parental push that goes with them. That it involves students' cramming by rote-memorizing has turned preparation for the Scholastic Aptitude Tests (SAT or, in the United Kingdom, SATs) into a highly lucrative commercial field. Many private companies and organizations now offer test-preparation and revision books in the form of easily remembered chunks (or "Bitesize," as the BBC revision website puts it).

Little wonder that pupils are turned off in droves. Surveys confirm that large majorities of pupils in school are bored most of the time, lack confidence, feel there is no point in working hard at school, do not enjoy school, and do not feel valued. It is hardly surprising that so many end up believing that they simply do not have the genes or brains to learn very much. In America, 9 percent of school-age children are now diagnosed with attention deficit hyperactivity disorder. There might be a connection. However, the process is being exposed for what it is in two contradictory ways.

THE DRIVE FOR ACHIEVEMENT

One of these is the intensified pressure on schools for improving attainments, as noted above. It is partly a reaction to growing social inequalities and the need to reinforce the equal opportunities ideology. And it is partly a fear of falling behind the developing economies, particularly those of East Asia. The superiority of the latter in the PISA ratings, as mentioned earlier in chapter 10, has led to many attempts, on both sides of the Atlantic to emulate them through school reforms.

It has resulted in more performance league tables, more test-focused teaching, more pressure on children, more homework, and more pressure on parents to get their children into the "best" schools. In the United States, the Race to the Top program, has increased the use of standardized testing for student, teacher, and administrator evaluations. These rank and label students, as well as teachers and administrators, according to the results of tests that are widely known to be imperfect.

In the United Kingdom, the schools minister has announced (August 2015) an extension of the "Chinese method of maths learning." It involves highly disciplined practice with the fine details of arithmetical processes (e.g., solving an equation) that pupils will then be able to repeat mechanistically, step by step, but it is largely devoid of meaningful content or social relevance.

Many educators are complaining about the narrow form of learning now becoming pervasive and the authoritarian atmosphere that it fosters in schools. They are warning that the intensification of schooling to be stressful "exam factories" is damaging the mental health of children. For example, in 2014, nearly one hundred educators from around the world signed a letter to Andreas Schleicher, head of PISA, attacking the OECD's PISA rankings and saying that the next round of tests should be canceled. They complain about the shift to "short-term fixes," about "alliances with multi-national for-profit companies," about the way the testing focus "harms our children and impoverishes our classrooms, as it inevitably involves more and longer batteries of multiple-choice testing, more scripted 'vendor'-made lessons, and less autonomy for teachers. It also further increases pupil and teacher stress levels" (letter published in the *Guardian*, May 6, 2014).

The view is reinforced by a letter from a Head of Year in a U.K. state school, who says that "real problems lie in the nationwide drive to train our young people to pass examinations and the demotion of their humanity below the data they generate. A proper, relevant, curriculum is decades away because of political dogma and the need for schools to find all means to maintain their pass rates. I would argue that our children are not being 'educated' at all" (letter to the *Observer*, July 5, 2015).

The kind of "attainments" fostered by the new regimes are illustrated in the U.S. Success Charter schools—publicly funded independent schools that have continued to expand across the United States, with supporters seeing them as a way of boosting standards in state education. The schools put great effort into teaching and motivating students to take tests. Discipline and sanctions for students are strict. Parents are quickly called in when there are problems.

The burden on teachers is enormous. They work long hours, which is difficult for those with children of their own. So many of the teachers are young recent graduates. And performance is closely monitored for student success. It means rapid promotion for teachers when students perform well. Otherwise, the teacher may be demoted to teaching assistant, or even removed if their performance does not improve. One consequence is that many teachers quit, complaining about the harsh atmosphere. By following a similar path, the United Kingdom is now experiencing record numbers of teachers leaving the profession.[11]

This illustrates where the intensification of the ideology of potential, as something "within" pupils, to be drawn out by pressure and "stretching," has already taken us in schools. I predict that it will only increase the inequalities that pupils already enter schools with. Indeed, research commissioned by the National Union of Teachers in the United Kingdom has already found that accountability measures such as league tables fail to reduce social class gaps in attainment and instead result in higher levels of stress. "Despite the government focus on reducing gaps, including pupil premium payments, the attainment gap at GCSE level [i.e., sixteen year olds] between pupils eligible for free school meals and those who are not has remained at about 27 percentage points throughout the last decade," the researchers found.[12]

Halley Potter, who studies charter schools at the Century Foundation, said that "Success Academy's strong test scores tell us that they have a strong model for producing good test scores," but she also says, "The conclusions that can be made from tests are limited."[13] That, it seems, is consistent with limited goals.

PSEUDO-ACHIEVEMENT

The other way in which the hidden curriculum of schooling is being exposed is simple. If school attainment is, indeed, an index of each child's learning potential, then we would expect a transfer into subsequent domains of life and learning. We have to remember that a test score is not just a "measure" of an individual's current status but also of his/her whole social background. If that background does not change or changes only a little into adulthood, then some correlation with future status is inevitable. In spite of that, relationships between the venerated school attainments and performance in later life are weak or difficult to demonstrate.

For example, we would expect that students with good grades or test scores would also perform better in higher education and the world of work. Across numerous studies, however, evidence for this is very thin. Every "measure" of ability or potential turns out to be a very poor predictor of performance outside the narrow confines of school learning.

Take, for example, the prediction of university performance from high school exam results. In the United States, the latter have consisted of SAT scores and more recently the Graduate Record Examination. Research has reported only small correlations, usually under 0.3, meaning that 90 percent of performance variation in higher education is not related to high school performance. A review, published in 2012, of thirteen years of previous research found only moderate average associations with university performance.[14] And of course even that may be due to noncognitive attributes, such as self-confidence and self-efficacy beliefs, and other social background factors, as explained earlier.[15] Indeed, the review by Michelle Richardson and colleagues, just mentioned, revealed a large correlation "for performance self-efficacy, which was the strongest correlate (of 50 measures)."[16]

Similar conclusions have been reached with the United Kingdom's A-level grades (at the end of high school). They have always been uncertain predictors of performance at university, as admissions tutors have long realized. Studies in the 1980s found that A-level grades have little predictive value for performance at university, either in medicine or in nonmedical subjects.[17] A study in the mid-1990s indicated that A-level grades accounted for only 8 percent of the variation in final degree performance on average. A study by King's College London a few years ago confirmed that, when picking the candidates with the best potential, universities might just as well toss a coin. The 2012 report by Richardson and colleagues, mentioned above, found that "in U.K. data, a small correlation was observed between A level points and university GPA ($r = 0.25$), again reflecting previous findings." An enquiry commissioned by the U.K. government noted that "many respondents were of the opinion that A levels do not seem to have a strong relationship to actual performance in higher education—and that those with the best grades can perform disappointingly (and vice versa)."[18]

In the United States, associations between high school and college grades have been a focus of much debate. One aspect of this has been inconsistencies between the different measures used as predictors. For example, as reported by the College Board in 2008, the correlation between SAT scores and high school grade point average (HSGPA) is only 0.28. As the Board says, "This finding suggests that the SAT and HSGPA may measure different aspects of academic achievement," though it does not suggest what those aspects may be.

Even predictions of high school grades for first-year college grades have never been very high (0.3–0.6). Correlations with final college grades fall off to about 0.3–0.4, but this is after a number of statistical corrections that always involve some guesswork. Suggesting they might be "veils of deception," one review responded to them as follows: "The ETS [Education Testing Service] has recently released new estimates of validities of the GRE [Graduate Record Exam] for predicting cumulative graduate GPA. They average in the middle thirties—twice as high as those previously reported by a number of independent investigators. It is shown . . .

that this unexpected finding can be traced to a flawed methodology that tends to inflate multiple correlation estimates."[19]

At least part of such correlations is a foregone conclusion: to some extent college exams will be testing the same knowledge already learned at high school. But part of it will be the mutual reflection of noncognitive aspects of exam readiness: social-class-related anxiety levels, self-confidence, and self-efficacy beliefs, all of which are important performance variables (see chapter 3). In other words, not so much "less able"—the usual interpretation of test scores—as less "prepared."

This is why much of the debate surrounding the SAT as a college predictor has been about its fairness to working-class and female students, and why it has undergone frequent revisions (the next one being due in 2016). Calling the use of SAT scores for college admissions a "national scandal," Jennifer Finney Boylan, an English professor at Colby College, argued in the New York Times (March 6, 2014) that we need to look at the "complex portrait" of students lives: "what their schools are like; how they've done in their courses; what they've chosen to study; what progress they've made over time; how they've reacted to adversity." Similarly, Elizabeth Kolbert wrote in the New Yorker that "the SAT measures those skills—and really only those skills—necessary for the SATs."[20]

In sum, the so-called measures of potential seem to be of dubious merit. Just as worrying, however, is that in the conventional school curriculum, children are being cheated out of their futures in more ways than one. In lacking relevance and the deeper structures essential for learning, the conventional curriculum actually denies all children the learning they most need for fuller participation in a democratic society.

By relevance, I mean learning how their local and national economies work, how social institutions operate, the social and technical nature of economic production and services, how local and national administrations function, the nature of civic rights and responsibilities, the true nature of science, artistic creativity, historical movements, and so on. In sum, young people are being deprived of access to the broader understanding of our society, of the perspective and vision that comes with true learning, and, thereby, their true learning powers and development of potential.

Sadly, the "means to an end" mentality is widespread. In her book, The Smartest Kids in the World: And How They Got That Way, Amanda

Ripley largely extolls the "hard-work" Chinese and South Korean educational methods: "In an automated, global economy, kids needed to be driven; they need to know how to adapt, since they would be doing it all their lives."[21]

In my view there is a sad irony here. Is this the kind of potential we want our children to develop? Should we not be promoting in them the education and intelligence through which they can cultivate a more congenial world for all, rather than become the toilers in someone's "hamster wheel" (an image Ripley also uses)? Susan Engel puts the contrasting view that American schools have allowed "the pursuit of money to guide our educational practices," and in so doing "we have miseducated everyone."[22]

LITTLE CONNECTION WITH JOB PERFORMANCE

As just explained, we would expect correlations between school and higher education performance because they are partly testing the same learning rather than potential. Moreover, since educational level is a condition of level of entry to the job market, there is bound to be some association between educational attainment and occupational level. School or college performance, that is, automatically predicts *level* of occupation. A more important test of whether they reflect genuine potential, as opposed to a host of other background variables, is whether they predict job performance.

As explained in chapter 3, it has always been the case that data have to be stretched a great deal to demonstrate the existence of any relationship between test or exam performances and job performance. I described how investigators have resorted to almost any "test" as surrogates of potential (or "g") to find such a relationship. Among these have been educational attainment test scores, including simple reading scores. Whatever they use, the correlations have always been tiny, around 0.2. And the causes of that cannot be distinguished from noncognitive factors known to affect such correlations.

Likewise, surveys going back to the 1960s have routinely shown that neither school nor university grades are good predictors of performance in the world of work. In his review, mentioned above, J. Scott Armstrong has put the correlations, six or more years after graduation, as low as 0.05.[23] Higher-performing pupils do not tend to become "high-performing" adults. Conversely, the vast majority of high achievers in the real world, as adults, did not stand out in school.

A legitimate objection may be that correlations will be underestimates, because many of those who took the school exam are not present among those assessed for job performance—they have been selected out. In statistical terminology, this is called "restriction of range." However, as mentioned in chapter 3, even when corrections are made, correlations are small or confusing.

For example, excellent high school grades are required for entry to medical school. Given that medical studies involve a real accretion of related knowledge over time, we would expect there to be some correlation from exam to exam. That appears to be the case. When we turn to actual medical practice, however, the situation is different. A study in the United Kingdom (also mentioned in chapter 3) reports small and statistically nonsignificant correlations between A-level results and the Practical Assessment of Clinical Examination Skills several years later. Also, correlations between A-level scores and having been promoted to the Specialist Registrar (or senior doctor) grade were low (below 0.2) or not statistically significant.[24]

The most surprising result is the tenuous association between school-acquired knowledge with useful real-life knowledge. It has frequently been shown how high school and university students, steeped in objectified curriculum knowledge, have difficulty translating it to, and thus understanding, real-life practical situations in corresponding domains. In his book, *The Unschooled Mind*, Howard Gardner described the results of a large number of studies, on both sides of the Atlantic, as follows: "Perhaps most stunning is the case of physics. . . . Students who receive honours grades in college level physics are frequently unable to solve basic problems and questions encountered in a form slightly different from that on which they have been formally instructed and tested. . . . Indeed, in dozens of studies of this sort, young adults trained in science continue to exhibit the very same misconceptions and misunderstandings that one encounters in primary school children. . . . Essentially the same situation has been encountered in every scholastic domain in which inquiries have been conducted."[25]

Among employers there is now considerable uncertainty about whether college grades predict potential in a job. Many will agree—and surveys suggest—that GPA is a useful indicator of who can "hit the ground running." But, again, the correlation may be due to noncognitive factors. J. Scott Armstrong has claimed that the relationship between grades and job performance is low and is becoming lower in recent studies. He cites research from a variety of occupations that suggests that those with good college grades did no better in the job than those without.[26]

Others are also skeptical. For example, Trudy Steinfeld, executive director of the Wasserman Center for Career Development at New York University, has deplored the fix on grade scores as longer-term predictors of job performance: "Nobody even cares about G.P.A. after a few years," she says.[27] That is also the view of Laszlo Bock, a vice president of human resources at Google. In an interview with the *New York Times* (June 13, 2013) he said, "One of the things we've seen from all our data crunching is that GPAs are worthless as a criteria for hiring, and test scores are worthless—no correlation at all except for brand-new college grads, where there's a slight correlation. Google famously used to ask everyone for a transcript and GPAs and test scores, but we don't anymore, unless you're just a few years out of school. We found that they don't predict anything. . . . What's interesting is the proportion of people without any college education at Google has increased over time as well."[28]

They are not alone. As reported by the BBC (January 18, 2016), publisher Penguin Random House decided that job applicants will no longer be required to have a university degree. The firm wants to have a more varied intake of staff, because there is no clear link between holding a degree and performance on a job. This announcement follows a series of other companies dropping academic requirements for applicants. For example, Ernst and Young has scrapped its former threshold for certain A-level and degree requirements and is removing all academic and education details from its application process. PriceWaterhouseCooper has also announced that it would stop using A-levels grades as a threshold for selecting graduate recruits.

As with IQ then, which similarly has little if any association with job performance (see chapter 3), educational grading of individuals seems to have little intrinsic value. Instead the function of both of them is the ideological one of legitimizing and maintaining a specific vision of human potential and, through that, the class structure of society. That is what the expensive hunt for "genes" and "brain structures" related to education is really about.

That is not to say that education per se is not important. Of course, over thirteen or more years of schooling, we would expect something to stick. And, indeed, that is the case. Many, if not most, people are grateful to schooling for their learning of many things. Many from poor backgrounds have found inspiration in exposure to knowledge resources, especially in developing countries.

When schooling is an instrument primarily for the identification of supposed potential, however, it is suppressive of learning anything deeper than exam fodder. Then it is usually only identifying social background. This is why the appeal to brain sciences, or to genetic "chips," in order to drive children more forcibly yet onto such a treadmill is so worrying.

In contrast, when alternative goals and approaches are attempted, learners can be highly motivated and schooling highly rewarding. Let us now have a look at some possibilities.

ALTERNATIVE EDUCATION

Most attempts to find alternatives to a crushing curriculum recognize sources of potential other than genes. They realize how potential emerges in individuals as dynamical systems functioning in dynamical coalitions. Individuals are not encapsulated learning machines. As a review by Yu Yuan and Bill McKelvey explains, "While the *acquisition metaphor* of traditional learning theories stresses the individual mind and what goes 'into it,' the *participation metaphor* of the situated theory of learning shifts the focus to the evolving bonds between the individual and others."[29]

Berliner and Glass quote John Dewey, who suggested, quite logically, that the best preparation for social life was to actually engage in social life.

Dewey was also less interested in what a child learned, the official school outcomes, than that the child learned how to learn.[30] This would seem of increasing importance in preparation for a changing but democratic society.

After all, human learning evolved for knowledge abstraction in socially meaningful contexts. Concepts are learned because they are intrinsic to goals shared globally in a group, as well as between novice and expert, teacher and learner. Countless complex social rules and customs are acquired in such a supportive framework. It is the kind of learning-for-cooperation that our uniquely huge human brains evolved for. Parents do not send their children to school to learn how to speak.

When people, and children, really want to learn something for a socially relevant activity, they do so, with their "good enough" systems, very quickly and easily. The widely used example is that of human language, consisting of highly abstract grammar and other rules of use that the vast majority of infants have acquired before starting school. Indeed, adult experts in a domain, including practicing mathematicians, scientists, and engineers, will often say how they found the subjects difficult in school, but learning later flourished in real contexts.

Taking school math as an example, David Carraher and Analucia Schliemann have pointed out (as I also noted above) that the emphasis in school math is on the computational rules and seldom on the meaning of the process. In contrast, informal math (as in young street traders) preserves meaning, and is used and expressed in a diversity of ways for different problems. The math knowledge may thus be quite different across the two situations and thus transfer to novel problems with different ease.

Practical math was the form in which it was passed across generations long before formal schooling. In a modern world, we need both sides of the coin, the practical and the abstract. But in the typical curriculum, it has become too one sided. Schooled individuals are coached in rules, but the search for rules can be a source of improved performance when meaning plays a more important role, Carraher and Schliemann say. A teaching approach that combines both aspects would obviously approximate, rather more closely, math knowledge as it has evolved among mathematicians and is used in science and engineering and other practical goals.[31]

This is why there have been many explorations of alternatives to the traditional curriculum and its hidden agenda. Many efforts exist in schools to make lessons more realistic, and curriculum reviews and innovations abound. But these, all too often, consist of made-up "projects," involving some sort of simulation of real activities, and pretend problems. Though laudatory, they are still divorced from real social contexts.

The point is to find ways through which wide curriculum objectives—currently encapsulated in "subjects" like physics or biology—could be found in meaningful contexts. In one scheme (reported a few years ago by Sally Goodman in the journal *Nature*), high school students became engaged with university research teams to build three-dimensional models of anthrax toxins in school. Researchers took time out to discuss genuine problems with the students during lab visits. And the scientists were rewarded with really helpful models: "an excellent tool for testing hypotheses," as one of them put it.

In another scheme in Milan (reported by Giulio Pavesi and colleagues in the journal of the *European Molecular Biology Organisation* in 2008), high-school students became involved with university researchers in genetic analyses. They developed much deeper understanding of connections between the theoretical and the practical. In the collaborative context, also, both students *and* their teachers developed a far more fertile understanding of the science topic, as well as the inspiration to continue.

In yet another scheme, students helped build equipment for experiments related to particle physics. And in a project in Germany, students spent two days a week at a research center, where they were taught by their own teachers alongside researchers. In Britain, Nuffield-sponsored projects have invited students to work with researchers over the summer on a one-to-one basis, even resulting in scholarly publications.

In all these cases, students suddenly felt they were part of a team and were rewarded with the creativity and excitement of genuine research. Connecting the theoretical with the practical fostered genuine learning that students will probably not forget. Students from all social backgrounds became more self-confident, and more challenging in classes. They also encouraged a tremendous groundswell of interest among peers. Researchers noted significant improvement in their own ability to communicate their work (and thus, perhaps, to think about it more clearly and explicitly).

Such students, so far, are a fortunate few. This is no doubt because such real learning projects are subversive to the other aims of education. But they need not be so special—if only we could drop the ideology of innate potential and the exam factory model of schooling.

In such real-life problems, all the aims and objectives of any acceptable school curriculum could be worked out. The above experiments need to be extended to workplaces and institutions of all kinds. Association with such meaningful contexts would not only develop abstract concepts in a grounded way, but also engender economic sense, a sense of worthwhileness about activities in schools, as well as civic identity and responsibility.

Such ideas are now being explored around the world as radical solutions to the current dismal state of education. For example, the Innovation Unit in London is working with schools on some experimental approaches. On their website, they refer to research from the United States showing that "Learning Through REAL Projects" (REAL being an acronym for Rigorous, Engaging, Authentic Learning) has significant impact on pupil development and engagement. "REAL Projects allow teachers to formulate lessons and activities around a single complex enquiry, and require students to produce high quality outputs with real-world application . . . students acquire subject knowledge systematically as part of the process of producing outputs, and are assessed for the quality of the work produced."[32]

In another document, they say that prescription for a twenty-first-century education system must include the following ingredients:

Students learn through "meaningful projects . . . of genuine value (to themselves, to the community, or to a client)."

School is a "base camp" for enquiry not merely "the place you go to acquire knowledge" but also "for enquiries that will take students into their communities, and online."

Learners collaborate in their learning, rather than "consuming" it.

Education takes advantage of digital technologies and helps students become both digitally literate and digitally adept.[33]

I see no reason why these goals could not be achieved in various ways involving local communities. Plenty of genuine problems are available from the practical contexts of local producers, practitioners, and residents. They can be presented to school students such as to require thought, knowledge, research, and practical action in the detached context of a school or college. The news agent may have a delivery organization problem, the regional council a reporting problem, the plastics factory a chemistry problem, the steelworks some physics problems, the health center a health education problem, the shirt factory a design problem, the farmer all kinds of botanical and zoological problems, and so on.

All the aims and objectives of any accepted curriculum could be worked out in such contexts. Teachers would have the far more challenging task of scaling the problem to a suitable level, identifying curriculum objectives in them, and then organizing the resources to achieve them. In addition, parents, communities, and business managers would need to have much greater sense of responsibility for what goes on in their local schools. And at least some parts of workplaces would need to be safe for students to visit, move around in, and learn. This would not only help develop abstract concepts in a grounded way, but also engender economic sense, a sense of worthwhileness about activities in schools, as well as civic identity and responsibility.

Above all, such schemes would help avoid the semi-enforced digestion of prepackaged, "dead" skills and knowledge. They would transform the curriculum from the slog of motivation and persistence it largely is and that currently does so much damage to people's real potentials.

Meanwhile, the myth of our age—the idea that school performance is a measure of children's true learning potential—must be purged. Parents should protest against it and its subversive social function. They should refuse to engage with a game that is cheating most of our children of their self-belief and their learning potentials, while creating the inequality that politicians pretend to be mystified about.

There will be much resistance to such ideas. They threaten the current ideology of individual differences in potential. However, human history is replete with cases in which envisaged constraints on potential have been overcome by lifting the ideology that defines it. I need only mention the recoveries from slavery, the sudden discovery of abilities of women to run economic production in times of war, and advances of working-class organizations in times of extreme oppression. Indeed, without potential in precisely that form, there would, quite simply, have been no human history.

Strangely enough, much of the pressure for more inclusive, cooperative forms of organization is now coming from commerce and industry. For example, organization management specialist Charles Handy has spoken of the benefits of "cognitive enfranchisement" of workers at all levels of production. In their 50 Myths, Berliner and Glass say that "modern companies, including modern factories, have begun to realize the potential of collaboration and have incorporated this method of work in to their daily routines."[34]

Michael Norton has more recently summarized the "decreasing motivation and labor productivity, impairing decision making, and increasing ethical lapses" that arise with increasing inequality in workplaces. He describes how behavioral research supports the benefits of what he calls "unequality."[35] Jodie Berg speaks of a "symbiotic vision" that arises from alignment of personal and organizational consciousness: "When alignment is felt through the sense of the greater purpose, there is a deep, almost spiritual, commitment to making the world a better place and helping the organization contribute to that."[36]

The organization Psychologists against Austerity have also earmarked the general psychological outcomes of living in a healthy, well balanced society and economy as:[37]

Agency
Security
Connection
Meaning
Trust

These are all aspects of the natural state of humans existing in the cultural dynamics they were evolved to assimilate. But they are thwarted by those who will maintain a gross imbalance of social intelligence and, thereby, limits to human potential.

12

SUMMARY AND CONCLUSIONS

uman potential—genes, biology, brains, intelligence and education, and so on—forms a big subject in a big area, and we have now covered a lot of ground. So a summary of some sort might be helpful. My Preface offered a brief outline of intent across the chapters. Here I summarize—albeit almost as briefly—how it has been achieved.

Opening comments in chapter 1 described the importance of the subject of human potential in human affairs, but how understanding of it remains backward. For decades, discourse has mixed ideology and science and successfully contributed to a culture of inequality that retards rather than promotes the sum total of human potential.

The chapter attributed much of the problem to the vagueness of fundamental concepts. This was illustrated by the current concepts of intelligence, the gene, and the brain. In spite of the enthusiastic application of the new technologies—chiefly gene sequencing and brain scanning—I showed how the same vagueness has muddied the results and also produced the nonreplications that plague the area. More recent responses illustrate how the hype crumbles under critical analysis because of such lack of substance. Some foretaste of the more substantive criticism to follow was offered, and the chapter ends with some foreshadowing of new perspectives now emerging.

The next two chapters critically analyzed the two mainstays of the current—and, indeed, older—claims about genes, environments, and intelligence. First is the remarkable, almost theological, concept of the gene on which current claims about "genes for intelligence" are based: remarkable because, even at first glance, it has little correspondence with real genes. Chapter 2 then described the many assumptions that have to be incorporated into statistical models to uphold a particular picture of genes and intelligence. These were exposed one by one and show the final picture to be a rather weird concoction of unlikely, even bizarre, conjectures. The data—most prominently from twin studies—are easily explained by the clear falsity of such assumptions.

Chapter 3 described something similar for intelligence testing, as seen in the IQ test. Remarkably, such tests have no test validity in the way that ordinary scientific and medical instruments have—that is, we do not know what they measure. In the absence of a proper scientific model of intelligence it is, of course, impossible to know. Instead, I show how the tests are constructed on a far more subjective basis. That is to assume in advance which people have more or less intelligence—and then construct the test to agree with it. The rest of the chapter examined attempts to gloss over that subjectivity; challenged the many defenses of it; and described the various ripples through, and aftermaths in, our societies, their institutions, and their real people.

Chapter 4 introduced the alternative new view. It attempted to bring together many recent findings and theoretical advances in a coherent picture. It explained that a dynamic systems view of living things is needed, because the standard picture of the genes—as fixed codes for form and variation—would be useless in the rapidly changing environments that most organisms encounter. Instead what is needed are adaptable systems that can develop suitable forms and variations in the course of life. The chapter explained that living systems can do this by assimilating the informational structure in the environment as experienced. Hence we got "intelligent systems" even at the origins of life, but more spectacularly in their evolution into more complex forms in increasingly changeable environments.

Such systems create potential, rather than merely expressing it; they generate far wider, but more useful, variations than could be produced by the standard genes-plus-environments model. The chapter described the many implications of these new foundations for understanding potential. For example, in intelligent systems, we would expect to find little relationship between variation in genes and variation in forms and behaviors—just as research finds. Another is that intelligent systems are a much more exciting feature of evolution than fixed adaptations to fixed niches, the preoccupation of neo-Darwinism. Yet another is that this dynamical perspective offers, at last, the foundations of a genuine theory of intelligence and its evolution.

Chapter 5 described developmental systems as an evolutionary extension of the same dynamic logic: that is, the regulation of components in single cells has been extended to regulate interactions among cells in order to deal with more changeable environments. Many illustrations were given of the role of that logic in transforming that original "speck" of matter into bodies and brains. A vast variety of cells thus emerges; each type in just the right place at just the right time, derived from the structural information between them and utilizing genes from a common genome.

So development itself is an adaptable intelligent process, not a fixed program. That adaptability is seen in the "canalization" of some avenues of development and the extreme plasticity of others (especially in brains). In development, genes are used as servants, according to current needs, not its masters: functional potentials are not merely expressed; they are made.

Intelligent developmental systems thus add a new dynamical level—an intercellular one—to those already evolved in the single cell. The coordination of numerous players as a team, in an ever-changing environment, is what we call physiology. So chapter 5 also described how physiology amplifies and extends developmental functions and becomes a new intelligent system—a new level of intelligence—dealing with environmental change on a life-long basis. Its dynamic processes coordinate harmonious responses from its many parts, whether these are cell or tissue processes, such as hormone production, or gross body movements. Thus they create the major life transitions (e.g., metamorphosis in reptiles and insects, or puberty in mammals). And they constantly recalibrate the system as needed.

The chapter also explained how the conventional notion of individual differences in mental ability was based on intuitions from physiology. However, when individuals are measured for biomarkers of physiological function, it is found that the vast majority operate perfectly normally within very wide ranges. That is, whatever the variation of detail, the whole is "good enough" for doing what it needs to do.

Variation in intelligence is conceived as some kind of "brain power." Yet brain studies have failed to construct an overall, well-integrated, theory of what the brain is for. Chapter 6 explained how brains further evolved from physiology as even more intelligent systems in more changeable environments. It explains how their primary function is to abstract (statistical) structural information from rapidly changing environments. The information is assimilated as dynamical attractors (or structural grammars) and used to constantly create novel responses. These are analogous to those operating in cells at the molecular level. Only now they operate across cells (neurons), with a specialized form of communication, deeper information, and many times faster. Much experimental work was described in support of that view.

How cognition is related to brain activity was taken up in chapter 7. Among psychologists, there is little agreement about the true nature of cognition, its evolutionary origins, and its forms and variations. The chapter showed why the dominant models have been less than convincing. I explained how their drawbacks stem from misunderstandings of the nature of experience and the true functions of cognition.

Accordingly, the chapter explained how cognition is an emergent phenomenon of the nonlinear dynamical processes of the brain, dealing in statistical patterns, or structures and not in specific contents like symbols, features, or images. The patterns reflect the deeper correlations among interacting variables, not shallow associations. They are structures that can be constantly updated, and they create novel constructs out of variable and sparse "data," imparting great predictability for interpretation and action.

It is important to realize how those constructs are now *cognitive* entities, not merely neural ones. As multidimensional, spatiotemporal discharges, they enter into a new emergent level of regulation with one another, creating new properties of life in the process. They are different from the brain in the same way that molecules become different from their constituent atoms. One of those properties is that of reflective abstraction, which discovers deeper structure in the world. It emerges from coalitions of attractors that take the system far beyond current experience, permitting more intelligent action on it than simple associations ever could. The chapter explained how key components of cognition—

perception, conceptual categorization, learning, knowledge, memory, and thinking—are properties of attractor coalitions at successive levels of emergence.

The evolved intelligent systems that create such cognition in brains are brought into still higher coalitions in social groups. Chapter 8 reviewed research on the kinds of "epicognition" that result, from swarm intelligence in bees and ants, through bird flocks and fish schools, to the variety of groups in mammals. It showed how the same logic of evolving dynamic systems has been most useful in describing these interactions in cognition and behavior.

The chapter also questioned the popular idea that the evolution, brains, and social cognition of apes somehow represents the pathway to humans. This notion seems to have arisen from weak ecological analyses. Evolving as social hunters, true cooperation was demanded far more intensively in humans than in even the most "social" primate groups, presenting far greater demands on cognitive systems.

This deeper ecological analysis was accordingly taken up in chapter 9. It explained how the need for intense cooperative action between brains created a new intelligent system—the socio-cognitive or cultural one— that sets humans apart from apes and other mammals. I showed how the reflective abstraction through this new attractor landscape has produced such enormous creativity of cognition and behavior—so much so that humans, unlike all other species, are freed from specific environments (or niches).

The realization that humans are obligatory users of "cultural tools" and develop from the outside in, as much as the inside out, has important implications for intelligence (IQ) testing. Claims to be measuring brain or cognitive "power," independently of cultural background, are seriously misleading. I also suggest that if human intelligence systems really were a simple quantitative trait, like height or weight—as behavioral geneticists suggest—then humans would still be confined to a single niche like the apes.

All of that has profound consequences for understanding social inequalities and for intervention programs aiming to ameliorate them. Chapter 10 considered the "models" of genetic and environmental interventionism that have figured in research and some of the programs

deduced from them. It explained that these have been less than successful because they assume a linear causal model, as in a "horticultural"
view of the child. A dynamical model, however, places the problem not
in a particular social class (and all its supposed deficiencies) but in the
class system as a whole. This view is supported by increasing numbers of
studies that show the inhibiting and distorting effects of class structure
on the development of individuals' potential.

Education is often described as the cultivator of potential and the "engine of social justice": an institution in which equal opportunities allow
natural potentials to be expressed and find their own innate limits.
Chapter 11 showed that such descriptions are misleading. Schooling largely
functions by simply reproducing the class structure of society across
generations. And it does that by false attributions of learning ability.
First, children are labeled as soon as they enter school, and a self-fulfilling
process swings in to encourage some, but discourage others, largely on
the basis of their class backgrounds. Second, a largely boring, irrelevant
curriculum is imposed on young people, making progress dependent on
motivation and self-confidence, which is in turn dependent on home
background and parental push.

The chapter argues that this hidden agenda of schooling is evident in
swathes of findings contradicting the institutional claims. First, achievement measures are poor predictors of later achievement either in subsequent education or the world of work (or for anything other than passing
exams). Second, most young people leave the institution with a poor
understanding of the world in which they live and will inherit—and even
of their own academic discipline. The chapter duly ended with possible
solutions, based on principles of cognition and learning described in the
previous chapters.

The positive messages from these chapters will hopefully help dispel
the fatalism, pessimism, and sheer "bad science" engendered by genetic
and brain reductionist approaches. These have done more to perpetuate
an ideology of inequality than to challenge it. The positive alternative is
to recognize that the dynamical biological constitutions of the vast majority of humans will be "good enough" for participation at all levels of
social existence. The challenge is to recognize and dispel the ideologies
and other forces that prevent the realization of the positive alternative.

NOTES

PREFACE

1. E. F. Keller, "From Gene Action to Reactive Genome," *Journal of Physiology* 592 (May 2014): 2423.

1. PINNING DOWN POTENTIAL

1. Open Science Collaboration, "Estimating the Reproducibility of Psychological Science," *Science* 349 (August 28, 2015): 943–945, 943, http://dx.doi.org/10.1126/science .aac4716.
2. For a brief overview, see E. Rhodes, "Replication: Is the Glass Half Full, Half Empty, or Irrelevant?" *Psychologist*, March 9, 2016.
3. D. Sarawitz, "Reproducibility Will Not Cure What Ails Science," *Nature* 525 (September 2015): 159.
4. F. Galton, *Inquiry into Human Faculty and Its Development* (London: Macmillan, 1883), 199.
5. D. A. P. Delzell and C. D. Poliak, "Karl Pearson and Eugenics: Personal Opinions and Scientific Rigor," *Science and Engineering Ethics* 19 (September 2013): 1057–1070.
6. L. M. Terman, "Feeble Minded Children in the Public Schools of California," *School and Society* 5 (1917): 161–165, 162.
7. A. Cohen, *Imbeciles: The Supreme Court, American Eugenics, and the Sterilization of Carrie Buck* (New York: Penguin, 2015); T. C. Leonard, *Illiberal Reformers: Race, Eugenics and American Economics in the Progressive Era* (Princeton, N.J.: Princeton University Press, 2016). Cohen points to a 1913 *New York Times* article headlined "Social Problems Have Proven Basis of Heredity." How little things have changed. (Thanks to Jay Joseph for this snippet.)
8. M. C. Fox and A. Mitchum, "Confirming the Cognition of Rising Scores," *PLoS One* 9 (May 2014): e95780.

9. I. J. Deary, *Intelligence: A Very Short Introduction* (Oxford: Oxford University Press, 2000).

10. For references and full discussion, see K. Richardson and S. H. Norgate, "Does IQ Measure Ability for Complex Cognition?" *Theory and Psychology* 24 (December 2015): 795–812.

11. Royal Society, *Brain Waves Module 2: Neuroscience: Implications for Education and Lifelong Learning* (London: Royal Society, 2011), 3.

12. S. Oyama, *The Ontogeny of Information* (Cambridge, Mass.: MIT Press, 1984), 31.

13. J. G. Daugman, "Brain Metaphor and Brain Theory," in *Philosophy and the Neurosciences*, ed. W. Bechtel et al. (Oxford: Blackwell, 2001), 23–36.

14. L. S. Gottfredson, "What If the Hereditarian Hypothesis Is True?" *Psychology, Public Policy, and Law* 11 (May 2005): 311–319.

15. A. W. Toga and P.M. Thompson, "Genetics of Brain Structure and Intelligence," *Annual Review of Neuroscience* 28 (July 2005): 1–23, 17.

16. See E. Yong, "Chinese Project Probes the Genetics of Genius: Bid to Unravel the Secrets of Brainpower Faces Scepticism," *Nature* 497 (May 2013), 297–299.

17. M. R. Johnson et al., "Systems Genetics Identifies a Convergent Gene Network for Cognition and Neurodevelopmental Disease," *Nature Neuroscience* 19 (January 2016): 223–232.

18. C. A. Rietveld, S. E. Medland, J. Derringer, J. Yang, T. Esko, et al., "GWAS of 126,559 Individuals Identifies Genetic Variants Associated with Educational Attainment," *Science* 340 (June 2013): 1467–1471. See also N. M. Davies et al., "The Role of Common Genetic Variation in Educational Attainment and Income: Evidence from the National Child Development Study," *Scientific Reports* 5 (November 2015): 16509, doi: 10.1038/srep16509.

19. K. Asbury and R. Plomin, *G Is for Genes* (London: Wiley, 2014), 12.

20. G. Davies, A. Tenesa, A. Payton, J. Yang, S. E. Harris, et al., "Genome-Wide Association Studies Establish That Human Intelligence Is Highly Heritable and Polygenic," *Molecular Psychiatry* 16 (October 2011): 996–1005, 996.

21. This is what Eric Turkheimer calls Jay Joseph over the latter's exhaustive critique of twin studies; E. Turkheimer, "Arsonists at the Cathedral," *PsychCritiques* 60(40) (October 2015): 1–4, doi: http://dx.doi.org/10.1037/ a0039763; and J. Joseph, *The Trouble with Twin Studies* (London: Routledge, 2014).

22. E. Turkheimer, "Commentary: Variation and Causation in the Environment and Genome," *International Journal of Epidemiology* 40 (June 2011): 598–601; E. Turkheimer, "Arsonists at the Cathedral."

23. S. J. Gould, *The Mismeasure of Man* (New York: Norton, 1981), 272.

24. O. Zuk et al. (2012). "The Mystery of Missing Heritability: Genetic Interactions Create Phantom Heritability," *Proceedings of the National Academy of Sciences* 109 (January 2012): 1193–1198.

25. P. Wilby, "Psychologist on a Mission to Give Every Child a Learning Chip," *Guardian* (February 18, 2014), www.theguardian.com/education/2014/feb/18/psychologist-robert -plomin-says-genes-crucial-education.

26. E. Turkheimer, "Commentary," 600.

27. A. C. Love, "Reflections on the Middle Stages of Evo-Devo," *Biological Theory* 1 (January 2007): 94–97, 94.

28. L.C. Mayes and M. Lewis, *The Cambridge Handbook of Environment in Human Development* (Cambridge: Cambridge University Press, 2012), 1; D. Goldhaber, *The Nature Nurture Debates: Bridging the Gaps* (Cambridge: Cambridge University Press, 2012), 8.

29. K. Asbury and R. Plomin, *G Is for Genes*, 102–103.

30. G. Claxton and S. Meadows, "Brightening Up: How Children Learn to Be Gifted," in *Routledge Companion to Gifted Education*, ed. T. Balchin, B. Hymer, and D. Mathews (London: Routledge, 2008), 3–9, 5.

31. R. J. Haier, interview with the National Institute for Early Education Research (December 2008), http://nieer.org/publications/richard-j-haier-reading-young-minds-unlock-their-possibilities.

32. J. J. Yang, U. Yoon, H. J. Yun, K. Im, Y. Y. Choi, et al., "Prediction for Human Intelligence Using Morphometric Characteristics of Cortical Surface: Partial Least Squares Analysis," *Neuroscience* 29 (April 2013): 351–361.

33. P. A. Howard-Jones W. Holmes, S. Demetriou, C. Jones, C., O. Morgan, et al., "Neuroeducational Research in the Design and Use of a Learning Technology," *Learning, Media and Technology* 40 (September 2014): 1–20, 1.

34. Royal Society, *Brain Waves Module 2: Neuroscience*, 3.

35. British Psychological Society, www.bps.org.uk/events/neuroscience-coaching.

36. P. A. Howard-Jones, "Neuroscience and Education: Myths and Messages," *Nature Reviews Neuroscience* 15 (October 2015): 817–824, 818.

37. M. Carandini, "From Circuits to Behaviour: A Bridge Too Far?" in *The Future of the Brain: Essays by the World's Leading Neuroscientists*, ed. G. Marcus and J. Freeman (Princeton, N.J.: Princeton University Press, 2014).

38. C. Bennett, A. A. Baird, M. B. Miller, and G. L. Wolford, "Neural Correlates of Interspecies Perspective Taking in the Post-Mortem Atlantic Salmon: An Argument for Multiple Comparisons Correction," *Journal of Serendipitous and Unexpected Results* 1 (2010): 1–5, www.improbable.com/ig/winners/#ig2012.

39. D. M. Barch and T. Yarkoni, "Introduction to the Special Issue on Reliability and Replication in Cognitive and Affective Neuroscience Research," *Cognitive, Affective and Behavioral Neuroscience* 13 (December 2013): 687–689, 687.

40. R. E. Nisbett, J. Aronson, C. Blair, J. Dickens, J. Flynn, et al., "Intelligence: New Findings and Theoretical Developments," *American Psychologist* 67 (February–March 2012): 130–159, 130.

41. M. Rutter and A. Pickles, "Annual Research Review: Threats to the Validity of Child Psychology and Psychiatry," *Journal of Child Psychology and Psychiatry* 57 (March 2016): 398–416, 406.

42. J. S. Bowers, "The Practical and Principled Problems with Educational Neuroscience," *Psychological Review* (March 2016): http://dx.doi.org/10.1037/rev0000025.

43. S. Rose, "50 Years of Neuroscience," *Lancet* (February 14, 2015), 599.

44. J. Roiser quote on p. 285. A version of the G. Marcus article appears in print on June 28, 2015, on page SR12 of the New York edition.

45. For discussion, see S. J. Schwartz, S. O. Lilienfeld, A. Meca, and K. C. Sauvigné, "The Role of Neuroscience Within Psychology: A Call for Inclusiveness over Exclusiveness," *American Psychologist* 71 (January 2016): 52–70.

46. O. James, *Not in Your Genes* (London: Vermillion, 2016).

47. R. M. Lerner, "Eliminating Genetic Reductionism from Developmental Science," *Research in Human Development* 12 (October 2015): 178–188, 185; see also R. M. Lerner and J. B. Benson, "Introduction: Embodiment and Epigenesis: A View of the Issues," *Advances in Child Development and Behavior* 44 (2013): 1–20.

2. PRETEND GENES

1. C. Burt (1959). "Class Differences in Intelligence," *British Journal of Statistical Psychology* 12 (May1959): 15–33.

2. R. A. Fisher, "On the Correlation Between Relatives on the Supposition of Mendelian Inheritance," *Transactions of the Royal Society of Edinburgh* 52 (1918): 399–433, 433.

3. R. A. Fisher, "Limits to Intensive Production in Animals," *Journal of Heredity* 4 (September 1951): 217–218. For critical analysis of the difficulties, see J. Tabery, *Beyond Versus: The Struggle to Understand the Interaction of Nature and Nurture* (Cambridge, Mass.: MIT Press, 2014).

4. L. J. Kamin, *The Science and Politics of IQ* (New York: Erlbaum, 1974). See also S. Rose, R. Lewontin, and L. J. Kamin, *Not in Our Genes: Biology, Ideology and Human Nature* (New York: Random House, 1985).

5. C. Burt and M. Howard, "The Multifactorial Theory of Inheritance and Its Application to Intelligence," *British Journal of Statistical Psychology* 8 (November 1956): 95–131.

6. R. Rust and S. Golombok, *Modern Psychometrics: The Science of Psychological Assessment*, 3rd ed. (New York: Routledge, 2014).

7. G. Buzsáki and K. Mizuseki, "The Log-Dynamic Brain: How Skewed Distributions Affect Network Operations," *Nature Reviews Neuroscience* 15 (August 2014): 264–278, 264.

8. J. Daw, G. Guo, and K. M. Harris, "Nurture Net of Nature: Re-evaluating the Role of Shared Environments in Academic Achievement and Verbal Intelligence," *Social Science Research* 52 (July 2015): 422–439, 422.

9. O. Zuk, E. Hechterra, S. R. Sunyaeva, and E. S. Landerr, "The Mystery of Missing Heritability: Genetic Interactions Create Phantom Heritability," *Proceedings of the National Academy of Sciences, USA* 109 (January 2012): 1193–1198, 1193.

10. S. Wright, "Gene Interaction," in *Methodology in Mammalian Genetics*, ed. W. J. Burdette (San Francisco: Holden-Day, 1956), 159–92, 189; H. Shao, L. C. Burragea, D. S. Sinasaca, A. E. Hilla, S. R. Ernesta, et al., "Genetic Architecture of Complex

Traits: Large Phenotypic Effects and Pervasive Epistasis," *Proceedings of the National Academy of Sciences, USA* 105 (December 2008): 19910–19914.

11. T. Bouchard, "IQ Similarity in Twins Reared Apart: Findings and Response to Criticisms," in *Intelligence, Heredity and Environment*, ed. R. J. Sternberg and E. Grigerenko (Cambridge: Cambridge University Press, 1997), 126–162, 145.

12. R. Plomin and I. J. Deary, "Genetics and Intelligence Differences: Five Special Findings," *Molecular Psychiatry* 20 (September 2015): 98–108.

13. W. Johnson, M. McGue, and W. G. Iacono, "Genetic and Environmental Influences on the Verbal-Perceptual-Image Rotation (VPR) Model of the Structure of Mental Abilities," *Intelligence* 35 (2007): 542–562, 548.

14. S. W. Omholt, "From Beanbag Genetics to Feedback Genetics: Bridging the Gap Between Regulatory Biology and Quantitative Genetics Theory," in *The Biology of Genetic Dominance*, ed. R. A. Veitia (Austin, Tex.: Eurekah/Landis, 2014), 1.

15. T. J. Bouchard Jr., D. T. Lykken, M. McGue, N. L. Segal, and A. Tellegen, "Sources of Human Psychological Differences: The Minnesota Study of Twins Reared Apart," *Science* 250 (October 1990): 223–250, 223; T. J. Bouchard Jr. and M. McGue, "Familial Studies of Intelligence: A Review," *Science* 212 (May 1981): 1055–1059, 1055.

16. R. Plomin, "Nature and Nurture: Perspective and Prospective," in *Nature, Nurture, and Psychology*, ed. R. Plomin and G. E. McClearn (Washington, D.C.: American Psychological Association, 1993), 457–483, 458

17. See J. Joseph, *The Trouble with Twin Studies* (Basingstoke, U.K.: Routledge, 2014).

18. J. Joseph, *The Trouble with Twin Studies*.

19. J. Joseph, *The Trouble with Twin Studies*.

20. J. Joseph, *The Trouble with Twin Studies*.

21. Z. A. Kaminsky, T. Tang, S. C. Wang, C. Ptak, G. H. Oh, et al., "DNA Methylation Profiles in Monozygotic and Dizygotic Twins," *Nature Genetics* 41 (February 2009): 240–245.

22. D. M. Evans and N. G. Martin, "The Validity of Twin Studies," *GeneScreen* 1 (July 2000): 77–79.

23. J. Joseph, *The Gene Illusion: Genetic Research in Psychiatry and Psychology Under the Microscope* (Ross-on-Wye, U.K.: PCCS Books, 2003).

24. Y. Kovas, R. A. Weinberg, J. M. Thomson, and K. W. Fischer, "The Genetic and Environmental Origins of Learning Abilities and Disabilities in the Early School Years," *Monographs of the Society for Research in Child Development* 72 (2007): vii–160, 6.

25. K. Richardson and S. H. Norgate, "The Equal Environments Assumption of Classical Twin Studies May Not Hold," *British Journal of Educational Psychology* 75 (September 2005): 1–13.

26. D. Conley, E. Rauscher, C. Dawes, P. K. Magnusson, and M. L. Siegal, "Heritability and the Equal Environments Assumption: Evidence from Multiple Samples of Misclassified Twins," *Behavior Genetics* 43 (September 2013): 415–426.

27. R. J. Sternberg, "For Whom the Bell Curve Tolls: A Review of the Bell Curve," *Psychological Science* 5 (1995): 257–261, 260.

28. E. Bryant, *Twins and Higher Multiple Births: A Guide to Their Nature and Nurture* (London: Hodder and Stoughton, 1992), 136.

29. E. Bryant, *Twins and Higher Multiple Births*, 136.

30. For full reference details, see K. Richardson and S. H. Norgate, "A Critical Analysis of IQ Studies of Adopted Children," *Human Development* 49 (January 2005): 319–335.

31. K. Richardson and S. H. Norgate, "A Critical Analysis of IQ Studies of Adopted Children."

32. K. R. Murphy, "The Logic of Validity Generalization," in *Validity Generalization: A Critical Review*, ed. K. R. Murphy (Hove, U.K.: Erlbaum, 2003), 1–30, 16.

33. Susan Dominus, "The Mixed-Up Brothers of Bogota," *New York Times* (July 9, 2015).

34. P. Christe, A.P Moller, N. Saino, and F. De Lope, "Genetic and Environmental Components of Phenotypic Variation in Immune Response and Body Size of a Colonial Bird, *Delichon urbica* (the House Martin)," *Heredity* 85 (July 2000): 75–83.

35. P. Schönemann, "Models and Muddles of Heritability," Genetica 99 (1997), 97–108, 105.

36. G. Davies, A. Tenesa, A. Payton, J. Yang, S. E Harris, et al., "Genome-Wide Association Studies Establish that Human Intelligence Is Highly Heritable and Polygenic," *Molecular Psychiatry* 16 (October 2012): 996–1005.

37. E. Charney, "Still Chasing Ghosts: A New Genetic Methodology Will Not Find the 'Missing Heritability,'" *Independent Science News* (September 19, 2013).

38. D. Conley, M. L. Siegal, B. W. Domingue, K. M. Harris, M. B. McQueen, and J. D. Boardman, "Testing the Key Assumption of Heritability Estimates Based on Genome-Wide Genetic Relatedness," *Journal of Human Genetics* 59 (June 2014): 342–345.

39. M. Trzaskowski, N. Harlaar, R. Arden, E. Krapohl, K. Rimfeld, et al., "Genetic Influence on Family Socioeconomic Status and Children's Intelligence," *Intelligence* 42 (January–February 2014): 83–88.

40. Y. Kim, Y. Lee, S. Lee, N. H. Kim, J. Lim, et al., "On the Estimation of Heritability with Family-Based and Population-Based Samples," *BioMed Research International* 2015 (August 2015): Article ID 671349, www.ncbi.nlm.nih.gov/pubmed/26339629.

41. S. K. Kumar, M. W. Feldman, D. H. Rehkopf, and S. Tuljapurkar, "Limitations of GCTA as a Solution to the Missing Heritability Problem," *Proceedings of the National Academy of Sciences, USA* (December 2015): doi/10.1073/pnas.1520109113.

42. E. Charney, "Still Chasing Ghosts."

43. D. Conley et al., "Heritability and the Equal Environments Assumption," 415–426, 419.

3. PRETEND INTELLIGENCE

1. R. J. Sternberg and D. Kauffman (eds.), *Cambridge Handbook of Intelligence* (Cambridge: Cambridge University Press, 2011).

2. F. Galton, *Hereditary Genius* (London: Macmillan, 1869), 37.

3. E. Hunt, "On the Nature of Intelligence," Science 219 (Jan 1983): 141–146, 141.

4. G. A. Miller, *Psychology: The Science of Mental Life* (Harmondsworth, U.K.: Penguin, 1962), 313.

5. G. A. Miller, *Psychology*, 315.

6. L. M. Terman, "Feeble Minded Children in the Public Schools of California," *School and Society* 5 (June 1917): 161–165, 163. For a stark illustration of the kind of eugenics program that followed, see T. C. Leonard, *Illiberal Reformers: Race, Eugenics and American Economics in the Progressive Era* (Princeton, N.J.: Princeton University Press, 2016).

7. D. MacKenzie, "Karl Pearson and the Professional Middle Class," *Annals of Science* 36 (1979): 125–145, 137.

8. C. Spearman, *Human Abilities* (London: Macmillan, 1927).

9. R. E. Nisbett, J. Aronson, C. Blair, W. Dickens, J. Flynn, et al., "Intelligence: New Findings and Theoretical Developments," *American Psychologist* 67 (February–March 2012): 130–159, 131.

10. R. L. Thorndike and E. P. Hagen, *Measurement and Evaluation in Psychology and Education* (New York: Wiley, 1969), 325.

11. R. E. Nisbett et al., "Intelligence," 131.

12. R. D. Hoge and T. Colodarci, "Teacher-Based Judgments of Academic Achievement: A Review of Literature," *Review of Educational Research* 59 (Fall 1989): 297–313.

13. For a fuller account and all references on this topic, see K. Richardson and S. Norgate, "Does IQ Really Predict Job Performance?" *Applied Developmental Science* 19 (January 2015): 153–169.

14. K. R. Murphy, "The Logic of Validity Generalization," in *Validity Generalization: A Critical Review*, ed. K. R. Murphy (Hove, U.K.: Erlbaum, 2003), 16.

15. L. Gottfredson, "The Evaluation of Alternative Measures of Job Performance," in *Performance Assessment for the Workplace, Volume II: Technical Issues*, ed. Commission on Behavioral Social Sciences and Education (New York: National Academy Press, 1991), 75–125, 75.

16. R. M. Guion, *Assessment, Measurement, and Prediction for Personnel Decisions* (Hillsdale, N.J.: Lawrence Erlbaum, 2011).

17. J. A. Hartigan and A. K. Wigdor (eds.), *Fairness in Employment Testing: Validity Generalization, Minority Issues and the General Aptitude Test Battery* (Washington, D.C.: National Academic Press, 1989), 150.

18. E. Byington and W. Felps, "Why Do IQ Scores Predict Job Performance? An Alternative, Sociological Explanation," *Research in Organizational Behavior* 30 (January 2010): 175–202.

19. I. C. McManus, K. Woolf, J. Dacre, E. Paice, E. and C. Dewberry, "The Academic Backbone: Longitudinal Continuities in Educational Achievement from Secondary School and Medical School to MRCP(UK) and the Specialist Register in UK Medical Students and Doctors," *BMC Medicine* 11(November 2013): 242.

20. A. R. Jensen, "Individual Differences in Mental Ability," in *Historical Foundations of Educational Psychology*, ed. J. A. Glover and R. R. Ronning (New York: Plenum Press, 1987), 61–88, 82.

21. D. K. Detterman, "What Does Reaction Time Tell Us About Intelligence?" in *Speed of Information-Processing and Intelligence*, ed. P. A. Vernon (Westport, Conn.: Ablex, 1987), 177–200.

22. L. S. Gottfredson, "Why *g* Matters: The Complexity of Everyday Life," *Intelligence* 24 (1997): 79–132, 79. For references and full discussion, see K. Richardson and S. H. Norgate, "Does IQ Measure Ability for Complex Cognition?" *Theory and Psychology* 24 (December 2015): 795–812.

23. L. S. Gottfredson, "Why g Matters: The Complexity of Everyday Life," *Intelligence* 24 (1997): 79–132, 94.

24. U. Goswami, "Analogical Reasoning in Children," in *Children's Learning in Classroom and Laboratory Contexts*, ed. J. Campione et al. (New York: Routledge, 2007), 55–70.

25. P. A. Carpenter, M. A. Just, and P. Shell, "What One Intelligence Test Measures: A Theoretical Account of the Processing in the Raven Progressive Matrices Test," *Psychological Review* 97 (January 1990): 404–431.

26. R. E. Nisbett et al., "Intelligence,"131.

27. M. W. Eysenck, *Psychology: An International Perspective* (Hove, U.K.: Psychology Press, 2004), 371.

28. For references and full discussion, see K. Richardson and S. H. Norgate, "Does IQ Measure Ability for Complex Cognition?"

29. C. M. Walker and A. Gopnik, "Toddlers Infer Higher-Order Relational Principles in Causal Learning," *Psychological Science* 25 (January 2014): 161–169.

30. S. J. Ceci and J. K. Liker, "A Day at the Races: A Study of IQ, Expertise, and Cognitive Complexity," *Journal of Experimental Psychology: General* 115 (July 1986): 255–266.

31. S. Scribner, "Knowledge at Work," in *Mind & Social Practice: Selected Writings of Sylvia Scribner*, ed. E. Tobach et al. (Cambridge: Cambridge University Press, 1997), 308–318.

32. J. R. Flynn, "IQ Gains Over Time: Towards Finding the Causes," in *The Rising Curve*, ed. U. Neisser (Washington, D.C.: American Psychological Association, 1998), 25–66. See also J. R. Flynn, *Are We Getting Smarter? Rising IQ in the Twenty-First Century* (Cambridge: Cambridge University Press, 2013), 61.

33. R. E. Nisbett, *The Geography of Thought: Why We Think the Way We Do* (New York: Free Press, 2003), 203.

34. A. M. Abdel-Khalek and J. Raven, "Normative Data from the Standardization of Raven's Standard Progressive Matrices in Kuwait in an International Context," *Social Behaviour and Personality: An International Journal* 34 (February 2006): 169–180, 171.

35. M. Desert Préaux and R. Jund, "So Young and Already Victims of Stereotype Threat: Socio-economic Status and Performance of 6 to 9 Years Old Children on Raven's Progressive Matrices," *European Journal of Psychology of Education* 24 (June 2009): 207–218.

36. For example, I. J. Deary, "Intelligence, Health and Death," *Psychologist* 18 (October 2005): 610–613.

37. For all references and further discussion about this topic, see K. Richardson and S. H. Norgate, "Does IQ Measure Ability for Complex Cognition?"

38. See K. Richardson and S. H. Norgate, "Does IQ Measure Ability for Complex Cognition?"

4. REAL GENES, REAL INTELLIGENCE

1. I. Prigogine and G. Nicolis, *Exploring Complexity: An Introduction* (New York: W. H. Freeman, 1998).

2. S. Kauffman, *At Home in the Universe* (Oxford: Oxford University Press, 1991). See also W. J. Zhang, *Selforganizology: The Science of Self-Organization* (Hackensack, N.J.: World Scientific, 2016).

3. See K. Baverstock and M. Rönkkö, "Epigenetic Regulation of the Mammalian Cell," *PLoS One* 3 (June 2008): e2290. See also K. Baverstock and M. Rönkkö, "The Evolutionary Origin of Form and Function," *Journal of Physiology* 592 (May 2014): 2261–2265.

4. N. Lane, "Why Are Cells Powered by Proton Gradients?" *Nature Education* 3 (June 2010): 18.

5. In S. Mazur, "Replace the Modern Synthesis (Neo-Darwinism): An Interview with Denis Noble," *Huffington Post* (July 9, 2014), www.huffingtonpost.com/suzan-ma zur/replace-the-modern-sythes_b_5284211.htm.

6. P. Lyon, "The Cognitive Cell: Bacterial Behavior Reconsidered," *Frontiers in Microbiology* 6 (April 2015): article 264, p. 265.

7. I. Tagkopoulos, Y-C Liu, and S. Tavazoie, "Predictive Behavior Within Microbial Genetic Networks," *Science* 320 (June 2008): 1313–1317.

8. H. V. Westerhoff, A. N. Brooks, E. Simionides, R. García-Contreras, F. He, et al., "Macromolecular Networks and Intelligence in Micro-organisms," *Frontiers in Microbiology* 5 (July 2014): 379.

9. F. J. Bruggerman, W. C. van Heeswijk, F. C. Boogerd, H. V. Westerhoff, et al., "Macromolecular Intelligence in Micro-organisms," *Biological Chemistry* 381 (September–October 2000): 965–972, 965.

10. For full references, see K. Richardson, "The Evolution of Intelligent Developmental Systems," in *Embodiment and Epigenesis: Theoretical and Methodological Issues in Understanding the Role of Biology Within the Relational Developmental System*, ed. R. M. Lerner and J. B. Benson (London: Academic Press, 2014), 127–160; P. Ball, P. (2008). "Cellular Memory Hints at the Origins of Intelligence," *Nature* 451 (2008): 385.

11. N. Carey, *The Epigenetics Revolution* (London: Icon Books, 2011), 42.

12. C. Adrain and M. Freeman, "Regulation of Receptor Tyrosine Kinase Ligand Processing," *Cold Spring Harbor Perspectives in Biology*, 6 (January 2014): a008995.

13. C. Niehrs, "The Complex World of WNT Receptor Signaling," *Nature Reviews Molecular Cell Biology* 13 (December 2012): 767–779, 767.

14. I. Tagkopoulos et al., "Predictive Behavior Within Microbial Genetic Networks," 1313.

15. B. J. Mayer, "The Discovery of Modular Binding Domains: Building Blocks of Cell Signaling," *Nature Reviews Molecular Cell Biology* 16 (September 2015): 691–698, 691.

16. N. R. Gough, "A Coincidence Detector with a Memory," *Science Signalling* 5 (2012): ec48, stke.sciencemag.org/content/5/211/ec48.

17. S. Berthoumieux, H. de Jong, G. Baptist, C. Pinel, C. Ranquet, et al., "Shared Control of Gene Expression in Bacteria by Transcription Factors and Global Physiology of the Cell," *Molecular Systems Biology* 9 (January 2013): 634.

18. K. R. Nitta, A. Jolma, Y. Yin, E. Morgunova1, T. Kivioja, et al., "Conservation of Transcription Factor Binding Specificities Across 600 Million Years of Bilateria Evolution," *eLife* (March 2015): http://dx.doi.org/10.7554/eLife.04837.

19. J. B. Brown, N. Boley, R. Eisman, G. E. May, M. H. Stoiber, et al., "Diversity and Dynamics of the *Drosophila* Transcriptome," *Nature* 512 (August 2014): 393–339, 393, doi:10.1038/nature12962.

20. D. G. Dias and K. J. Ressler, "Parental Olfactory Experience Influences Behavior and Neural Structure in Subsequent Generations," *Nature Neuroscience* 17 (March 2014): 89–96. For reviews of these effects, see J. D. Sweatt, M. J. Meanyer, E. J. Nestler, and S. Akbarian, (eds.), *Epigenetic Regulation in the Nervous System* (New York: Elsevier, 2012).

21. S. Alvarado, R. Rajakumar, E. Abouheif, and M. Szyf, "Epigenetic Variation in the *Egfr* Gene Generates Quantitative Variation in a Complex Trait in Ants," *Nature Communications* 6 (March 2015): 10.1038/ncomms7513.

22. The Noble remark is from S. Mazur, "Replace the Modern Synthesis (Neo-Darwinism)," http://www.huffingtonpost.com/suzan-mazur/replace-the-modern-sythes_b_5284211 .html; J. S. Mattick, "Rocking the Foundations of Molecular Genetics," *PNAS* 109 (October 2012): 16400–16401, 16400.

23. M.-W. Ho, *Non-random Directed Mutations Confirmed*, ISIS Report 07/10/13 (October 2013): www.i-sis.org.uk/Nonrandom_directed_mutations_confirmed.php.

24. M. W. Kirschner and J. C. Gerhart, *The Plausibility of Life: Resolving Darwin's Dilemma* (New Haven, Conn.: Yale University Press, 2014), 3.

25. J. A. Shapiro, "How Life Changes Itself: The Read–Write (RW) Genome," *Physics of Life Reviews* 10 (July 2013): 287–323, 287.

26. R. Lickliter, (2013). "The Origins of Variation: Evolutionary Insights from Developmental Science," in *Embodiment and Epigenesis: Theoretical and Methodological Issues in Understanding the Role of Biology Within the Relational Developmental System*, ed. R. M. Lerner and J. B. Benson (London: Academic Press, 2014), 173–203, 193. See also M. W. Ho, *Evolution by Natural Genetic Engineering*, ISIS Report 02/06/14 (June 2014): www.i-sis.org.uk/Evolution_by_Natural_Genetic_Engineering.php.

27. M.-W. Ho, "How Mind Changes Genes Through Meditation," ISIS Report 21/05/14 (May 2014): www.i-sis.org.uk/How_mind_changes_genes_through_meditation.php.

28. I. V. Razinkov, B. L. Baumgartner, M. R. Bennett, L. S. Tsimring, and J. Hasty, "Measuring Competitive Fitness in Dynamic Environments," *Journal of Physical Chemistry* 17 (October 2015): 13175–13181, dx.doi.org/10.1021/jp403162v. See also A. Rossi, Z. Kontarakis, C. Gerri, H. Nolte, S. Hölper, et al., "Genetic Compensation Induced by Deleterious Mutations but Not Gene Knockdowns," *Nature* 524 (August 2015): 230–233.

29. A. Wagner and J. Wright, "Alternative Routes and Mutational Robustness in Complex Regulatory Networks," *BioSystems* 88 (March 2007): 163–172, 163.

30. H. F. Nijhout, J. A. Best, and M. C. Reed, "Using Mathematical Models to Understand Metabolism, Genes, and Disease," *BMC Biology* 13 (September 2015): 79.

31. P. M. Visscher, D. Smith, S. J. G. Hall, and J. L. Williams, "A Viable Herd of Genetically Uniform Cattle," *Nature* 409 (January 2001): 303.

5. INTELLIGENT DEVELOPMENT

1. Quoted in P. Griffiths and J. Tabery, "Developmental Systems Theory: What Does It Explain, and How Does It Explain It?" *Advances in Child Development and Behavior* 44 (May 2013): 65–95, 171; S. Dominus, "The Mixed-Up Brothers of Bogotá," *New York Times* (July 9, 2015).

2. D. H. Ford and R. M. Lerner, *Developmental Systems Theory: An Integrative Approach* (Newbury Park, Calif.: Sage, 1992); G. Greenberg and T. Partridge, "Biology, Evolution, and Development," in *Cognition, Biology, and Methods*, vol. 1 of *Handbook of Life-Span Development*, ed. W. F. Overton (Hoboken, N.J.: Wiley, 2010), 115–148.

3. L. Wolpert, "Positional Information Revisited," *Development Supplement* 1989 (1989): 3–12, 10.

4. There are numerous excellent books and papers on these processes. See, for example, J. Jaeger and A. Martinez-Arias, "Getting the Measure of Positional Information," *PLoS Biology* 7 (March 2009): e1000081, doi:10.1371/journal.pbio.1000081.

5. Y. Komiya and R. Habas, "Wnt Signal Transduction Pathways," *Organogenesis* 4 (April–June 2008): 68–75, 68.

6. A. H. Lang, H. Li, J. J. Collins, and P. Mehta, "Epigenetic Landscapes Explain Partially Reprogrammed Cells and Identify Key Reprogramming Genes," *PLoS Computational Biology* 10 (August 2014): e1003734.

7. E. J. Kollar and C. Fisher, "Tooth Induction in Chick Epithelium: Expression of Quiescent Genes for Enamel Synthesis," *Science* 207 (February 1980): 993–995, 993.

8. C. H. Waddington, *The Strategy of the Genes* (London: Allen and Unwin, 1957); P. Griffiths and J. Tabery, "Developmental Systems Theory."

9. See chapter 4. See also, for example, A. Manu, S. Surkova, A. V. Spirov, V. V. Gursky, H. Janssens, et al., "Canalization of Gene Expression and Domain Shifts in the *Drosophila* Blastoderm by Dynamical Attractors," *PLoS Computational Biology* 5 (March 2009): e1000303, doi: 10.1371/journal.pcbi.1000303.

10. O. S. Soyer and T. Pfeiffer, "Evolution Under Fluctuating Environments Explains Observed Robustness in Metabolic Networks," *PLoS Computational Biology* 6 (August 2010): e1000907, doi: 10.1371/journal.pcbi.1000907.

11. S. A. Kelly, T. M. Panhuis, and A. M. Stoehr, "Phenotypic Plasticity: Molecular Mechanisms and Adaptive Significance," *Comprehensive Physiology* 2 (April 2012): 1416–1439.

12. G. Gottlieb, "Experiential Canalization of Behavioral Development: Theory," *Developmental Psychology* 27 (1991): 4–13, 9.

13. See contributions in A. Love (ed.), *Conceptual Change in Science: Scientific and Philosophical Papers on Evolution and Development* (New York: Springer, 2015).

14. R. Lickliter, "The Origins of Variation: Evolutionary Insights from Developmental Science," in *Embodiment and Epigenesis: Theoretical and Methodological Issues in Understanding the Role of Biology Within the Relational Developmental System*, ed. R. Lerner and J. Benson (London: Academic Press, 2014), 173–203, 193.

15. M.-W. Ho, "Development and Evolution Revisited," in *Handbook of Developmental Science, Behavior and Genetics*, ed. K. Hood et al. (New York: Blackwell, 2009), 61–109.

16. See the EuroStemCell website: "iPS Cells and Reprogramming: Turn Any Cell of the Body into a Stem Cell," eurostemcell.org (last updated September 29, 2015).

17. S. Rose, *Lifelines: Life Beyond the Gene* (New York: Oxford University Press, 2003), 17.

18. M. Joëls and T. Z. Baram, "The Neuro-Symphony of Stress," *Nature Reviews: Neuroscience* 10 (June 2009): 459–466, 459.

19. M. J. Wijnants, (2014). "Presence of 1/f Scaling in Coordinated Physiological and Cognitive Processes," *Journal of Nonlinear Dynamics* 2014 (February 2014): article 962043, http://dx.doi.org/10.1155/2014/962043.

20. B. J. West, "Fractal Physiology and the Fractional Calculus: A Perspective," *Frontiers in Physiology 1: Fractal Physiology* 1 (October 2010): article 12.

21. A. L. Goldberger, L. A. N. Amaral, J. M. Hausdorff, P. Ch. Ivanov, C.-K. Peng, et al., "Fractal Dynamics in Physiology: Alterations with Disease and Aging," *Proceedings of the National Academy of Sciences, USA* 99 (suppl. 1, February 2002): 2466–2472, 2471.

22. C. Darwin, *The Power of Movement in Plants* (London: John Murray, 1880), 572. Thanks to Keith Baverstock, personal communication, February 14, 2015, for this quote.

23. R. Karban, *Plant Sensing and Communication* (Chicago: University of Chicago Press, 2015), 1; A. Barnett, "Intelligent Life: Why Don't We Consider Plants to Be Smart?" *New Scientist*, May 2015, 30.

24. A. M. Johnstone, S. D. Murison, J. S. Duncan, K. A. Rance, and J. R. Speakman, "Factors Influencing Variation in Basal Metabolic Rate Include Fat-Free Mass, Fat Mass, Age, and Circulating Thyroxine but Not Sex, Circulating Leptin, or Triiodothyronine," *American Journal of Clinical Nutrition* 82 (November 2005): 941–948.

25. A. M. Johnstone et al., "Factors Influencing Variation in Basal Metabolic Rate."

26. For example, Z. Boratynski, E. Koskela, T. Mappes, and E. Schroderus, "Quantitative Genetics and Fitness Effects of Basal Metabolism," *Evolutionary Ecology* 27 (March 2013): 301–314.

6. HOW THE BRAIN MAKES POTENTIAL

1. R. Plomin, J. C. DeFries, V. S. Knopik, and J. M. Neiderhiser, *Behavioral Genetics*, 6th ed. (New York: Worth Publishers, 2013), 63.

2. J. Flynn, *Intelligence and Human Progress* (New York: Academic Press, 2013), 63; J. Flynn, interview in the *Independent*, June 5, 2014.

3. E. Goldberg, *The New Executive Brain* (Oxford: Oxford University Press, 2009), 4.

4. R. del Moral, A. M. Brandmaier, L. Lewejohann, I. Kirste1, M. Kritzler, et al., "From Genomics to Scientomics: Expanding the Bioinformation Paradigm," *Information* 2 (July 2011): 651–671, 661; J. S. Allen, *The Lives of the Brain: Human Evolution and the Organ of Mind* (Cambridge, Mass.: Harvard University Press, 2012), 6.

5. W. H. Warren and R. E. Shaw, "Events and Encounters as Units of Analysis for Ecological Psychology," in *Persistence and Change: Proceedings of the First International Conference on Event Perception*, ed. W. H. Warren and R. E. Shaw (Hillsdale, N.J.: Lawrence Erlbaum, 1985), 1–28, 6.

6. R. H. Masland, "The Neuronal Organization of the Retina," *Neuron* 76 (October 18, 2012). p. 266–280, 266; D. D. Hoffman, *Visual Intelligence: How We Create What We See* (New York: W. W. Norton, 1998).

7. H. R. Barlow, "The Knowledge Used in Vision and Where It Comes From," *Philosophical Transactions of the Royal Society* 35, Series B (1997): 1141–1147, 1141.

8. J. S. Lappin, D. Tadin, and E. J. Whittier, "Visual Coherence of Moving and Stationary Image Changes," *Vision Research* 42 (June 2002): 1523–1534.

9. K. Richardson and D. S. Webster, "Recognition of Objects from Point-Light Stimuli: Evidence for Covariation Hierarchies in Conceptual Representation," *British Journal of Psychology* 87 (July 1996): 567–591.

10. D. Mackay, "Vision: The Capture of Optical Covariation," in *Visual Neuroscience*, ed. J. D. Pettigrew, K. J. Sanderson, and W. R. Levick (Cambridge: Cambridge University Press, 1986), 365–373, 370.

11. N.-L. Xu, M. T. Harnett, S. R. Williams, D. Huber, D. H. O'Connor, et al., "Nonlinear Dendritic Integration of Sensory and Motor Input During an Active Sensing Task," *Nature* 492 (December 2014): 247–251, 247; G. Buzsáki and K. Mizuseki, "The Log-Normal Brain: How Skewed Distributions Affect Network Operations," *Nature Reviews Neuroscience* 15 (April 2014): 264–278.

12. T. S. Lee, T. Stepleton, B. Potetz, and J. Samonds, "Neural Encoding of Scene Statistics for Surface and Object Inference," in *Object Categorization: Computer and Human Vision Perspective*, ed. S. Dickinson et al. (Cambridge: Cambridge University Press, 2007), 451–474.

13. See A. Hyvärinen and J. Hurri, *Natural Image Statistics: A Probabilistic Approach to Early Computational Vision* (New York: Springer, 2009).

14. S. Onat, D. Jancke, and P. König, "Cortical Long-Range Interactions Embed Statistical Knowledge of Natural Sensory Input: A Voltage-Sensitive Dye Imaging Study," *F1000Research* 2 (February 2013): doi: 10.3410/f1000research.2-51.v1.

15. S. J. Blackmore, "Three Experiments to Test the Sensorimotor Theory of Vision," *Behavioural and Brain Sciences* 24 (2001): 977.

16. V. Michalski, R. Memisevic, and K. Konda, "Modeling Deep Temporal Dependencies with Recurrent Grammar Cells," *Advances in Neural Information Processing Systems* 27 (May 2014): 1925–1933, 1926.

17. S. Bao, "Perceptual Learning in the Developing Auditory Cortex," *European Journal of Neuroscience* 41 (March 2015): 718–724, 718.

18. J. A. Garcia-Lazaro, B. Ahmed, and J. W. H. Schnupp, "Emergence of Tuning to Natural Stimulus Statistics Along the Central Auditory Pathway," *PLoS One* 6 (August 2011): e22584, doi: 10.1371/journal.pone.0022584.

19. I. Nelken, "Processing of Complex Stimuli and Natural Scenes in the Auditory Cortex," *Current Opinion in Neurobiology* 14 (July 2004): 474–480, 474.

20. G. A. Calvert and R. Campbell, "Reading Speech from Still and Moving Faces," *Journal of Cognitive Neuroscience* 15 (January 2003): 57–70; K. G. Munhall and J. N. Buchan, "Something in the Way She Moves," *Trends in Cognitive Sciences* 8 (February 2004): 51–53.

21. M. S. Grubb and I. D. Thompson, "The Influence of Early Experience on the Development of Sensory Systems," *Current Opinion in Neurobiology* 14 (August 2004): 503–512.

22. B. E. Stein, T. R. Stanford, and B. A. Rowland, "Development of Multisensory Integration from the Perspective of the Individual Neuron," *Nature Reviews Neuroscience* 15 (July 2014): 520–535, 520.

23. E. A. Phelps and J. E. LeDoux, "Contributions of the Amygdala to Emotion Processing: From Animal Models to Human Behavior," *Neuron* 48 (September 2005): 175–187; L. Pessoa and R. Adolphs, "Emotion and the Brain: Multiple Roads Are Better Than One," *Nature Reviews Neuroscience* 12 (November 2011): 425.

24. L. Pessoa, *The Cognitive-Emotional Brain: From Interactions to Integration* (Cambridge, Mass.: MIT Press, 2013).

25. I. Deary, "Intelligence," *Annual Review of Psychology* 63 (2012): 453–482, 465. (For fMRI scan illustrations, see the Wikipedia article "Functional Magnetic Resonance Imaging.")

26. R. J. Haier, The Great Courses, "The Intelligent Brain," www.thegreatcourses.co.uk /courses/the-intelligent-brain.html.

27. R. J. Davidson and B. S. McEwan, "Social Influences on Neuroplasticity: Stress and Interventions to Promote Well-Being," *Nature Neuroscience* 15 (April 2012): 689–695.

28. M. Rutter and A. Pickles, "Annual Research Review: Threats to the Validity of Child Psychology and Psychiatry," *Journal of Child Psychology and Psychiatry* 57 (March 2016): 398–416.

29. H. Okon-Singer, T. Hendler, L. Pessoa, and A. J. Shackman, "The Neurobiology of Emotion-Cognition Interactions: Fundamental Questions and Strategies for Future Research," *Frontiers in Human Neuroscience* 9 (February 2015): 1–14, 8, 10.3389/ fnhum.2015.00058.

30. C. Thomas, F. Q. Yec, M. O. Irfanoglu, P. Modia, K. S. Saleem, et al., "Anatomical Accuracy of Brain Connections Derived from Diffusion MRI Tractography Is Inherently Limited," *Proceedings of the National Academy of Sciences*, USA 111 (June 2014): 16574–16579.

31. M. Rutter and A. Pickles, "Annual Research Review: Threats to the Validity of Child Psychology and Psychiatry," *Journal of Child Psychology and Psychiatry* 57 (March 2016): 398–416, 406.

32. K. Martinez, S. K. Madsen, A. A. Joshi, S. H. Joshi, F. J. Román, et al., "Reproducibility of Brain-Cognition Relationships Using Three Cortical Surface-Based Protocols: An Exhaustive Analysis Based on Cortical Thickness," *Human Brain Mapping* 36 (August 2015): 3227–3245, 3227.

33. C. I. Bargmann and E. Marde, "From the Connectome to Brain Function," *Nature Methods*, 10 (June 2013): 483–490, 484.

34. H. G. Schnack, H. E. van Haren, R. M. Brouwer, A. Evans, S. Durston, et al., "Changes in Thickness and Surface Area of the Human Cortex and Their Relationship with Intelligence," *Cerebral Cortex* 25 (June 2015): 1608–1617, 1609.

35. R. Haier, R. Colom, D. H. Schroeder, C. A. Condon, C. Tang, et al., "Gray Matter and Intelligence Factors: Is There a Neuro-*g*? *Intelligence* 37 (January 2009): 136–144, 136.

36. G. Buzsáki and K. Mizuseki, "The Log-Dynamic Brain: How Skewed Distributions Affect Network Operations," *Nature Reviews Neuroscience* 15 (April 2014): 264–278, 264.

37. D. M. Barch and T. Yarkoni, "Introduction to the Special Issue on Reliability and Replication in Cognitive and Affective Neuroscience Research," *Cognitive, Affective and Behavioral Neuroscience* 13 (December 2013): 687–689, 687.

38. C. I. Bargmann and E. Marde, "From the Connectome to Brain Function," 488.

39. M. Hawrylycz, C. Dang, C. Koch, and H. Zeng, "A Very Brief History of Brain Atlases," in *The Future of the Brain: Essays by the World's Leading Neuroscientists*, ed. G. Marcus and J. Freeman (Princeton, N.J.: Princeton University Press, 2014), 3–16, 11.

40. S. R. Quartz and T. J. Sejnowski, "The Neural Basis of Cognitive Development: A Constructivist Manifesto," *Behavioral and Brain Sciences* 20 (December 1997): 537–596, 537.

41. H. Lee, J. T. Devlin, C. Shakeshaft, L. H. Stewart, A. Brennan, et al., "Anatomical Traces of Vocabulary Acquisition in the Adolescent Brain," *Journal of Neuroscience* 27 (January 2007): 1184–1189.

42. A. May, "Experience-Dependent Structural Plasticity in the Adult Human Brain," *Trends in Cognitive Sciences* 15 (October 2011): 475–482, 475. See also M. Lövdén, E. Wenger, J. Mårtensson, U. Lindenberger, and L. Bäckman, "Structural Brain Plasticity in Adult Learning and Development," *Neuroscience and Biobehavioral Reviews* 37 (June 2013): 2296–2310.

43. A. May, "Experience-Dependent Structural Plasticity," 475.

44. J. Freund, A. M. Brandmaier, L. Lewejohann, I. Kirstel, M. Kritzler, et al., "Emergence of Individuality in Genetically Identical Mice," *Science* 340 (May 2013): 756–759, 756.

7. A CREATIVE COGNITION

1. R. Plomin, J. C. DeFries, V. S. Knopik, and J. M. Neiderhiser, *Behavioral Genetics*, 6th ed. (New York: Worth Publishers, 2013), 187–188.

2. S. J. Shettleworth, "Animal Cognition and Animal Behaviour," *Animal Behaviour* 61 (February 2001): 277–286, 278.

3. M. W. Eysenck and M. T. Keane, *Cognitive Psychology: A Student's Handbook* (Hove, U.K.: Erlbaum, 2015), 31.

4. M. Giurfa, *Animal Cognition* (Cold Spring Harbor, N.Y.: Cold Spring Harbor Press, 2009), 281.

5. R. Adolphs, "The Unsolved Problems of Neuroscience," *Trends in Cognitive Science* 19 (April 2015):173–175; W. James, *Principles of Psychology*, vol. 1 (New York: Dover Publications, 1890), 6.

6. S. Pinker, *How the Mind Works* (Harmondsworth: Penguin, 1999), 21.

7. T. Stone and M. Davies, "Theoretical Issues in Cognitive Psychology," in *Cognitive Psychology*, 2nd ed., ed. N. Braisby and A. Gellatly (Oxford: Oxford University Press, 2012), 639 (emphasis in original); also S. Bem and H. L. de Jong, *Theoretical Issues in Psychology: An Introduction* (London: Sage, 2012).

8. R. Wallace and D. Wallace, *A Mathematical Approach to Multilevel, Multiscale Health Interventions* (London: Imperial College Press, 2013), 8.

9. For a full discussion, see K. Richardson, *Models of Cognitive Development* (Hove, U.K.: Psychology Press, 2000).

10. E. L. Ardiel and C. H. Rankin, "An Elegant Mind: Learning and Memory in *Caenorhabditis elegans*," *Learning & Memory* 17 (April 2010): 191–201.

11. M. Giurfa, "Learning and Cognition in Insects," *Cognitive Science* 6 (March 2015): 10.1002/wcs.1348.

12. E. L. Ardiel and C. H. Rankin, "An Elegant Mind."

13. M. Giurfa, "Learning and Cognition in Insects."

14. E. Başar (ed.), *Chaos in Brain Functions* (New York: Springer, 2012); K. Clancy, "Your Brain Is on the Brink of Chaos: Neurological Evidence for Chaos in the Nervous System Is Growing," *Nautilus* (July 10, 2014): nautil.us/issue/15/turbulence/your-brain-is-on-the-brink-of-chaos.

15. W. J. Freeman, *Societies of Brains* (Hillsdale, N.J.: Erlbaum, 1995), 67. See also W. J. Freeman, *How Brains Make Up Their Minds* (London: Weidenfeld and Nicolson, 1999).

16. R. Gregory, "The Blind Leading the Sighted: An Eye-Opening Experience of the Wonders of Perception," *Nature* 430 (August 2004): 836.

17. M. O. Ernst, "The 'Puzzle' of Sensory Perception: Putting Together Multisensory Information," in *Proceedings of the 7th International Conference on Multimodal Interfaces* (New York: ACM Press, 2005), 1.

18. L. Pessoa, "On the Relationship Between Emotion and Cognition," *Nature Reviews Neuroscience* 9 (February 2008): 148–158, 148.

19. M. Johnson, "Embodied Understanding," *Frontiers in Psychology* 6 (June 2015): article 875, http://dx.doi.org/10.3389/fpsyg.2015.00875.

20. W. J. Freeman, "Noise-Induced First-Order Phase Transitions in Chaotic Brain Activity," *International Journal of Bifurcation and Chaos* 9 (November 1999): 2215–2218, 2218.

21. J. Almeida D. He, Q. Chen, B. Z. Mahon, F. Zhang, et al., "Decoding Visual Location from Neural Patterns in the Auditory Cortex of the Congenitally Deaf," *Psychological Science* 26 (September 2015): 1771–1782, 10.1177/0956797615598970.

22. J. Elman, E. Bates, A. Karmiloff-Smith, M. Johnson, D. Parisi, and K. Plunkett, *Rethinking Innateness: A Connectionist Perspective on Development* (Cambridge, Mass.: MIT Press, 1996), 359.

23. E. M. Pothos, "The Rules versus Similarity Distinction," *Behavioral and Brain Sciences* 28 (February 2003): 1–49, 26.

24. E. Turkheimer, "Commentary: Variation and Causation in the Environment and Genome," *International Journal of Epidemiology* 40 (June 2011): 598–601, 598.

25. P. Carruthers, "Evolution of Working Memory," *Proceedings of the National Academy of Sciences, USA* 110 (June 2013): 10371–10378.

26. P. Carruthers, "Evolution of Working Memory," 10371.

27. P. Carruthers, "Evolution of Working Memory," 10371–10372.

28. A. D. Baddeley, *Working Memory* (Oxford: Oxford University Press, 1986).

29. S. M. Jaeggi, B. Studer-Luethi, M. Buschkuehl, Y-F. Su, J. Jonides, et al., "The Relationship Between *n*-Back Performance and Matrix Reasoning—Implications for Training and Transfer," *Intelligence* 38 (June 2010): 625–635, 626.

30. P. Carruthers, "Evolution of Working Memory," 10372.

8. POTENTIAL BETWEEN BRAINS: SOCIAL INTELLIGENCE

1. K. L. Visick and C. Fuqua, "Decoding Microbial Chatter: Cell-Cell Communication in Bacteria," *Journal of Bacteriology* 187 (August 2005): 5507–5519, 5512.

2. D. E. Jackson and F. Ratnieks, "Communication in Ants," *Current Biology* 16 (August 2006): R570–R574, R571.

3. M. Mossaid, S. Garnier, G, Theralauz, and D. Helbing, "Collective Information Processing and Pattern Formation in Swarms, Flocks, and Crowds," *Topics in Cognitive Science* 1 (July 2009): 469–497, 471.

4. C. Detrain and J.-L. Deneubourg, "Self-Organized Structures in a Superorganism: Do Ants 'Behave' Like Molecules?" *Physics of Life Reviews* 3 (January 2006): 162–187, 165.

5. C. Detrain and J.-L. Deneubourg, "Self-Organized Structures in a Superorganism: Do Ants 'Behave' Like Molecules?" *Physics of Life Reviews* 3 (January 2006): 162–187, 172.

6. I. D. Couzin, "Collective Cognition in Animal Groups," *Trends in Cognitive Sciences* 13 (September 2009): 36–43, 36.

7. N. E. Leonard, "Multi-Agent System Dynamics: Bifurcation and Behavior of Animal Groups," plenary paper, IFAC Symposium on Nonlinear Control Systems (NOLCOS), Toulouse, France, September 2013.

8. I. D. Chase and K. Seitz, "Self-Structuring Properties of Dominance Hierarchies: A New Perspective," *Advances in Genetics* 75 (January 2011): 51–81.

9. I. D. Chase and K. Seitz, "Self-Structuring Properties of Dominance Hierarchies," 51.

10. A. Cavagna, A. Cimarelli, I. Giardina, G. Parisi, R. Santagati, et al., "Scale-Free Correlations in Starling Flocks," *Proceedings of the National Academy of Sciences, USA* 107 (June 2010): 11865–11870.

11. A. Cavagna et al., "Scale-Free Correlations in Starling Flocks," 11865.

12. W. Bialek, A. Cavagne, I. Giardina, T. Mora, O. Pohl, et al., "Social Interactions Dominate Speed Control in Poising Natural Flocks Near Criticality," *Proceedings of the National Academy of Sciences* 111 (May 2014): 7212–7217, 7216.

13. M. Watve, "Bee-Eaters (*Merops orientalis*) Respond to What a Predator Can See," *Animal Cognition* 5 (December 2002): 253–259.

14. J. M. Thom and N. S. Clayton, "Re-caching by Western Scrub-Jays (*Aphelocoma californica*) Cannot Be Attributed to Stress," *PLoS One* 8 (January 2013): e52936, doi:-10.1371/journal.pone.0052936.

15. R. Bshary, W. Wickler, and H. Fricke, "Fish Cognition: A Primate's Eye View," *Animal Cognition* 5 (March 2002): 1–13, 5.

16. S. Creel and N. M. Creel, *The African Wild Dog: Behavior, Ecology, and Conservation* (Princeton, N.J.: Princeton University Press, 2002).

17. D. Sol, S. Bacher, S. M. Reader, and L. Lefebvre, "Brain Size Predicts the Success of Mammal Species Introduced into Novel Environments," *American Naturalist* 172 (July 2008, supplement): S61–S71.

18. R. I. Dunbar, "The Social Brain Hypothesis and Its Implications for Social Evolution," *Annals of Human Biology* 36 (September 2009): 562–572, 563.

19. N. K. Humphrey, "The Social Function of Intellect," in *Growing Points in Ethology*, ed. P. P. G. Bateson and R. A. Hinde (Cambridge: Cambridge University Press, 1976), 303–317.

20. See S. F. Brosnan, L. Salwiczek, and R. Bshary, "The Interplay of Cognition and Cooperation," *Philosophical Transactions of the Royal Society* B 365 (August 2010): 2699–2710.

21. A. Abbott, "Animal Behaviour: Inside the Cunning, Caring and Greedy Minds of Fish," *Nature* 521 (May 2015): 412–414, 413.

22. C. J. Charvet and B. L. Finlay, "Embracing Covariation in Brain Evolution: Large Brains, Extended Development, and Flexible Primate Social Systems," in *Progress in Brain Research*, vol. 195, ed. M. A. Hoffman and D. Falk (Amsterdam: Elsevier, 2012), 71–87.

23. S. M. Reader, Y. Hager, and K. N. Laland, "The Evolution of Primate General and Cultural Intelligence," *Philosophical Transaction of the Royal Society* B 366 (March 2011): 1017–1027.

24. C. B. Stanford, "How Smart Does a Hunter Need to Be?" in *The Cognitive Animal: Empirical and Theoretical Perspectives on Animal Cognition,* ed. M. Beloff, C. Allen, and G. M. Burghardt (Cambridge, Mass.: MIT Press, 2002), 399–403, 401.

25. A. Strandburg-Peshkin, D. R. Farine, I. D. Couzin, and M. C. Crofoot, "Shared Decision-Making Drives Collective Movement in Wild Baboons," *Science* 348 (June 2015): 1358–1361, 1358.

26. C. Detrain and J.-L. Deneubourg, "Self-Organized Structures in a Superorganism," 164.

27. A. F. Bullinger, J. M. Burkart, A. P. Melis, and M. Tomasello, "Bonobos, *Pan paniscus,* Chimpanzees, *Pan troglodytes,* and Marmosets, *Callithrix jacchus,* Prefer to Feed Alone," *Animal Behaviour* 85 (January 2013): 51–60.

28. S. Harder, T. Lange, G. F. Hansen, M. Væver, and S. Køppe, "A Longitudinal Study of Coordination in Mother-Infant Vocal Interaction from Age 4 to 10 Months," *Developmental Psychology* 51 (December 2015): 1778–1790.

29. K. Gillespie-Lynch, P. M. Greenfield, H. Lyn, and S. Savage-Rumbaugh, "Gestural and Symbolic Development Among Apes and Humans: Support for a Multimodal Theory of Language Evolution," *Frontiers in Psychology* (October 2014): 10.3389/fpsyg.2014.01228.

30. M. Tomasello, "The Ultra-Social Animal," *European Journal of Social Psychology* 44 (April 2014): 187–194, 187; M. Tomasello, *A Natural History of Human Thinking* (Cambridge, Mass.: Harvard University Press, 2014).

31. P. Kropotkin, *Mutual Aid: A Factor of Evolution* (New York: New York University Press, 1972), 7.

9. HUMAN INTELLIGENCE

1. K. N. Laland, J. Odling-Smee, and S. Myles, "How Culture Shaped the Human Genome: Bringing Genetics and the Human Sciences Together," *Nature Reviews Genetics* 11 (February 2010): 137–148, 138.

2. M. V. Flinn, D. C. Geary, and C. V. Ward, "Ecological Dominance, Social Competition, and Coalitionary Arms Races: Why Humans Evolved Extraordinary Intelligence," *Evolution and Human Behavior* 26 (January 2005): 10–46, 15.

3. F. M. Menger, *The Thin Bone Vault* (Singapore: World Scientific, 2009), 143.

4. E. O. Wilson, *The Meaning of Human Existence* (New York: Liveright, 2014), 28.

5. U. Hasson, A. A. Ghazanfar, B. Galantucci, S. Garrod, and C. Keysers, "Brain-to-Brain Coupling: A Mechanism for Creating and Sharing a Social World," *Trends in Cognitive Sciences* 16 (February 2011): 114–121, 114.

6. R. Tallis, *Aping Mankind* (Durham, U.K.: Acumen, 2011), 11. Now Abingdon, U.K.: Routledge, 2011.

7. C. Stringer, *The Origin of Our Species* (Harmondsworth, U.K.: Penguin, 2012); L. Gabora and A. Russon, "The Evolution of Human Intelligence," in *The Cambridge Handbook of Intelligence,* ed. R. Sternberg and S. Kaufman (Cambridge: Cambridge University Press, 2011), 328–350.

8. S. C. Antón, R. Potts, L.C. Aiello, "Evolution of Early *Homo:* An Integrated Biological Perspective," *Science* 345 (July 2014): 1236828-0, doi: 10.1126/science.12368280.

9. L. Gabora and A. Russon, "The Evolution of Human Intelligence," in *The Cambridge Handbook of Intelligence,* ed. R. Sternberg and S. Kaufman (Cambridge: Cambridge University Press, 2011), 328–350, 332.

10. E. Discamps and C. S. Henshilwood, "Intra-Site Variability in the Still Bay Fauna at Blombos Cave: Implications for Explanatory Models of the Middle Stone Age Cultural and Technological Evolution," *PLoS One* 10 (December 2016): e0144866, doi: 10.1371/journal.pone.0144866; University of Bergen, "Humans Evolved by Sharing Technology and Culture," *ScienceDaily* (April 2016): www.sciencedaily.com /releases/2016/02/160202121246.htm.

11. M. Tomasello, "The Ultra-Social Animal," *European Journal of Social Psychology* 44 (April 2014): 187–194.

12. J. Prado-Martinez, P. H. Sudmant, J. M. Kidd, H. Li, J. L. Kelley, et al., "Great Ape Genetic Diversity and Population History," *Nature* 499 (August 2014): 471–475.

13. "Science & Music: Bountiful Noise" (editorial), *Nature* 453 (May 8, 2008): 134, doi:10.1038/453134a.

14. O. Sacks, *Musicophilia: Tales of Music and the Brain* (New York: Vintage Books, 2011), 266, cited by H. Hennig, "Synchronization in Human Musical Rhythms and Mutually Interacting Complex Systems," *Proceedings of the National Academy of Sciences, USA,* 11 (September 2014): 12974–12979.

15. J. Sänger, V. Müller, and U. Lindenberger, "Intra- and Interbrain Synchronization and Network Properties When Playing Guitar in Duets," *Frontiers in Human Neuroscience* 29 (November 2012): http://dx.doi.org/10.3389/fnhum.2012.00312.

16. U. Hasson et al., "Brain-to-Brain Coupling," 114.

17. L. Gabora and A. Russon, "The Evolution of Human Intelligence," 341

18. B. L. Fredrickson, K. M. Grewen, S. B. Algoe, A. M. Firestine, J. M. G. Arevalo, et al., "Psychological Well-Being and the Human Conserved Transcriptional Response to Adversity," *PLoS One* 10 (March 2015): e0121839, doi: 10.1371/journal.pone.0121839.

19. M. Bond, "Reflections of the Divine: Interview with Desmond Tutu," *New Scientist* 29 (April 2006): 1163.

20. P. Zukow-Goldring, "A Social Ecological Realist Approach to the Emergence of the Lexicon," in *Evolving Explanations of Development,* ed. C. Dent-Read and P. Zukow-Goldring (Washington, D.C.: American Psychological Association, 1997), 199–250, 210.

21. Quoted in O. Lourenço, "Piaget and Vygotsky: Many Resemblances, and a Crucial Difference," *New Ideas in Psychology* 30 (December 2012): 281–295, 282.

22. L. S. Vygotsky, "The Genesis of Higher Mental Functions," in *The Concept of Activity in Soviet Psychology,* ed. J. V. Wertsch (New York: Sharpe, 1981), 144–188, 160.

23. L. S. Vygotsky, "The Genesis of Higher Mental Functions," 163.

24. A. May, "Experience-Dependent Structural Plasticity in the Adult Human Brain," *Trends in Cognitive Sciences* 15 (October 2011): 475–482.

25. D. C. Park and C.-M. Huang, "Culture Wires the Brain: A Cognitive Neuroscience Perspective," *Perspectives on Psychological Science* 5 (August 2010): 391–400.

26. T. Ingold, "The Social Child," in *Human Development in the Twenty-First Century: Visionary Ideas from Systems Scientists*, ed. A. Fogel et al. (Cambridge: Cambridge University Press, 2007), 112–118, 117.

27. As reported by Tom Heyden, "When 100 People Lift a Bus," *BBC News Magazine*, June 4, 2015.

28. C. Geertz, *The Interpretation of Culture* (New York: Basic Books, 1973), 68.

29. D. Kahneman, *Thinking, Fast and Slow* (New York: Farrar, Straus & Giroux, 2011).

30. L. Malafouris, *How Things Shape the Mind: A Theory of Material Engagement* (Cambridge, Mass.: MIT Press, 2015).

31. L. Gottfredson, "Unmasking the Egalitarian Fiction," Duke Talent Identification Program (Durham, N.C.: Duke University, 2006), 10, Tip.duke.edu.

32. J. Freeman, "Giftedness in the Long Term," *Journal for the Education of the Gifted* 29 (June 2006): 384–403.

33. M. Howe, *The Psychology of High Abilities* (London: Macmillan, 1999), 5.

34. J. Freeman, "Giftedness in the Long Term," 392.

35. S. Bergia, "Einstein and the Birth of Special Relativity," in *Einstein: A Centenary Volume*, ed. A. P. French (London: Heinemann, 1979); A. E. Einstein, *Ideas and Opinions*, ed. and comp. C. Seelig (New York: Crown Publications, 1982).

36. G. Claxton and S. Meadows, "Brightening Up: How Children Learn to Be Gifted," in *Routledge Companion to Gifted Education*, ed. T. Balchin, B. Hymer, and D. Matthews (London: Routledge, 2008), 3–7, 3.

10. PROMOTING POTENTIAL

1. G. M. Howe, S.R. Beach, G.H. Brody, and P.A. Wyman, "Translating Genetic Research into Preventive Intervention: The Baseline Target Moderated Mediator Design," *Frontiers in Psychology* (January 2016): doi: 10.3389/fpsyg.2015.01911.

2. I. Pappa, V. R. Mileva-Seitz, M. J. Bakermans-Kranenburg, H. Tiemeier, and M. H. van Ijzendoorn, "The Magnificent Seven: A Quantitative Review of Dopamine Receptor d4 and Its Association with Child Behavior," *Neuroscience and Biobehavioral Reviews* 57 (October 2015): 175–186, 175.

3. P. Wilby, "Psychologist on a Mission to Give Every Child a Learning Chip," *Guardian* (February 18, 2014), https://www.theguardian.com/education/2014/feb/18/psychologist-robert-plomin-says-genes-crucial-education.

4. For critical discussion, see H. Rose and S. Rose, *Genes, Cells and Brains: The Promethean Promises of the New Biology* (London: Verso, 2012).

5. S. Hsu, "Super-Intelligent Humans Are Coming: Genetic Engineering Will One Day Create the Smartest Humans Who Have Ever Lived," *Nautilus*, no. 18 (October 16, 2014).

6. J. M. Goodrich and D. C. Dolinoy, "Environmental Exposures: Impact on the Epigenome," in *Epigenetics: Current Research and Emerging Trends*, ed. B. P. Chadwick (Norfolk, U.K.: Caister Academic Press, 2012), 330–345.

7. For example, C. G. Victora, B. L. Hort, C. Loret de Mola, L. Quevedo, R. T. Pinheiro, et al., "Association Between Breastfeeding and Intelligence, Educational Attainment, and Income at 30 Years of Age: A Prospective Birth Cohort Study from Brazil," *Lancet Global Health* 3 (January 2015): 199–205.

8. E. B. Isaacs, B. R. Fischl, B. T Quinn, W. K Chong, D. G Gadian, and A. Lucas, "Impact of Breast Milk on IQ, Brain Size and White Matter Development," *Pediatric Research* 67 (April 2010): 357–362.

9. B. M. Kar, S. L. Rao, and B. A. Chandramouli, "Cognitive Development in Children with Chronic Protein Energy Malnutrition," *Behavior and Brain Functions* 4 (July 2008): 31.

10. J. R. Galler and L. R. Barrett, "Children and Famine: Long-Term Impact on Development," *Ambulatory Child Health* 7 (June 2001): 85–95, 85.

11. E. B. Isaacs et al., "Impact of Breast Milk on IQ."

12. S. L. Huffman, R. K. A. Harika, A. Eilander, and S. J. Osendarp, "Essential Fats: How Do They Affect Growth and Development of Infants and Young Children in Developing Countries? A Literature Review," *Maternal and Child Nutrition* 3 (October 2011, supplement): 44–65, 44.

13. A. D. Stein, "Nutrition in Early Life and Cognitive Functioning," *American Journal of Clinical Nutrition* 99 (November 2014): 1–2, 1.

14. A. D. Stein, "Nutrition in Early Life," 1.

15. R. Kumsta, J. Kreppner, M. Kennedy, N. Knights, E. Sonuga-Barke, and E. & M. Rutter, "Psychological Consequences of Early Global Deprivation: An Overview of Findings from the English & Romanian Adoptees Study," *European Psychologist* 20 (April 2015): 138–151, 138.

16. M. Weinstock, "The Potential Influence of Maternal Stress Hormones on Development and Mental Health of the Offspring," *Brain, Behavior, and Immunity* 19 (July 2005): 296–308.

17. R. Bogdan and A. R. Hariri, "Neural Embedding of Stress Reactivity," *Nature Neuroscience* 15 (November 2012): 1605–1607.

18. P. La Marca-Ghaemmaghami and U. Ehlert, "Stress During Pregnancy: Experienced Stress, Stress Hormones, and Protective Factors," *European Psychologist* 20 (January 2015): 102–119.

19. M. R. Rosenzweig and E. L. Bennett, "Psychobiology of Plasticity: Effects of Training and Experience on Brain and Behavior," *Behavioural Brain Research* 78 (June 1996): 57–65, 57.

20. For a review, see C. A. Nelson, N.A. Fox, and C.H. Zeanah, *Romania's Abandoned Children: Deprivation, Brain Development, and the Struggle for Recovery* (Cambridge, Mass.: Harvard University Press, 2014).

21. M. Lövdén, E. Wenger, J. Mårtensson, U. Lindenberger, and L. Bäckman, "Structural Brain Plasticity in Adult Learning and Development," *Neuroscience and Biobehavioral Reviews* 37 (November 2013): 2296–2310.

22. See also B. M. Caldwell and R. H. Bradley, *Home Observation for Measurement of the Environment: Administration Manual* (Tempe, Ariz.: Family & Human Dynamics Research Institute, Arizona State University, 2003), fhdri.clas.asu.edu/home.

23. R. Plomin and D. Daniels, "Why Are Children in the Same Family So Different from Each Other?" *Behavior and Brain Sciences* 10 (March 1987): 1–16.

24. E. Turkheimer and M. Waldron, "Nonshared Environment: A Theoretical, Methodological, and Quantitative Review," *Psychological Bulletin* 126 (January 2000): 78–108, 78.

25. L. C. Mayes and M. Lewis, *The Cambridge Handbook of Environment in Human Development* (Cambridge: Cambridge University Press, 2012), 1.

26. A. W. Heim, *The Appraisal of Intelligence* (London: Methen, 1954), 154.

27. For example, D. R. Topor, S. P. Keane, T. L. Shelton, and S. D. Calkins, "Parent Involvement and Student Academic Performance: A Multiple Mediational Analysis," *Journal of Prevention & Intervention in the Community* 38 (July 2010): 183–197.

28. C. Koede and T. Techapaisarnjaroenkit, "The Relative Performance of Head Start," *Eastern Economic Journal* 38 (February 2012): 251–275, 251.

29. L. F. Cofer, "Dynamic Views of Education," in *Human Development in the Twenty-First Century*, ed. A. Fogel et al. (Cambridge: Cambridge University Press, 2008), 128–135, 133.

30. See Every Student Succeeds Act, 2010, U.S. Department of Education, Washington, D.C., www.ed.gov/esea; and U.S. Department of Education, "Equity of Opportunity," 2016, Washington, D.C., www.ed.gov/equity.

31. U.S. Department of Education, "The Threat of Educational Stagnation and Complacency: Remarks of U.S. Secretary of Education Arne Duncan at the release of the 2012 Program for International Student Assessment (PISA)," December 13, 2013, Washington, D.C., www.ed.gov/news/speeches/threat-educational-stagnation-and-complacency.

32. Michael Gove's Speech to the Policy Exchange on Free Schools, June 20, 2011, https://www.gov.uk/government/speeches/michael-goves-speech-to-the-policy-exchange-on-free-schools.

33. "The Tories and Scotland," *Herald* (Glasgow), April 25, 2009.

34. B. Lott, "The Social Psychology of Class and Classism," *American Psychologist* 67 (November 2012): 650–658, 650.

35. American Psychological Association, *Report of the APA Task Force on Socioeconomic Status*, Washington, D.C., 2006, 7, http://www.apa.org/pi/ses/resources/publications/task-force-2006.pdf; F. Autin and F. Butera, "Editorial: Institutional Determinants of Social Inequalities," *Frontiers in Psychology* 6 (January 2016): http://dx.doi.org/10.3389/fpsyg.2015.02027.

36. A. Reuben, Gap Between Rich and Poor 'Keeps Growing,'" BBC News, May 21, 2015, www.bbc.com/news/business-32824770.

37. E. N. Wolff and M. Gittleman, "Inheritances and the Distribution of Wealth: Or, Whatever Happened to the Great Inheritance Boom?" Bureau of Labor Statistics, Working Paper 445, January 2011, doi: 10.1007/s10888-013-9261-8.

38. M. I. Norton, "Unequality: Who Gets What and Why It Matters," *Policy Insights from the Behavioral and Brain Sciences* 1 (October 2014): 151–155.

39. Y. Kim and M. Sherraden, "Do Parental Assets Matter for Children's Educational Attainment? Evidence from Mediation Tests," *Children and Youth Services Review* 33 (June 2011): 969–979.

40. S. Loughnan, P. Kuppens, J. Allik, K. Balazs, S. de Lemus, et al., "Economic Inequality Is Linked to Biased Self-Perception," *Psychological Science* 22 (October 2011): 1254–1258.

41. P. K. Smith, N. B. Jostmann, A. D. Galinsky, and W. W. van Dijk, "Lacking Power Impairs Executive Functions," *Psychological Science* 19 (May 2008): 441–447, 446.

42. C. L. Odgers, "Income Inequality and the Developing Child: Is It All Relative?" *American Psychologist* 70 (November 2015): 722–731, 722.

43. K. D. Vohs, "The Poor's Poor Mental Power," *Science* 341 (August 2013): 968–970, 969.

44. J. T. McGuire and J. W. Kable, "Rational Temporal Predictions Can Underlie Apparent Failures to Delay Gratification," *Psychological Review* 120 (April 2013): 395–410.

45. A. K. Shah, S. Mullainathan, and E. Shafir, "Some Consequences of Having Too Little," *Science* 338 (November 2012): 682–685, 682.

46. A. Chirumbola and A. Areni, "The Influence of Job Insecurity on Job Performance and Absenteeism: The Moderating Effect of Work Attitudes," *Journal of Industrial Psychology* 31 (October 2005): 65–71, 65.

47. A. Bandura, C. Barbaranelli, G. V. Caprara, and C. Pastorelli, "Multifaceted Impact of Self-Efficacy Beliefs on Academic Functioning," *Child Development* 67 (June 1996): 1206–1222, 1206.

48. W. E. Frankenhuis and C. de Weerth, "Does Early-Life Exposure to Stress Shape, or Impair, Cognition?" *Current Directions in Psychological Science* 22 (October 2013): 407–412.

49. T. Schmader and W. M. Hall, "Stereotype Threat in School and at Work: Putting Science into Practice," *Policy Insights from the Behavioral and Brain Sciences* 1 (October 2014): 30–33, 30.

50. P. K. Piff, M. W. Kraus, S. Côté, B. H. Cheng, and D. Keltner, "Having Less, Giving More: The Influence of Social Class on Prosocial Behavior," *Journal of Personality and Social Psychology* 99 (November 2010): 771–784.

11. THE PROBLEMS OF EDUCATION ARE NOT GENETIC

1. F. Autin, A. Batruch, and F. Butera, "Social Justice in Education: How the Function of Selection in Educational Institutions Predicts Support for (Non)Egalitarian Assessment Practices," *Frontiers in Psychology* 6 (June 2015): 707, http://dx.doi.org/10.3389/fpsyg.2015.00707.

2. S. Bowles and H. Gintis, "The Inheritance of Economic Status: Education, Class and Genetics," in *Genetics, Behavior and Society*, ed. M. Feldman (New York: Elsevier, 2001), 4132–4141. See also S. Bowles and H. Gintis, *Schooling in Capitalist America: Educational Reform and the Contradictions of Economic Life* (Chicago: Haymarket Books, 2013).

3. See, for example, D. C. Berliner and G. V. Glass (eds.), *"50 Myths and Lies That Threaten America's Public Schools: The Real Crisis in Education* (New York: Teachers College Press, 2014), 63.

4. G. Claxton and S. Meadows, "Brightening Up: How Children Learn to Be Gifted," in *Routledge Companion to Gifted Education*, ed. T. Balchin et al. (London: Routledge, 2008), 3–9, 5.

5. B. Elsner and I. E. Isphording, "A Big Fish in a Small Pond," Institute for the Study of Labor Discussion Paper 9121 (Bonn: Institute for the Study of Labor, 2015).

6. J. Boaler, "Ability and Mathematics: The Mindset Revolution That Is Reshaping Education," *Forum* 55, no. 1 (2013): 143–152, 146.

7. D. K. Ginther and S. Kahn, "Comment on 'Expectations of Brilliance Underlie Gender Distributions Across Academic Disciplines,'" *Science* 349 (July 2015): 341–343. And also A. Cimpian and S.-J. Lesli, "Response to Comment on 'Expectations of Brilliance Underlie Gender Distributions Across Academic Disciplines,'" *Science* 349 (July 2015): 391–393, 391.

8. J. Robinson-Cimpian, S. T. Lubienski, C. M. Ganley, and Y. Copur-Gencturk, "Teachers' Perceptions of Students' Mathematics Proficiency May Exacerbate Early Gender Gaps in Achievement," *Developmental Psychology* 50 (April 2013): 1262–1281.

9. See, for example, D. C. Berliner and G. V. Glass (eds.), *50 Myths and Lies That Threaten America's Public Schools*, 239.

10. J. S. Armstrong, "Natural Learning in Higher Education," *Encyclopedia of the Sciences of Learning* (London: Springer, 2011), 2426–2433, 2426.

11. For example, H. Richardson, "Warning Over England's 'Teacher Brain Drain,'" BBC News (February 26, 2016).

12. National Union of Teachers, "Exam Factories? The Impact of Accountability Measures on Children and Young People" (July 4, 2015): 39, teachers.org.uk.

13. Quoted by K. Taylor, "At Success Academy Charter Schools, High Scores and Polarizing Tactics," *New York Times* (April 6, 2015). A version of this article appears in print on April 7, 2015, on page A1 of the New York edition.

14. M. Richardson, C. Abraham, and R. Bond, "Psychological Correlates of University Students' Academic Performance: A Systematic Review and Meta-Analysis," *Psychological Bulletin* 138 (March 2012): 353–387.

15. P. H. Schonemann and M. Heene, "Predictive Validities: Figures of Merit or Veils of Deception?" *Psychology Science Quarterly* 51 (August 2009): 195–215.

16. M. Richardson et al., "Psychological Correlates of University Students' Academic Performance," 353.

17. I. C. McManus and P. Richards, "Prospective Survey of Performance of Medical Students During Preclinical Years," *British Medical Journal* 293 (July 1986): 124–127.

18. M. Richardson et al., "Psychological Correlates of University Students' Academic Performance," 372; Office of Qualifications and Examinations Regulation (Ofqual), "Fit for Purpose? The View of the Higher Education Sector, Teachers and Employers on the Suitability of A Levels," (April 2012): 46, www.ofqual.gov.uk/files/2012-04-03-fit-for-purpose-a-levels.pdf.

19. P. H. Schonemann and M. Heene, "Predictive Validities," 195.

20. E. Kolbert, "Big Score," *New Yorker*, March 3, 2014.

21. A. Ripley, *The Smartest Kids in the World and How They Got That Way* (New York: Simon and Schuster, 2013), 192.

22. S. Engel, *The End of the Rainbow: How Educating for Happiness (Not Money) Would Transform Our Schools* (New York: New Press, 2015), 7.

23. J. S. Armstrong, "Natural Learning in Higher Education," in *Encyclopedia of the Sciences of Learning* (Heidelberg: Springer, 2011), 2426–2433.

24. I. C. McManus, K. Woolf, J. Dacre, E. Paice, and C. Dewberry, "The Academic Backbone: Longitudinal Continuities in Educational Achievement from Secondary School and Medical School to MRCP(UK) and the Specialist Register in UK Medical Students and Doctors," *BMC Medicine* 11 (November 2013): 242, doi:10.1186/1741-7015-11-242.

25. H. S. Gardner, *The Unschooled Mind* (New York: Basic Books, 2004), 3–4.

26. J. S. Armstrong, "Natural Learning in Higher Education."

27. Quoted by D. Koeppel, "Those Low Grades in College May Haunt Your Job Search," *New York Times* (December 31, 2006): HW1 of the New York ed.

28. A. Bryant, "In Head-Hunting, Big Data May Not Be Such a Big Deal," *New York Times*. A version of this article appeared in print on June 20, 2013, on page F6 of the New York edition.

29. Y. Yuan and B. McKelvey, "Situated Learning Theory: Adding Rate and Complexity Effects via Kauffman's NK Model," *Nonlinear Dynamics, Psychology, and Life Sciences* 8 (January 2004): 65–103, 70.

30. See, for example, D. C. Berliner and G. V. Glass (eds.), *50 Myths and Lies That Threaten America's Public Schools.*

31. A. D. Schleimann and D. W. Carraher, "The Evolution of Mathematical Reasoning: Everyday Versus Idealized Understandings," *Developmental Review* 22 (October 2002): 242–266.

32. Innovation Unit, "Redesigning Secondary Schools," http://innovationunit.org/real-projects.

33. See Innovation Unit, "Redesigning Secondary Schools."

34. See, for example, D. C. Berliner and G. V. Glass (eds.), *50 Myths and Lies That Threaten America's Public Schools*, 63.

35. M. I. Norton, "Unequality: Who Gets What and Why It Matters," *Policy Insights from the Behavioral and Brain Sciences* 1 (October 2014): 151–155, 151.

36. J. L. Berg, "The Role of Personal Purpose and Personal Goals in Symbiotic Visions," *Frontiers in Psychology* (April 2015): doi: 10.3389/fpsyg.2015.00443; see also S. Denning, "The Copernican Revolution in Management," *Forbes Magazine* (July 11, 2013): http://www.forbes.com/sites/stevedenning/2013/07/11/the-copernician-revolution-in-management/#49c4e8af63a0.

37. Psychologists Against Austerity, "The Psychological Impact of Austerity: A Briefing Paper" (2015): https://psychagainstausterity.wordpress.com/?s=the+psychological+impact+of+austerity.

INDEX

Page numbers in *italics* indicate images.